DA £58.50

Human Genome Evolution

The HUMAN MOLECULAR GENETICS series

Series Advisors

D.N. Cooper, *Institute of Medical Genetics, University of Wales College of Medicine, Cardiff, UK*

S.E. Humphries, *Division of Cardiovascular Genetics, University College London Medical School, London, UK*

T. Strachan, *Department of Human Genetics, University of Newcastle upon Tyne, Newcastle upon Tyne, UK*

Human Gene Mutation
From Genotype to Phenotype
Functional Analysis of the Human Genome
Molecular Genetics of Cancer
Environmental Mutagenesis
HLA and MHC: Genes, Molecules and Function
Human Genome Evolution

Forthcoming titles

Gene Therapy

Human Genome Evolution

Michael S. Jackson
Department of Human Genetics, University of Newcastle upon Tyne, Newcastle upon Tyne, UK

Tom Strachan
Department of Human Genetics, University of Newcastle upon Tyne, Newcastle upon Tyne, UK

G. Dover
Department of Genetics, University of Leicester, Leicester, UK

First published in 1996

A CIP catalogue record for this book is available from the British Library.

ISBN 1 859960 95 2

BIOS Scientific Publishers Ltd
9 Newtec Place, Magdalen Road, Oxford OX4 1RE, UK.
Tel. +44 (0) 1865 726286. Fax. +44 (0) 1865 246823
World-Wide Web home page: http://www.Bookshop.co.uk/BIOS/

DISTRIBUTORS

Australia and New Zealand
DA Information Services
648 Whitehorse Road, Mitcham
Victoria 3132

India
Viva Books Private Limited
4346/4C Ansari Road
New Delhi 110002

Singapore and South East Asia
Toppan Company (S) PTE Ltd
38 Liu Fang Road, Jurong
Singapore 2262

USA and Canada
BIOS Scientific Publishers
PO Box 605, Herndon,
VA 20172–0605

Typeset by Footnote Graphics, Warminster, UK.
Printed by Biddles Ltd, Guildford, UK.

Contents

Contributors

Armour, J.A.L. Department of Genetics, University of Leicester, University Road, Leicester LE1 7RH, UK (contact address: Department of Genetics, Queens Medical Centre, University of Nottingham, Nottingham NG7 2UH, UK)

Britten, R.J. Division of Biology, California Institute of Technology, 101 Dahlia Avenue, Corona del Mar, CA 92625, USA

Cooper, D.N. Institute of Medical Genetics, University of Wales College of Medicine, Heath Park, Cardiff CF4 4XN, UK

Ellis, N.A. Laboratory of Human Genetics, New York Blood Center, 303–319 East 66th Street, New York, NY 10021, USA

Hancock, J.M. Gene and Genome Evolution Group, MRC Clinical Sciences Centre, Royal Postgraduate Medical School, Hammersmith Hospital, Du Cane Road, London W12 0NN, UK

Hughes, A.L. Department of Biology and Institute of Molecular Evolutionary Genetics, The Pennsylvania State University, University Park, PA 16802, USA

Jobling, M. Department of Genetics, University of Leicester, University Road, Leicester LE1 7RH, UK

Krawczak, M. Institute of Medical Genetics, University of Wales College of Medicine, Heath Park, Cardiff CF4 4XN, UK

Patthy, L. Institute of Enzymology, Biological Research Center, Hungarian Academy of Sciences, Budapest H-1113, Hungary

Royle, N.J. Department of Genetics, University of Leicester, University Road, Leicester LE1 7RH, UK

Starmer, W.T. Department of Biology, Syracuse University, 130 College Place, Syracuse, NY 13244, USA

Stoneking, M. Department of Anthropology, Pennsylvania State University, University Park, PA 16802, USA

Warburton, P.E. Institute of Cell and Molecular Biology, Swann Building, King's Buildings, University of Edinburgh, Edinburgh EH9 3JR, UK

Willard, H.F. Department of Genetics and Center for Human Genetics, Case Western University School of Medicine, Cleveland, OH 44106, USA

Yokoyama, S. Department of Biology, Syracuse University, 130 College Place, Syracuse, NY 13244, USA

Abbreviations

ADA	adenosine deaminase gene
β_2-m	β_2-microglobulin
C1I	C1 inhibitor gene
CONSBI	conserved before insertion
DMD	Duchenne muscular dystrophy
DRPLA	dentatorubral pallidoluysian atrophy
EGF	epidermal growth factor
F9	factor IX gene
FISH	fluorescence *in situ* hybridization
5′ FR	5′ flanking region
FSH	follicle-stimulating hormone
GDP	guanosine diphosphate
GH-1	growth hormone 1
G6PD	glucose-6-phosphate dehydrogenase gene
GPA	G-protein α-subunits
GPCR	G-protein-coupled receptor
GTP	guanosine triphosphate
HBB	β-globin gene
hCG	human chorionic gonadotrophin
HERV	human endogenous retrovirus
HLA	human leukocyte antigen
HMG proteins	high mobility group proteins
HNPCC	hereditary nonpolyposis colorectal cancer
HPRT	hypoxanthine phosphoribosyltransferase gene
HVR	hypervariable region
IP3	inositol triphosphate
LINE	long interspersed repetitive sequence
LH	luteinizing hormone
LRP	LDL-receptor-related protein
MVR-PCR	minisatellite variant repeat PCR
MYHCB	myosin heavy chain β gene
OTC	ornithine transcarbamylase gene
PAR	pseudoautosomal region
PBR	peptide-binding region
PFGE	pulsed-field gel electrophoresis
PGK	phosphoglycerate kinase gene
PROC	protein C gene
RBI	retinoblastoma gene
RSF	relative simplicity factor
RTVL	retrovirus-like element
SCA1	spinocerebellar ataxia type 1
SRY	sex-determining region Y
SSM	significantly simple motif
STS	steroid sulphatase gene
TAP	transporter associated with antigen processing
TBP	TATA-binding protein
THE	transposable human element
t-PA	tissue plasminogen activator
USCE	unequal sister chromatid exchange
UTR	untranslated region
VNTR	variable number of tandem repeats
YAP	Y *Alu* polymorphic element

Preface

The title of this volume, *Human Genome Evolution*, should not imply a uniqueness to the human genome. On the contrary, all available information suggests that it is typical of mammalian genomes in terms of size, gene content, organization and evolution. What sets the human genome aside is the extent to which we have analysed it, due to our understandable fascination with our own species and the belief that an inventory of all human genes is of fundamental importance if we are to understand and influence genetic disease. This has been a major justification for the Human Genome Project, the ambitious global scientific collaboration which aims to identify, map and sequence all human genes. However, the resulting map, while invaluable as a reference, can convey little of the processes which have shaped our genome, or provide us with insight into our evolutionary history. The present volume is an attempt to provide some of this background.

Individual chapters in this book present detailed accounts of specific areas of research into human genome evolution which have proved particularly instructive. Collectively they provide what we would hope is a reasonably balanced and accurate overview of our current knowledge in a complex and rapidly changing field. Chapter 1 provides a review of the processes of gene mutation and evolution, illustrating both the underlying constraints which influence the spectrum of observed mutations and the breadth of mechanisms which give rise to natural variation within and between genes. A detailed account of the formation of novel genes by the reorganization of existing protein modules (exon-shuffling, Chapter 2) is followed by evolutionary analyses of two gene families (HLA, Chapter 3; G-protein-coupled receptors, Chapter 4), illustrating the power of phylogenetic approaches to infer the relationships, evolution and function of genes within multigene families. Chapters 5–9 focus primarily on non-genic, repetitive, components of the human genome and highlight the complexity and diversity of the mutational processes affecting these sequences. The usual suspects – unequal crossing over, gene conversion, slippage mechanisms and transposition – are all implicated and in certain cases can be distinguished from each other. Furthermore, it is becoming clear that these are not merely turnover mechanisms which blindly homogenize repetitive DNA within the genome; they can have profound and unexpected effects on the evolution of gene function. Finally, Chapters 10–12 discuss the organization and evolution of sex chromosomes and mitochondria, highlighting how these are currently being used to investigate the evolution and movement of early human populations.

Perhaps the most striking feature emerging from these chapters is the sheer dynamism of our genetic material, a dynamism which can often only be appreciated when appropriate data from a number of related species are available. Given the current rapid development of genome maps and of comparative genome analyses, it is likely that research in the coming years will provide both clarity and depth to our understanding of human genome evolution in particular, and molecular evolution in general.

Michael S. Jackson (*Newcastle upon Tyne*)
Tom Strachan (*Newcastle upon Tyne*)
G. Dover (*Leicester*)

1

Mutational processes in pathology and evolution

Michael Krawczak and David N. Cooper

1.1 Introduction

One of the most important properties of the genetic material is its capacity to organize its own faithful replication within the cell. If, however, no errors were ever made during the process of DNA replication, the evolution of complex genomes would have been impossible. Although studies of eukaryotic DNA replication *in vivo* have indicated that this process is extremely accurate, errors do nevertheless occur at a frequency of 10^{-9}–10^{-11} per incorporated nucleotide (Drake, 1970; Thacker, 1985).

The discovery in phage T4 of mutational 'hotspots' (Benzer, 1961) first indicated that the frequency of mutational lesions might be strongly influenced by the local DNA sequence environment. Moreover, this and other early work provided the first clear evidence of endogenous mechanisms of mutation (e.g. DNA replication and repair) as distinct from the hitherto better-characterized exogenous causes such as radiation or chemical mutagens. Considerable evidence for endogenous mechanisms of mutagenesis has since emerged for the genomes of higher eukaryotes. Thus, nucleotide substitutions may result either from chemical (e.g. deamination of 5-methylcytosine or depurination: Coulondre *et al.*, 1978; Loeb and Preston, 1986), physical (e.g. DNA slippage: Kunkel and Soni, 1988) or enzymatic (e.g. post-replicative mismatch repair or exonucleolytic proofreading: Loeb and Kunkel, 1982; Modrich, 1987) mechanisms. Since the efficiency of all these processes is known to be sequence-dependent, it is hardly surprising that the distribution of single base-pair substitutions is non-random, thereby giving rise to 'hotspots' of base substitution in eukaryotic genomes.

Gene deletions also occur non-randomly in prokaryotes and viruses. For example, spontaneous deletions in the *lac*I gene of *Escherichia coli* were frequently found to be flanked by direct repeat sequences, which was suggestive

Human Genome Evolution, edited by M. Jackson, T. Strachan and G. Dover.

of DNA slippage as a mutational mechanism (Schaaper *et al.*, 1986). To date, an enormous amount of experimental data has accumulated which has left no doubt that repeat-mediated transient misalignment during DNA replication is also a very important source of deletions in eukaryotes, including humans (reviewed by Cooper and Krawczak, 1993).

Mutation rates calculated for specific human genetic diseases vary quite widely (Kondrashov and Crow, 1993; Vogel and Motulsky, 1986), which implies underlying differences in the relative mutability of gene sequences in the germline. In addition to gene deletions and single base-pair substitutions, however, which embody the vast bulk of mutations causing human inherited disease (Cooper and Krawczak, 1993), mutation rates comprise all types of lesion, including insertions and duplications. The observed variation, therefore, that all types of gene lesion have in common is that they are highly non-random in terms of both frequency and location (Cooper and Krawczak, 1993).

Mutational processes which occur at high frequency and cause human genetic disease can be assumed to have had a disproportionate impact on the evolution of present-day genomes. For this reason, this chapter concentrates on mutational hotspots in the human genome in the belief that these could shed light upon the mutational mechanisms underlying both gene pathology and evolution.

1.2 Single base-pair substitutions

1.2.1 CpG suppression in the vertebrate genome and its origin

In eukaryotic genomes, 5-methylcytosine (5mC) occurs predominantly in CpG dinucleotides comprising cytosine (C) and guanine (G), the majority of which appear to be methylated. Methylation of C results in a high level of mutation due to the propensity of 5mC to undergo deamination to form thymine (T). Deamination of 5mC is unlikely to occur during the enzymatic replication of the methylation pattern, which appears to be a high-fidelity process. Indeed, 5mC deamination probably occurs with the same frequency as the deamination of C to uracil (U), but uracil DNA glycosylase activity in eukaryotic cells is able to recognize and excise U. Thymine, however, being a 'normal' DNA base, is thought to be both less readily detectable and removable by cellular DNA repair mechanisms.

One consequence of the hypermutability of 5mC is the paucity of CpG in the genomes of many eukaryotes (Setlow, 1976), the heavily methylated vertebrate genomes exhibiting the most extreme 'CpG suppression' (Bird, 1980; Schorderet and Gartler, 1992). Spatially, the distribution of CpG is also non-random: about 1% of vertebrate genomes consist of a fraction that is rich in CpG and which accounts for about 15% of all CpG dinucleotides (reviewed by Bird, 1986). In contrast to most of the scattered CpG dinucleotides, these

'CpG islands' represent unmethylated domains and in many cases appear to coincide with transcribed regions (Larsen *et al.*, 1992). CpG islands, which are predominantly found in early-replicating (R-band) as opposed to late-replicating (G-band) regions of the genome (Craig and Bickmore, 1994), are considered to be the last remnants of the unmethylated domains that once dominated the vertebrate genome. The evolution of the heavily methylated vertebrate genome has been accompanied by a progressive loss of CpG dinucleotides as a direct consequence of their methylation in the germline.

A first estimate of the rate at which 5mC is deaminated and fixed as T *in vivo* was arrived at by extrapolation from *in vitro* data (Cooper and Krawczak, 1989). To this end, the deamination rate of 5mC measured under laboratory conditions in single-stranded DNA was modified so as to be consistent with the observed spectrum of point mutations found to have caused human genetic disease. The rate estimate of 1.66×10^{-16} sec^{-1}, which was later confirmed by the analysis of a much larger sample following a different analytical approach (Cooper and Krawczak, 1993), turned out to be consistent with the evolutionary pattern of CpG substitution exhibited by β-globin gene and pseudogene sequences in humans, chimpanzees and macaques (Cooper and Krawczak, 1989). Mathematical modelling allowed a detailed study of the dynamics underlying the CpG suppression currently found in the bulk DNA of vertebrate genomes (Cooper and Krawczak, 1989). The evolutionary changes in frequency of C and G mononucleotides and of CpG dinucleotides, as deduced retrospectively from the employed model, are depicted in *Figure 1.1*. It can be inferred that the timespan required to create the currently observed CpG frequency (1%) was 50–100 million years. On the other hand, the process of CpG loss must have lasted for approximately 450 million years in order for the mononucleotide frequencies to have attained their present

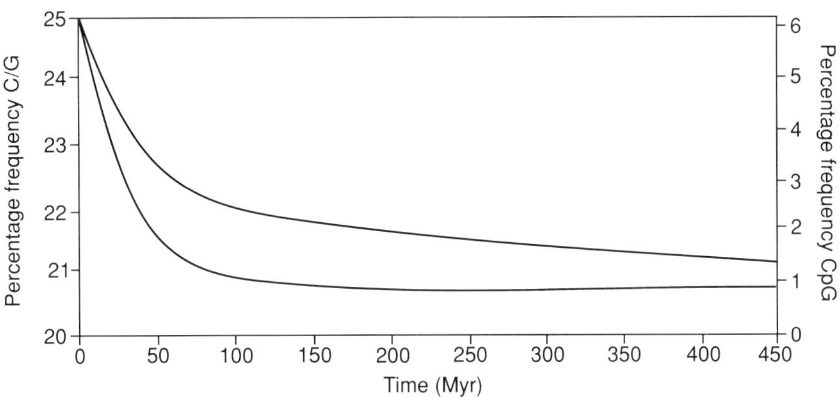

Figure 1.1. Time-dependent frequencies of C/G mononucleotides (upper line) and CpG dinucleotides (lower line) in vertebrate genomes. The curves represent theoretical expectations based upon mathematical modelling. Myr, million years. Data from Cooper and Krawczak (1989).

levels. This timespan corresponds closely to the estimated time since the emergence and adaptive radiation of the vertebrates, and therefore coincides with the probable advent of heavily methylated genomes. Thus, these data are consistent with patterns of vertebrate gene methylation being comparatively stable over relatively long periods of evolutionary time.

1.2.2 The CpG dinucleotide and human genetic disease

At the level of the gene, a high rate of C→T transitions was first reported by Vogel and Röhrborn (1965) in a study of the mutations responsible for haemoglobin variants in humans. Further studies (Vogel and Kopun, 1977) confirmed the existence of this phenomenon. Many sequence comparison studies in eukaryotes (reviewed by Cooper and Krawczak, 1993) have now shown that deamination of 5mC to T in CpG dinucleotides is specifically associated with a high frequency of C→T and G→adenosine (A) transitions, depending on which strand the 5mC is mutated. (G→A transitions arise as a result of a 5mC→T transition on the 'reverse' DNA strand, creating a G:T mismatch, followed by miscorrection of G→A on the 'forward' strand.) A high frequency of polymorphism has also been detected in the human genome by restriction enzymes containing CpG in their recognition sequences (Barker *et al.*, 1984).

CpG has been found to be a hotspot for mutation in a wide range of different human genes (reviewed by Cooper and Krawczak, 1993). Of a total of 3532 base changes so far reported to cause human genetic disease (*Tables 1.1* and *1.2*), 32.6% involve a CpG dinucleotide, and 83.0% (27.0% of the total) of these were either C→T or G→A transitions. These data imply that the rate at which CpG mutates to either TpG or CpA is 6–7 times higher than the mutation rate per nucleotide. Methylation-mediated deamination of 5mC therefore represents a major cause of gene mutability leading to human genetic disease.

The proportion of human gene mutations compatible with a model of methylation-mediated deamination (CG→TG or CA) is, however, very much

Table 1.1. Number of reported[a] point mutations in CpG dinucleotides causing human genetic disease consistent with methylation-mediated deamination of 5-methylcytosine

Mutation	Number
CG→TG	529
CG→CA	424
Total	953

[a]Reported up to November 1995 (Krawczak and Cooper, 1996).

Table 1.2. Number of reported[a] point mutations causing human genetic disease other than CG→TG or CG→CA transitions

Initial nucleotide	Nucleotide resulting from single base-pair change				
	T	C	A	G	Totals
T	–	300	124	140	564
C	314	–	175	170	659
A	89	75	–	255	419
G	285	197	455	–	937
Totals	688	572	754	565	2579

[a] Reported up to November 1995 (Krawczak and Cooper, 1996).

an average figure and provides no information on individual genes. When these values are recalculated on a gene-specific basis, considerable variation becomes apparent. The frequency of CG→TG and CG→CA mutations may be smaller than 10% [*HBB* (4%), *HPRT* (5%), *PALB* (9%)] or larger than 50% [*MYHCB* (67%), *ADA* (63%), *RB1* (57%)]. In the assumed absence of a detection bias (Cooper and Krawczak, 1990), we may surmise that this variation is due either to differences in germline DNA methylation and/or relative intragenic CpG frequency (itself dependent upon germline DNA methylation). Imbalances in the propensity to transition at CpG dinucleotides appear to exist even between the two DNA strands. For instance, in the human *HPRT* gene, Skandalis *et al.* (1994) have noted a significant strand bias in mutations recovered at CpG sites. Arginine residues are encoded by the triplet CpGpA at nucleotides 151–153 and 508–510 of this gene and mutations caused by C→T transitions in these triplets have been recovered from both strands. Mutation of the non-transcribed strand at positions 151 and 508 (to TpGpA) yield nonsense codons, whereas the same transition on the transcribed strand at positions 152 and 509 (to CpApA) yield conservative Arg→Gly substitutions. A statistically significant bias towards most *HPRT* mutations (17/19) being recovered from the non-transcribed strand, although partly explicable in terms of a higher detection likelihood for nonsense mutations, suggests a different rate of CG→TG transition for both strands.

CpG hypermutability in inherited disease implies that the CpG sites in question are methylated in the germline, thereby rendering them prone to 5mC deamination. However, on its own, CpG hypermutability still represents very indirect evidence for CpG methylation. More direct evidence that 5mC deamination underlies mutations is provided by the observation that several C residues known to have undergone a germline mutation in the genes for low-density lipoprotein receptor (*LDLR*, causing hypercholesterolaemia) and p53 (*TP53*, causing various types of tumour) are indeed methylated in the germline (Rideout *et al.*, 1990). As yet, however, this is one of only a few studies to have attempted to correlate CpG hypermutability directly with DNA methylation for specific CpG dinucleotides; more such studies on other genes are urgently needed.

The pattern of germline CpG mutation in the factor IX (*F9*) gene was found to be indistinguishable between Asians (mostly of Korean origin) and Caucasians (Bottema *et al.*, 1990), which argues both for an endogenous origin of these mutations and, indirectly, for the absence of population-specific methylation patterns. Analysis of patterns of DNA methylation exhibited by a variety of other DNA sequences also failed to provide evidence for frequent differences in methylation between individuals from different ethnic backgrounds (Behn-Krappa *et al.*, 1991).

A random sample of CpG transitions in the *F9* gene causing haemophilia B were considered by Green *et al.* (1990): 82% occurred at nine sites that corresponded to functionally critical and evolutionarily conserved amino acid residues, whereas the remaining 18% of transitions occurred at 11 poorly conserved sites. This disproportionately high number of CpG mutations at conserved residues was explained by selection retaining CpG dinucleotides at these locations, despite the attendant high risk of deleterious mutation. Green *et al.* (1990) concluded that consideration of the ratio of CpG to non-CpG mutation might, therefore, lead to overestimation of the intrinsic mutability of the CpG dinucleotide. However, our demonstration that similar values for the relative mutability of CpG may be derived from both clinical and evolutionary studies (Cooper and Krawczak, 1989) argues strongly against this postulate. Furthermore, Koeberl *et al.* (1990) have shown that all *F9* missense mutations, not merely those which occur in CpG dinucleotides, tend to be located in evolutionarily conserved residues. Thus, the puzzle is not so much why CpG dinucleotides are retained at functionally important sites but rather why these sites are methylated, since it is their modification that is ultimately responsible for their hypermutability.

Using a multi-domain molecular model of factor IXa constructed by comparative methods, we have studied the nature and location of over 200 missense mutations in the *F9* gene causing haemophilia B (Wacey *et al.*, 1994). Missense mutations affecting highly conserved amino acid residues were shown to be 15–20 times more likely to result in haemophilia B than those affecting non-conserved residues. However about 25% of this increase in likelihood of clinical observation could be attributed to the magnitude of the amino acid exchange. Missense mutations in structurally conserved residues were found to be 2.1 times more likely to come to clinical attention than those in structurally variable residues. The severity of the clinical phenotype correlated with both the extent of the evolutionary sequence conservation of the residue at the site of mutation and the magnitude of the amino acid exchange. These observations imply a hierarchy of sequence/structure conservation in the factor IX protein. Moreover, the likelihood that a given factor IX mutation will come to clinical attention is a complex function of the sequence characteristics of the *F9* gene, the nature of the amino acid substitution, its precise location and immediate environment within the protein molecule, and its resulting effects on the structure and function of the protein.

1.2.3 Other mechanisms responsible for CpG deamination

Steinberg and Gorman (1992) have suggested that CpG deamination may perhaps result from endogenous enzymatic activity. They found that 70% of their (independent) mouse lymphoma cell mutants possessed a specific CGG→TGG substitution, converting Arg334 to tryptophan in the gene encoding the protein kinase regulatory subunit. In 5% of these mutants, a second mutation (CGT→TGT) was found converting Arg332 to cysteine. The co-occurrence of these two mutations at such a high frequency is very hard to explain on the basis of two independent, yet spatially localized, spontaneous methylation-mediated, deamination events. Instead, it argues for some type of enzymatic mechanism. Such a mechanism could involve a deaminase, as has been postulated by Cambareri *et al.* (1989) to account for the high frequency of CpG transitions in duplicated genomic sequences of *Neurospora crassa* (although no such activity has yet been purified). Though intriguing, the above observation requires further work for its implications to be substantiated. Its relevance to human gene mutation is doubtful since: (i) it does not alter the fact that CpG is a hotspot for mutation; (ii) there are no known examples, including CpG dinucleotides, of pathological base changes in humans that occur with such a high frequency; and (iii) although a few isolated examples of a double mutation have been reported as causes of human genetic disease [e.g. in the gene for β-amyloid (*APP*) in Alzheimer's disease (Mullan *et al.*, 1992)], these do not involve CpG dinucleotides. Multiple spontaneous mutation of the gene for adenosine phosphoribosyltransferase (*APRT*) in a human carcinoma cell line has been observed (Harwood *et al.*, 1991). However, no preference for mutation in CpG sites was apparent and the cause was put down to error-prone DNA synthesis.

Shen *et al.* (1992a) have shown that DNA methyltransferase is capable of inducing C→T transitions directly. Deamination occurs enzymatically and is sensitive to the concentration of the methyl donor, S-adenosylmethionine.

One alternative to an endogenous deaminase could be the mispairing of 5mC with an A base, which, if A was stably incorporated, would increase the frequency of G→A transitions. However, no evidence for such a mechanism has been found from studies of a eukaryotic DNA polymerase α *in vitro* (Shen *et al.*, 1992b).

1.2.4 Mutations in human gene pathology and evolution

In the early days of genetics, spontaneous variation was viewed simply in terms of its role as the evolutionary fuel for speciation. The first to draw parallels with disease was probably the British geneticist William Bateson who, at the turn of the century, maintained that disease presents "a discontinuity closely comparable with that of many variations". Indeed, he speculated that the "problem of species [may well] be solved by the study of pathology; for the likeness between variation and disease goes far to support the view which Virchow has forcibly expressed, that 'every deviation from the type of the parent animal must have its foundation on a pathological accident'"(Bateson, 1894, 1913). Couched in modern terms, single base-pair substitutions in human gene pathology and evolution may be viewed as two sides of the same coin. This appealing supposition has recently been corroborated by an in-depth comparison of mutations causing inherited disorders with mutations in non-coding DNA that have become fixed during the evolutionary divergence of human and other mammalian genomes (Krawczak and Cooper, 1996). Of course, the latter date back at least several million years and their survival is assumed to be independent of natural selection. By contrast, disease-associated substitutions are of fairly recent origin in comparison with the evolutionary timescale. This difference notwithstanding, the two mutational spectra could be expected to show some similarities, assuming that the underlying molecular mechanisms of mutation have not changed substantially during mammalian evolution.

Figure 1.2 shows that the relative base substitution rates observed in the context of human genetic disease (Cooper and Krawczak, 1993; Krawczak and Cooper, 1996) are remarkably similar to those arrived at by Hess *et al.* (1994) in an extensive analysis of human gene–pseudogene alignments. The same pattern of similarity emerged when allowance was made for the DNA sequence environment of the site of mutation by the use of nearest neighbour-dependent mutation rates. The only notable difference between parts (a) and (b) in *Figure 1.2* is a slight under-representation among pseudogene mutations of C→T and G→A substitutions compared with those observed in human genetic disease. This can be explained by either a lower level of germ-line methylation or, more likely, a deficiency of the 5mC-containing CpG

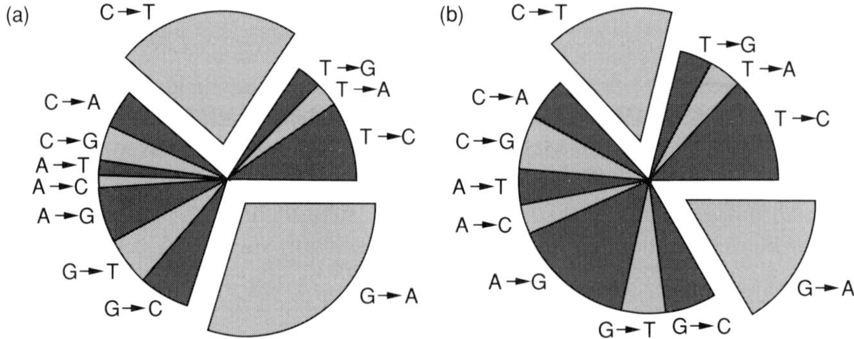

Figure 1.2. Relative mononucleotide substitution rates estimated from (a) human genetic disease data and (b) human gene–pseudogene alignments. Data from Krawczak and Cooper (1996).

dinucleotides ('CG suppression'; Bird, 1980) in non-coding DNA sequences.

That relative mutation rates exhibit a strong positive correlation between disease-associated and evolutionarily fixed mutations (Krawczak and Cooper, 1996) might appear surprising in the light of other experimental results. For example, Hanawalt (1990) has shown that induced pyrimidine dimers and interstrand DNA crosslinks *in vitro* are repaired with a substantially higher efficiency in active genes than in non-coding regions. Although this type of lesion was specific to particular exogenous chemical mutagens and irradiation, the idea of a system which is generally more effective at removing endogenous mutations from coding DNA as opposed to non-coding DNA is attractive. This is because efficient DNA repair should only have conferred a substantial selective advantage in coding regions. However, the comparison of evolutionary and disease data (Krawczak and Cooper, 1996) has demonstrated that the *relative* contribution (via variable efficiency) of different DNA repair pathways to the generation of mutations is unlikely to differ much between intragenic and intergenic sequences.

To what extent is the likelihood of generation of a mutation related to its phenotypic consequences? When codon substitutions causing genetic disease were categorized according to whether they were neutral or whether they changed the hydrophobicity or polarity of the encoded amino acid residue, it emerged that neutral changes are more likely to arise by mutation than non-neutral substitutions (Krawczak and Cooper, 1996). One possible explanation for this disparity is that selection has operated on the cellular DNA repair machinery in such a way as to optimize the removal of non-neutral mutations. If non-neutral changes were more likely to result in a disadvantageous (disease) phenotype than neutral substitutions, then any repair bias operating against these changes at the DNA level would have had a selective advantage.

The hypothesis of a mutational repair bias was further supported by a negative correlation observed in humans between the likelihood of generating a given amino acid substitution and its likelihood of coming to clinical attention (Krawczak and Cooper, 1996). The extent of the correlation, however, differed dramatically between different types of substitution. A significant decrease in mutation generation likelihood was only associated with an increased likelihood of clinical observation for substitutions which affected hydrophilic and polar residues. Various explanations may be advanced to account for these limitations. First, substitutions of hydrophilic and polar amino acid residues could have resulted in more severe and/or variable consequences for the phenotype than other types of substitution during most of the time when the DNA repair mechanisms were evolving. Secondly, it has been observed that, during evolution, the effect of an amino acid change in the hydrophobic protein core is often compensated for by another change in the immediate vicinity (Schirmer, 1979). Thus, any selective pressure acting upon the single organism to avoid substitutions at hydrophobic core residues may have been balanced by the requirement for evolutionary fixation of a second compensatory mutation. Finally, the majority of adaptive events shaping the eukaryotic DNA repair process will have occurred in organisms other than humans. It therefore follows that, were current clinical observation likelihoods to be specific to humans, this could have obscured any correlation between mutation generation likelihoods and phenotypic consequences at hydrophobic or non-polar residues.

If mutations in human genes are biased against amino acid replacements with a high current probability of resulting in a disease phenotype, the question arises as to whether such a bias might also be reflected in the evolutionary history of protein sequences. To explore this possibility, a quantity termed the 'relative evolutionary acceptability' was calculated for each possible amino acid substitution from the data of Collins and Jukes (1994). These authors had reported the numbers and types of amino acid mismatches deduced from the alignment of 337 pairs of human–rodent cDNA sequences. Indeed, clinical observation likelihoods were found to be negatively correlated with evolutionary acceptability values. This implies that the more likely a given amino acid substitution is to result in a disease phenotype (at least in contemporary humans), the less often has it been tolerated during the evolution of human and rodent protein sequences.

In summary, it is evident that the evolutionary requirement to avoid a deleterious phenotype has left its footprints in the mechanisms of mutation generation. In this context, the most promising target for selection would appear to be the intracellular DNA repair mechanism. Although the effect of a given amino acid replacement upon protein structure is known to be heavily dependent upon its precise location within the tertiary structure of the molecule (Alber, 1989; Pakula and Sauer, 1989; Wacey *et al.*, 1994), some basic rules which relate local causes and consequences may nevertheless be perceived (De Filippis *et al.*, 1994). If the efficiency of mutation removal is directed by the immediate DNA sequence context of a lesion, it may be that this has facilitated the avoidance of hazardous amino acid replacements during evolution.

1.2.5 Equilibrium of synonymous codon substitutions

Single base-pair substitutions which do not change the encoded amino acid can be assumed to be comparatively free of selectional constraints (Creighton, 1993). Although it cannot be excluded that these 'synonymous' or silent changes influence gene expression via effects on mRNA translation (e.g. alterations in RNA secondary structure, activation of a cryptic splice site, or local imbalances in the tRNA reservoir of a cell), the probability of survival and ultimate population fixation should be much higher for silent mutations than for missense mutations (Nei, 1987). This view is consistent with the fact that the vast majority of evolutionarily stable base substitutions in coding regions of human genes have taken place at the wobble positions of degenerate codons (Wilbur, 1985).

There are 19 groups of triplets which encode the same amino acid such that the constituent triplets of each group can be replaced by each other via

Table 1.3. Euclidean distance between the vectors of current and equilibrium frequencies within degenerate codons of human genes

Encoded amino acid	Codon group	Euclidean distance
Glu	GAA GAG	0.083
Lys	AAA AAG	0.088
Asp	GAT GAC	0.164
Asn	AAT AAC	0.173
Tyr	TAT TAC	0.196
His	CAT CAC	0.222
Pro	CCT CCC CCA CCG	0.240
Thr	ACT ACC ACA ACG	0.275
Gln	CAA CAG	0.294
Ser	TCT TCC TCA TCG	0.295
Cys	TGT TGC	0.311
Gly	GGT GGC GGA GGG	0.314
Ala	GCT GCC GCA GCG	0.318
Phe	TTT TTC	0.367
Arg	CGT CGC CGA CGG AGA AGG	0.371
Ser	AGT AGC	0.416
Leu	TTA TTG CTT CTC CTA CTG	0.431
Val	GTT GTC GTA GTG	0.444
Ile	ATT ATC ATA	0.540

single base-pair substitution (*Table 1.3*). If mutations that cause an amino acid exchange are ignored, the mutation dynamics within each group of codons can be modelled by a simple system of linear equations involving the relative rates of different single base-pair substitutions. With evolutionary time, this system will approach an equilibrium state and the equilibrium codon frequencies within each group can be determined by solving this system of equations. As can be inferred from *Table 1.3*, the actual frequencies within degenerate codons are still some distance from equilibrium in humans.

When a similar analysis is performed for other vertebrate species with known codon usage (Wada *et al.*, 1991), it turns out that the human is not the species closest to its own equilibrium. In 17/19 codon groups, the toad *Xenopus laevis* is nearest equilibrium whereas humans and rodents form a second group of species, all ranking equally low. This class is followed by chickens, dogs, cows, pigs, rabbits and sheep in order of distance from equilibrium, with chickens closest and sheep furthest. There is thus some correlation between distance from equilibrium and generation time and, with the exception of humans, the ranking of species reflects the total synonymous substitution rates estimated by Bulmer *et al.* (1991). These data are therefore consistent with a model of DNA sequence evolution which, after species divergence, allows ancestral gene sequences to approach equilibrium codon usage faster in one species than in another.

If current codon usage in different species is indeed the result of the divergent evolution of common ancestral sequences progressing at different absolute (albeit equal relative) substitution rates, then one species should always be closer to equilibrium than the other in *all* codon groups. This, however, is not the case. For example, hamsters are closer to equilibrium than dogs for the tyrosine (TAT, TAC) and histidine (CAT, CAC) encoding triplets, but not for glutamine (CAA, CAG). Furthermore, the codon frequencies for lysine (AAA, AAG), aspartic acid (GAT, GAC) and glutamic acid (GAA, GAG) in *X. laevis* are on opposite sites of the equilibrium when compared with all other vertebrates.

One may therefore surmise that codon usage has not evolved in a strictly uniform way. Although it cannot be excluded that relative mutation rates differ between species, thereby resulting in different equilibria, the fact that *Xenopus* and rodents are very close to an equilibrium that is itself based upon human genetic disease data argues strongly against this objection. Thus, it is more likely that species divergence has been accompanied by substantial changes in codon usage, suddenly allowing some species to overtake or be overtaken with respect to distance from equilibrium. This would also explain the theoretical finding that different synonymous codon usage in different vertebrate species is not explicable alone by differential absolute DNA repair efficiency (Eyre-Walker, 1994).

Ikemura and Wada (1991) were able to demonstrate through an analysis of approximately 2000 human gene sequences that codon usage differs dramatically between different genomic regions. The major proportion of GC-rich genes was observed in T-bands whereas AT-rich genes were located mainly in G-bands. Furthermore, the average percentage of G+C at the third position of codons was found to be related to the quinacrine dullness and the mitotic chiasma density of a particular chromosome. Since species divergence was almost always characterized by gross structural chromosomal rearrangements, and since different genes are known to evolve at different absolute rates even in the same species (Bulmer *et al.*, 1991), the piece-wise reconstitution of new genomes from their common ancestors is likely to have altered codon frequencies substantially.

1.3 Deletions

1.3.1 Gross gene deletions

Gene deletions have so far been found to be responsible for more than 150 different inherited conditions in humans (Cooper and Krawczak, 1993). The non-randomness of human gene deletion is readily apparent at two distinct levels. First, in some X-linked recessive conditions of similar incidence, the frequency of gene deletion does not correlate with the size and complexity of the underlying gene. Secondly, 'hotspots' for deletion breakpoints have been reported within several human genes including those encoding the Duchenne muscular dystrophy protein (*DMD*; Wapenaar *et al.*, 1988), growth hormone-1 (*GH-1*; Vnencak-Jones and Phillips, 1990), *LDLR* (Langlois *et al.*, 1988) and α-globin (*HBA-1*; Nicholls *et al.*, 1987). These observations are consistent with deletional events in human genes being at least partly sequence-directed, with the frequency of occurrence reflecting underlying structural differences between genes.

Gross gene deletions may arise through a number of different recombinational mechanisms including homologous unequal recombination between either gene sequences or repetitive elements, and non-homologous recombination between sequences with little or no obvious homology. Perhaps the best example of repetitive-DNA-mediated gene deletion is provided by *Alu* sequences. *Alu* sequences are short, primate-specific interspersed repeated elements (SINES) of about 300 bp in length; they are thought to be degenerate forms of 7SL RNA that have been reverse-transcribed and integrated into the genome. Up to 750 000 copies are thought to exist in the human genome with an average spacing of 4 kb (see Chapter 9; Hwu *et al.*, 1986; Kariya *et al.*, 1987). Copy-number polymorphism between higher primate species is evident in specific gene regions (e.g. the gene for α-globin), implying the sequential insertion of *Alu* elements by retroposition into specific chromosomal sites (Bailey and Shen, 1993).

Alu sequences flanking deletion breakpoints have been noted in a considerable number of human genetic conditions (reviewed by Cooper and Krawczak, 1993). Deletions mediated by *Alu* sequences are of essentially three types: (i) recombination takes place between one *Alu* repeat and a non-repetitive DNA sequence which may or may not possess sequence homology with the *Alu* repeat; (ii) recombination takes place between *Alu* sequences oriented in opposite directions; and (iii) recombination takes place between *Alu* sequences oriented in the same direction.

In terms of the proportion of known deletions, the most dramatic examples of the involvement of *Alu* sequences occur in the *LDLR* gene and the gene for C1–inhibitor (*C1I*). Almost all known deletion breakpoints in the *LDLR* gene occur within an *Alu* repeat sequence (of which there are a total of 21 within this gene). One of these lesions, a 5.5 kb deletion, involves the formation of a stem–loop structure mediated by inverted repeats on the same DNA strand and derived from the oppositely oriented *Alu* sequences (Lehrman *et al.*, 1987). For *LDLR* gene deletions bounded by *Alu* repeats of the same orientation, however, deletion is proposed to occur by meiotic (or mitotic) recombination between chromosomes misaligned at the highly homologous *Alu* sequences. In the vast majority of cases in which two *Alu* sequences have been implicated in the deletion event, they occur in the same orientation. We may therefore surmise that homologous unequal recombination between similarly oriented *Alu* family members, misaligned at meiosis, is probably the most common mechanism of *Alu*-mediated gene deletion in this instance. A similar situation pertains in the *C1I* gene, which possesses a total of 17 *Alu* repeats within a 17 kb region. Deletions and partial deletions of the *C1I* gene appear to account for 15–20% of the lesions which cause type-1 hereditary angioneurotic oedema, and a much higher proportion of these than expected occur within *Alu* sequences. Clustering of breakpoints is evident: Stoppa-Lyonnet *et al.* (1991) have shown that 5/5 deletions or duplications occurred within the first *Alu* sequence element preceding exon 4 of this gene. However, the breakpoints of these rearrangements were distributed over the entire *Alu* sequence element and were not themselves further clustered.

Thus, there are several examples of *Alu*-rich genes whose deletion breakpoints occur very frequently within members of this repeat family. There are, however, a number of other examples which suggest that the extent of association between deletion breakpoints and *Alu* sequences is very much dependent upon the gene under study. Henthorn *et al.* (1990) collated data on over 30 deletions in the β-globin cluster of genes but noted the presence of *Alu* sequences at only four breakpoints. These authors concluded that the occurrence of deletion breakpoints in *Alu* sequences within the β-globin gene region was not significantly different from that expected by chance alone. It could, however, be argued that this was due to the relative paucity of *Alu* sequences (only eight in 60 kb) within the gene cluster. The study of Kornreich *et al.* (1990) goes some way toward meeting this objection. These authors looked for an association between *Alu* sequences and deletion breakpoints in the gene encoding α-galactosidase A (*GLA*), deficiency of which causes Fabry disease. Although 12 *Alu* repeats are found in the 12 kb gene region, deletions were relatively infrequent (only 5/130 patients possessed a partial gene deletion). Moreover, only three breakpoints occurred within an *Alu* sequence and only one resulted from an *Alu–Alu* recombination. This is rather a low proportion since 30% of the *GLA* gene comprises *Alu* repeat sequence. Finally, no correlation has been found between the locations of deletion breakpoints and *Alu* sequences in the human *HPRT* gene (Monnat *et al.*, 1992). These authors suggested that this might be because the 130–210 bp *Alu* repeats in the *HPRT* gene rarely exhibited more than 30 bp sequence identity, much lower than the 200–300 bp sequence identity normally required to promote efficient intrachromosomal recombination in mammalian cells (Bollag *et al.*, 1989).

It must be made clear, however, that most gene deletions are not mediated by *Alu* repeat sequences. Indeed, although the 66 kb human growth hormone gene cluster contains some 48 *Alu* sequences, these do not appear to be the cause of the high frequency of clustered *GH–1* gene deletions causing familial growth hormone deficiency (Vnencak-Jones *et al.*, 1988). Vnencak-Jones and Phillips (1990) studied 10 such patients and showed that in nine of these the crossovers had occurred within two 99% homologous 594 bp regions flanking the *GH–1* gene. Since these two sequences are longer and probably more similar than any two *Alu* repeats in the vicinity, it is thought that they may serve as recombination targets in preference to *Alu* repeats.

Other types of repetitive sequence element may also mediate homologous unequal recombination. Approximately 90% of individuals with ichthyosis have a deletion at their locus for steroid sulphatase (*STS*) (Shapiro *et al.*, 1989). Yen *et al.* (1990) reported that 24/26 patients with an *STS* deletion had breakpoints clustered within and around a number of S232–type repetitive sequences flanking the *STS* gene, suggesting that the high frequency of deletion at this locus may be due to recombination between these sequences. The long terminal repeats of the endogenous retroviral RTVL-H family have also been found to mediate homologous unequal recombination events (Mager and Goodchild, 1989). In their study of some 30 deletions of the β-globin gene cluster, Henthorn *et al.* (1990) noted breakpoints within five long interspersed repeat elements (LINEs), transpositionally active sequences, some 6–7 kb in length, which are present in all mammals (Smit *et al.*, 1995). However, this was no higher than the random expectation figure and thus it is unnecessary to invoke an important role for LINE elements in the causation of deletions at this locus. Finally, sequence analysis of deletion breakpoints located within the deletion hotspot in intron 43 of the dystrophin (*DMD*) genes of two unrelated DMD patients has revealed the presence of a transposon-like element belonging to the transposable human element (THE-1) family (Pizzuti *et al.*, 1992).

Not surprisingly, gross gene deletions have also occurred during evolution. For example, the γ-globin gene in New World monkeys has been subjected to a 1.8 kb deletion which has removed most of exon 2, all of intron 2, exon 3 and much of the 3′ flanking region (Meireles *et al.*, 1995). Thus, in New World monkeys, γ_2-globin is the primary fetally expressed globin gene, whereas in Old World monkeys it is γ_1-globin.

1.3.2 Short gene deletions

Short human gene deletions (<20 bp) were analysed by Cooper and Krawczak (1993), who attempted to relate the presence of specific DNA sequence motifs in the vicinity of these lesions to possible mechanisms responsible for their generation. Many mechanistic models for deletion mutagenesis have been proposed but no one model alone appears to be sufficient to account for the spectrum of deletions observed in human genes. In many cases, slipped mispairing at the replication fork between homologous sequences in close proximity to one another on complementary DNA strands appears to be the causative mechanism. Slipped mispairing may occur, however, between direct repeat copies or through the formation of secondary structure intermediates potentiated by the presence of inverted repeats or 'symmetric' elements (Cooper and Krawczak, 1993; Krawczak and Cooper, 1991).

We have identified a consensus sequence, TGRRKM, which appears to be common to deletion 'hotspots' found in different human genes (Krawczak and Cooper, 1991). This consensus sequence is strikingly similar to the core motifs TGGGG and TGAGC found in the tandemly repeated immunoglobulin switch (Sμ) regions (Gritzmacher, 1989), and to putative arrest sites for DNA polymerase α which often contain a GAG motif. [Indeed, one 5 bp arrest site specifically mentioned by Weaver and DePamphilis (1982) fits perfectly with the first five positions of the deletion hotspot consensus sequence.] DNA polymerase arrest sites may fulfil one or more important biological function, such as the initial localization of the replication fork, the synchronization or termination of DNA replication, or the promotion of recombination at the replication fork. Secondary structure appears to be important too; it may be that the arrest of DNA polymerase α at the replication fork arises as a consequence of both the primary and secondary structure acting in concert. All these factors suggest the possible involvement of DNA polymerase α arrest sites in the generation of human gene deletions.

A more elaborate search for a variety of DNA sequence motifs was undertaken within 10 bp of the breakpoints associated with 776 deletions of less than 20 bp (update of database presented in Cooper and Krawczak, 1993). Observed frequencies of occurrence of such motifs are listed in *Table 1.4* and

Table 1.4 DNA sequence motifs potentially associated with short human gene deletions

Motif	Sequence	Observed frequency	Expected frequency
Vertebrate/plant topoisomerase I consensus cleavage sites	CAT	112	109
	CTY	247	220
	GTY	169	191
	RAT	202	222
DNA pol α pause site core motifs	GAG	167	170
	GCS	161	252
	ACG	54	105
Murine MHC deletion hotspot	CAGG	51	59
Vaccinia topoisomerase I cleavage consensus sequence	YCCTT	32	19
Polypyrimidine runs	YYYYY	218	170
Polypurine runs	RRRRR	280	237
Alternating purine–pyrimidine runs	RYRYR	152	180
Alternating pyrimidine–purine runs	YRYRY	136	172
Consensus Ig switch region	TGGGG	18	15
DNA pol-α frameshift hotspots	TCCCCC	5	4
	CTGGCG	3	4
DNA pol-β frameshift hotspots	ACCCWR	15	16
	TTTT	36	22
DNA pols-α/β frameshift hotspots	TGG*N*GT	12	11
	ACCCCA	3	4
Human deletion hotspot	TGRRKM	92	57
Murine parvovirus recombination hotspot	CTWTTY	11	8
Human fragile X breakpoint cluster	CGGCGG	5	2
Ig/TCR recombinase heptamer	CACAGTG	1	1
Chi and chi-like sequences	GCTGGTGG	1	1
	CCWCCWGC	0	2
Murine MHC recombination hotspot	CAGRCAGR	0	2
Ig/TCR recombinase nonamer	ACAAAAACC	0	0
Murine LTR recombination hotspot	TGGAAATCC	0	0

Motifs are taken from Monnat *et al.* (1992).
Ambiguous nucleotide symbols: R, A or G; Y, C or T; W, A or T; S, G or C; M, A or C; K, G or T; *N*, any base. Ig, immunoglobulin; MHC, major histocompatiblility complex; LTR, long terminal repeat; TCR, T-cell receptor.

are compared with randomly expected frequencies. *Table 1.4* shows that only two of the 29 sequence motifs appear to be over-represented in the vicinity of short human gene deletions: polypyrimidine runs of at least 5 bp (YYYYY) and the deletion hotspot consensus sequence TGRRKM. A detailed analysis of the YYYYY motifs showed that the maximum disparity occurred between the observed and expected frequency for the pentamer YTCYT (78 compared with 40). For the deletion hotspot consensus sequence, the significance of association with short gene deletions can be improved if M (i.e. A or C) at position 6 is replaced by A (observed and expected numbers of matches are 62 and 29, respectively). In contrast to our findings, however, Monnat *et al.* (1992) observed a significant association with *HPRT* deletion breakpoints only for vertebrate topoisomerase I cleavage sites with a CTY sequence.

Short gene deletions also occur during evolution, although at a tenth of the frequency of that of nucleotide substitutions (Saitou and Ueda, 1994). One example is the urate oxidase gene, which is found in baboons and rhesus monkeys but has been inactivated in gibbons by a 13 bp deletion (Wu *et al.*, 1992) and by a nonsense mutation in humans and chimpanzees (Yeldandi *et al.*, 1991). Since the loss of the urate oxidase gene has occurred twice as a result of two independent mutational events, it may be that its absence has conferred some selective advantage in higher primates.

1.4 Other mechanisms of mutation

1.4.1 Gene conversion

Gene conversion is the "modification of one of two alleles by the other" (Vogel and Motulsky, 1986). The end resulting change in the DNA sequence is very similar to that produced from a double crossing-over event. The difference between the two processes is, however, that in gene conversion the sequence change in the 'converted' allele is non-reciprocal, leaving the 'donor' allele physically unchanged. The process of gene conversion may involve the whole or only a part of a gene and occurs not only between allelic genes but also between highly homologous but non-allelic genes. Examples of the latter include conversions between the G-gamma (*HBG2*) and A-gamma (*HBG1*) globin genes (Powers and Smithies, 1986; Shen *et al.*, 1981; Stoeckert *et al.*, 1984) and between the α_1 (*HBA1*)- and α_2 (*HBA2*)- globin genes (Liebhaber *et al.*, 1981). The mechanism underlying gene conversion remains elusive but must presumably entail the close physical interaction between the homologous DNA sequences. Gene conversion has been invoked in instances where it is necessary to account for the association of the same disease-causing mutation with two or more different haplotypes [e.g. β^E-globin alleles in south-east Asian populations (Antonarakis *et al.*, 1982) and the β^S-globin mutation in African populations (Antonarakis *et al.*, 1984)].

Both Amor *et al.* (1988) and Matsuno *et al.* (1992) have noted the presence of *Chi*-like sequences (GCTGGGG, which are known to promote recombination both in *E. coli* and in mouse immunoglobulin genes; Smith, 1983) in the vicinity of regions of the cytochrome *CYP21A* or *CYP21B* and *HBB* genes. These authors speculated that these sequences might play a role in gene conversion events.

Gene conversion has had important effects on the evolution of multigene families (Schimenti *et al.*, 1994). It may act so as to homogenize the sequences of repeated genes (e.g. primate immunoglobulin Cα gene segments; Kawamura *et al.*, 1992) but may possibly also promote diversity by introducing sequence changes into different members of large gene families (e.g. genes of the major histocompatibility complex; Ohta, 1991).

1.4.2 Small insertions

The vast majority of mutations causing human genetic disease are point mutations and deletions (Cooper and Krawczak, 1993). However, a substantial proportion of the remainder involve short sequence duplications or insertions of novel bases, usually resulting in an alteration of the reading frame of the encoded protein, and most often causing termination of translation at some distance downstream. That insertional mutagenesis might be as intrinsically non-random as point mutations and gene deletions has been strongly suggested, for example, by the findings of Fearon *et al.* (1990), who reported 10 independent examples of DNA insertion within the same 170 bp intronic region of the *DCC* gene (deleted in colon cancer; a locus which has been proposed to play an important role in human colorectal neoplasia).

Cooper and Krawczak (1991) have studied the nature of short insertions in human genes causing inherited disease. Although the sample size was too small to detect mutational hotspots, two broad conclusions may nevertheless be stated:

(i) insertional mutation involving the introduction of <10 bp DNA sequence into a gene-coding region is not a random process and appears to be highly dependent upon the local DNA sequence context, and
(ii) mechanistic models which have explanatory value in the context of gene deletions also appear to be useful in accounting for the nature of gene insertions.

Small insertions have also occurred during evolution, although at a frequency half that of small deletions (Saitou and Ueda, 1994). For example, a 24 bp sequence found in the human glycophorin E (*GPE*) gene was inserted into the ancestral gene encoding glycophorin during primate evolution prior to the divergence of the gorilla from the lineage of the other great apes (Rearden *et al.*, 1993).

1.4.3 Insertion of transposable elements in pathology and evolution

Endogenous retroviral sequences comprise some 0.1–0.6% of the human genome (Leib-Mösch *et al.*, 1990). These retroviral elements are thought to have inserted themselves into the germline some 10–40 million years ago after infection with exogenous retroviruses (Shih *et al.*, 1991). They have persisted as endogenous proviruses although rendered replication-defective by multiple mutational events. A number of families of retroposon have been characterized in the genomes of human and other primates: some are confined to humans (e.g. SINE-R.C2; Zhu *et al.*, 1994); some are found distributed throughout the genomes of other primate species [e.g. RTVL-H (Goodchild *et al.*, 1993) and HERV-K (Steinhuber *et al.*, 1995)]; whereas still others are present in a wide range of mammals (e.g. LINE elements). Analysis of the copy number, distribution and sequence characteristics of such endogenous elements may provide important clues as to the evolutionary history and phylogeny of the various mammalian orders and suborders.

Mammalian transposable elements are invariably non-randomly distributed in the genome. LINE elements and some retroviral elements insert preferentially into chromosomal G-bands while *Alu* sequences appear to insert preferentially into R-bands (Wichman *et al.*, 1992). Since many transposable elements contain enhancer sequences, their transposition may serve to alter the pattern of host gene expression at or around the integration site. Indeed, once transposed, evolution may recruit such enhancers to play a role in the transcriptional regulation of a gene; for example, an HERV-E-derived enhancer is required for tissue-specific expression of the gene for human salivary amylase (*AMY1C*) (Ting *et al.*, 1992).

In a pathological context, a number of examples of gene inactivation through insertion of *Alu* or LINE elements into gene-coding sequences are known (reviewed by Cooper and Krawczak, 1993). Some of these sequences exhibit a preference for integration at AT-rich sequences (Kariya *et al.*, 1987; Kazazian *et al.*, 1988). Furthermore, the target sites of two LINE elements inserted into the factor VIII (*F8*) gene causing haemophilia A (Kazazian *et al.*, 1988) are 80% homologous to a 10-nucleotide motif (GAAGACATAC) present in one of the highly favoured retroviral insertion target sequences reported by Shih *et al.* (1988).

1.4.4 Gene duplications

Partial gene duplication has played an important role in evolution and is also a fairly common cause of human genetic disease (reviewed by Hu and Worton, 1992). Two distinct mechanisms are currently envisaged: homologous unequal recombination, either between homologous chromosomes or sister chromatids, and non-homologous recombination at sites with minimal homology.

Several possible topoisomerase I cleavage sites have been noted in the vicinity of the *F8* gene duplication described by Casula *et al.* (1990). This observation is interesting since topoisomerase activity has been implicated in several cases of non-homologous recombination (Bullock *et al.*, 1985). Other examples of an association between topoisomerase cleavage sites and gene duplications have been reported (Hu *et al.*, 1991; Kornreich *et al.*, 1990), including potential sites for topoisomerases I and II exactly coincident with the breakpoints of a duplication in the *DMD* gene (Hu *et al.*, 1991). The significance of these findings remains to be elucidated.

Several gene duplications have occurred during primate evolution. For example, the tandem duplication responsible for the creation of the γ_1- and γ_2-globin genes occurred prior to the divergence of Old World from New World monkeys (Fitch *et al.*, 1991). The 5.5 kb duplicated segment is bounded by two related LINE elements, suggesting that the duplication occurred via unequal homologous recombination (Fitch *et al.*, 1991). Perhaps the redundancy thus introduced created an opportunity for the γ-globin genes to evolve a fetal function to replace their original embryonic function.

The primate glycophorin genes A, B and E are thought to have emerged by duplication mediated in part by *Alu* sequences (Rearden *et al.*, 1993). *GPA* probably represents the ancestral gene and is present in all primates studied. The *GPB* gene is present in humans, chimpanzees and gorillas but not in orangutans and gibbons, whereas the *GPE* gene is present in humans and chimpanzees but intriguingly only in 7/16 gorillas tested (Rearden *et al.*, 1993). Duplications occurring during evolution have also originated by non-homologous recombination events. One example is the tandem duplication leading to the emergence of the complement *C4* gene and the cytochrome *CYP21* gene which occurred prior to the divergence of apes from the Old World monkeys (Horiuchi *et al.*, 1993).

The number and distribution of *Alu* sequences may often help to reconstruct the phylogeny of a gene duplication event, as in the case of the duplication of the apolipoprotein CI (*APOC1*) gene, timed at about 39 million years ago (Raisonnier, 1991).

1.4.5 Gene fusions

Gene fusions, such as that exemplified by haemoglobin Lepore, are an unusual cause of genome pathology (reviewed by Cooper and Krawczak, 1993). Such fusions are, however, thought to have had a more important role in evolution through the creation of novel genes encoding novel combinations of functional protein domains. The emergence of thyroglobulin via the ancient duplication of a carboxyesterase, followed by a gene fusion with a DNA sequence of unknown origin, is one example of the action of this process during evolution (Takagi *et al.*, 1991).

1.4.6 Inversions

Although relatively common in the evolution of the human karyotype (Dutrillaux, 1979), inversions are a highly unusual mutational mechanism in the context of inherited disease. The best-known example is that found recently in the *F8* gene: this rearrangement occurs in about 40% of severely affected patients and recurs at high frequency (Lakich *et al.*, 1993; Naylor *et al.*, 1993). The mechanism responsible is thought to be homologous, intra-chromosomal recombination between a gene (*F8A*) located in intron 22 of the *F8* gene and two additional homologues of the *F8A* gene situated 500 kb upstream of the *F8* gene.

1.5 Geographical dispersion of pathological mutations

For newly arisen mutations to become fixed in a population, it is obvious that they must become widely distributed geographically. Among those factors that influence the geographical distribution of such lesions, it is widely assumed that genetic drift and selection are the most important. We have studied the geographical distribution and prevalence of 256 pathological single base-pair substitutions (105 of them different) within the coding region of the human protein C (*PROC*) gene (Krawczak *et al.*, 1995). A significant positive correlation was observed between the likelihood of generation of a given mutation and its geographical dispersal. This relationship was attributed to the fact that, with few exceptions, high dispersal was only exhibited by CG→TG and CG→CA substitutions, compatible with methylation-mediated deamination. The statistical distribution of mutational likelihoods was as predicted on the basis of the *PROC* cDNA sequence alone, allowing for the redundancy of the genetic code. These findings suggested that genetic drift and lesion-specific selection have been of relatively minor importance in determining the mutational spectrum observed in the *PROC* gene. Furthermore, at least for protein C deficiency, most multiple reports of particular substitutions in different geographical locations appear to reflect recurrent mutation rather than identity by descent.

1.6 Mutations in tandemly repetitive DNA sequences

Tandemly repetitive DNA has been the subject of intense study over the past decade. Some families of repeats are associated with structural components of chromosomes (such as centromeres and telomeres; see Chapters 5 and 6). Others, such as minisatellites (Chapter 7), have been adopted as the tool of choice for identification and individualization purposes, while microsatellites have also been used in mapping studies (Chapter 8) and some simple repeats have been implicated in a variety of clinically important human diseases (e.g. some trinucleotide repeats). Since most of these are discussed in detail elsewhere in this book, we shall confine our discussion to two pathologically important phenomena: the association of mitotically unstable repetitive sequences with tumour development; and the associations of trinucleotide repeat expansions with pathological disorders.

1.6.1 Mitotic instability of repetitive DNA

Short oligonucleotide DNA probes, including $(CAC)_5$, $(CA)_8$ and $(GATA)_4$, have consistently failed to detect differences between microsatellite DNA from differentiated somatic tissues of the same individual. In tumours, however, many changes in banding patterns were observed (Nürnberg *et al.*, 1989). Similarly, Armour *et al.* (1989) showed that mutant minisatellite alleles are only present at very low levels in lymphoblastoid cell lines but occur at much higher frequencies in tumour cells (most notably in gastrointestinal adenocarcinomas). Since mutant alleles were found in tumours at a dosage greater than that of the wild-type allele, this was held to indicate that virtually all mutant tumour alleles were the clonal result of a single mutation. The incidence of somatic mutations in tumours varied from locus to locus, and the locus with the highest propensity to germline mutation also exhibited the most pronounced somatic instability.

A very strong association between $(CA)_n$ repeat instability and tumour development has been observed for hereditary nonpolyposis colorectal cancer (HNPCC), a syndrome characterized by familial predisposition to colorectal carcinoma and other cancers of the gastrointestinal, urological and female reproductive tracts (Aaltonen *et al.*, 1994). The defective gene in HNPCC was mapped to human chromosome 2p by linkage analysis, and genome-wide microsatellite instability is thought to result from mutations within that gene. Fishel *et al.* (1993) then identified a gene (*MSH2*) on chromosome 2p22–21 which exhibits sequence homology to a DNA mismatch repair gene (*MSH*) from *Saccharomyces cerevisiae*. Since both somatic and germline mutations of *MSH2* have been found to be associated with tumour development, this gene was implicated in the aetiology of HNPCC. The repeat instability associated with the *MSH2* mutation represents the footprint of defective mismatch repair correction. Slipped or mispaired replication intermediates cannot be repaired by the defective mismatch repair system, and therefore these replication errors tend to accumulate. It is not unlikely that polymorphisms in the *MSH2* gene and other mutator genes (*MLH1*, *PMS1*, *PMS2*) also contribute to inter-individual variation in the propensity of repeat sequences to mutate.

1.6.2 Triplet repeat expansion disorders

An insertional/duplicational mechanism of mutation with particular clinical importance involves the instability of certain trinucleotide repeat sequences (reviewed by Caskey *et al.*, 1992; Rousseau *et al.*, 1992; Sutherland and Richards, 1995; Willems, 1994). This mechanism was first reported in the context of fragile X syndrome, the most frequent cause of inherited mental retardation, which is associated with the expression of a fragile site on Xq27.3.

The gene underlying fragile X syndrome, *FMR1*, was isolated and identified through its association with a CpG island, by the presence within the *FMR1* gene of the fragile-site-induced breakpoint cluster, and by the greatly reduced levels of *FMR1* mRNA seen in patients with fragile X syndrome

(Oberlé *et al.*, 1991; Verkerk *et al.*, 1991; Yu *et al.*, 1991). A highly unusual CGG repeat sequence was found in the 5′ untranslated region of the brain-expressed *FMR1* gene (Verkerk *et al.*, 1991), length variation of which appeared to correlate with the expression of the fragile X phenotype (Oberlé *et al.*, 1991; Yu *et al.*, 1991). Indeed, while the bulk of normal healthy controls possess between 25 and 35 repeat copies (Fu *et al.*, 1991), phenotypically normal transmitting males exhibit a repeat copy number of between 45 and 200 ('premutation'), and affected males have 300 and sometimes in excess of 1000 copies ('full mutation'; Dobkin *et al.*, 1991; Fu *et al.*, 1991; Hirst *et al.*, 1991; Pergolizzi *et al.*, 1992; Rousseau *et al.*, 1991). The studies of Rousseau *et al.* (1991) and Fu *et al.* (1991) revealed how the *FMR1*-associated repeat cluster expands over successive generations. Alleles with less than 45 repeats showed no meiotic instability. Unaffected individuals bearing the premutation, however, exhibited a high probability of having either affected children or grandchildren. Expansion of premutations to full mutations only occurred during female meiotic transmission where alleles with 52–113 repeats expanded to the full mutation in 70% of cases. The expansion risk for alleles with >90 repeat copies was 100%.

Thus, in contrast to other apparently neutral microsatellites in humans (Valdes *et al.*, 1993), the probability of repeat expansion at the *FMR1* gene correlates with the repeat copy number of the premutation allele. The tendency is to increase repeat copy number rather than to decrease it; the odds ratio was given as 64:3 (Fu *et al.*, 1991). A correlation between informativity (i.e. heterozygosity) and repeat copy number has been observed for human $(CA)_n$ microsatellites by Weber (1990). Of over 100 such loci, those with a comparatively large number of repeats also possess the largest number of different alleles. This holds true, however, only if repetition is perfect. Whenever the run of repeats was interrupted, informativity was much lower than expected on the basis of the total number of repeats. Thus, Weber (1990) concluded that the longest run of uninterrupted CA or GT repeats is the best predictor of informativity. Similarly, Eichler *et al.* (1994) noted that the CGG run at the *FMR1* gene usually contains two interspersed AGG triplets according to the formula $(CGG)_{8-15}AGG(CGG)_{9-13}AGG(CGG)_n$. Instability of a premutation was observed only when one or both of these AGG trinucleotides was lacking, thereby creating a run of at least 33 pure CGG repeats.

Since the characterization of the molecular defect underlying fragile X syndrome, a total of 10 human disorders or phenotypes have been found to be associated with triplet repeat expansion (*Table 1.5*). These include four fragile sites (FRAXA at the *FMR1* gene, FRAXE, FRAXF and FRA16A), all of which are due to the expansion of CGG tracts to more than 200 repeats. At least for the X-chromosomal fragile sites, the region containing the expanded repeat is subject to delayed replication. Willems (1994) has pointed out that this is not surprising since fragile sites are induced by cell-culture conditions that inhibit DNA replication (e.g. folate deprivation).

A second category of triplet repeat expansion relates to five neurodegenera-

Table 1.5. Human genetic diseases and phenotypes associated with triplet repeat expansion

Disease or phenotype	Repeat motif	Localization		Repeat numbers in: Affected patients	Controls
Huntington's disease	CAG	4p16.3	ORF	16–70	9–34
SCA1	CAG	6p23	ORF	41–81	19–36
DRPLA	CAG	12pter–p12	ORF	49–75	7–23
Machado-Joseph disease	CAG	14q31.1	ORF	61–84	13–36
Kennedy disease	CAG	Xq11–q13	ORF	40–55	17–24
Myotonic dystrophy	CTG	19q13.3	3′ UTR	50–2000	5–37
FRA16A	CGG	16p13	?	>1000	16–49
FRAXA	CGG	Xq27–28	5′ UTR	>200	6–50 45–200[a]
FRAXE	CGG	Xq27–28	?	>200	6–25
FRAXF	CGG	Xq27–28	?	>200	12–26
Friedreich ataxia	GAA	9q13–q21.1	Intron	200–900	7–22

[a]Premutation.
DRPLA, dentatorubral pallidoluysian atrophy; SCA1, spinalcerebellar ataxia type 1.

tive disorders [Huntington's disease, spinocerebellar ataxia type 1 (SCA1), dentatorubral pallidoluysian atrophy (DRPLA), Machado-Joseph disease, and Kennedy disease] in which a CAG trinucleotide cluster is expanded within the open reading frame of a gene (reviewed by Sutherland and Richards, 1995). All five diseases are progressive, autosomal, have late onset and are characterized by similar neuropathological features (e.g. ataxia, chorea, dementia). The repeat number above which instability becomes apparent is approximately 35 in all five diseases, a figure which is strikingly close to the critical number of uninterrupted CCG repeats in fragile X syndrome. The upper limit for expansion appears to be 100 repeats. On the basis that the probability of expansion is related to repeat length, computer simulations have suggested that the CAG repeat distribution in the human gene for Huntington's disease may have originated with a primate ancestor and that the incidence of Huntington's disease will continue to increase (Rubinsztein *et al.*, 1994). However, given our current lack of a detailed understanding of the action of genes associated with trinucleotide repeats and the problems of ascertainment bias when comparing sequences identified in humans with paralogues from other species (Ellegren *et al*, 1995), the conclusions of these simulations must remain debatable.

Myotonic dystrophy, a progressive disorder of muscle weakness inherited as an autosomal dominant trait, is as yet the only representative of the third group of triplet disorders. Although its pathology and clinical appearance show some similarity to the CAG repeat disorder, the molecular mechanism underlying this disease is different. Expansion affects a CTG repeat sequence within the 3′ untranslated region of a putative serine–threonine protein kinase gene (*DMPK*), 500 bp upstream of the polyadenylation site (Aslanidis *et al.*,

1992; Brook *et al.*, 1992; Buxton *et al.*, 1992; Fu *et al.*, 1992; Harley *et al.*, 1992; Mahadevan *et al.*, 1992). The repeat is highly polymorphic, varying from 5 to 37 repeats in controls (Brunner *et al.*, 1992; Davies *et al.*, 1992). By contrast, minimally affected myotonic dystrophy patients possess at least 50 CTG repeats, whereas the more severe cases exhibit expansion of the repeat size up to 5 kb (Harley *et al.*, 1993; Lavedan *et al.*, 1993). The size of the CTG repeat correlates both with severity and age of onset. It is thought that the initial predisposing event(s) consisted of a transition from a $(CTG)_5$ allele to an allele with 19–30 repeats (Imbert *et al.*, 1993).

The number of triplet repeat disorders discovered will almost certainly increase, and studying them may provide explanations at the molecular level for a variety of intriguing phenomena in human genetics (Sutherland *et al.*, 1991). The process of triplet repeat expansion has been termed ‘dynamic mutation’ (Richards *et al.*, 1992), since the probability of mutation increases with repeat number and the transition of a normal allele into a clinically significant copy number can involve a number of steps. As to the mechanism of instability manifested by the repeats *in vivo*, slipped mispairing is one possibility. This idea is supported by the observed expansion of the repeat in multiples of 3 bp. Nevertheless, it has become clear that the precise mechanism of change in repeat copy number is complex and may differ from one repeat to another (Richards and Sutherland, 1994).

In some disorders (fragile X syndrome, myotonic dystrophy, Huntington’s disease), a founder effect has been noted in association studies with closely linked markers (Harley *et al.*, 1991; MacDonald *et al.*, 1992; Richards *et al.*, 1992), which argues in favour of *cis*-acting elements being involved in the mutation mechanism. Since the number of uninterrupted repeats was found to be critical for expansion (Chung *et al.*, 1993; Eichler *et al.*, 1994), the simplest explanation for a *cis*-acting component would be a return to perfect repetition, conferring instability upon the repeat sequence. Thus, Richards and Sutherland (1994) suggested that relatively small changes in the normal and premutation range of triplet repeat clusters result from replication slippage of Okazaki fragments still anchored at their 5′ end by a unique sequence flanking the repeat array. If the number of uninterrupted repeats exceeds a certain threshold, however, strand breakage becomes likely to occur at both ends within the repeat sequence. Then the Okazaki fragment is no longer anchored but can slip and slide freely during polymerization, thereby giving rise to much larger expansions.

Why this phenomenon appears to be confined to C*N*G trinucleotides (*N* being any base) has been explored by Kuryavyi and Jovin (1995). These authors proposed a novel structural model of DNA (and possibly RNA) called ‘triad-DNA’, which consists of an antiparallel double helix of base triads instead of the base pairs of conventional B-DNA. The triads involve both hydrogen-bonded and covalent interactions between a Watson–Crick G:C nucleotide pair and a third nucleotide, molecular constraints allowing this third base to be only A or G. Triads can be connected so as to generate a double helix in which two bases in one triplet are connected to a base from another triplet. Such a structure could provide a stable self-structured conformation for the extruded strand in a slippage event, serving to stabilize the intermediate in the pathological expansion of a triple repeat sequence. The complementary strand would be unpaired and could serve as a replication template maintaining the triplet repeat motif.

Two recent studies employing nuclear magnetic resonance techniques (Chen *et al.*, 1995; Gacy *et al.*, 1995) have revealed that the triplet arrays expanded in human genetic diseases can form single-stranded hairpin structures *in vitro*. In contrast to the theoretical considerations of Kuryavyi and Jovin (1995), however, the results of Chen *et al.* (1995) suggested that the propensity to hairpin formation of $(CCG)_n$ or $(CGG)_n$ clusters should be higher for the C-rich rather than the G-rich strand. The central cytosines of the CCG motifs will mismatch within the hairpin and confer some local flexibility upon it. Since C:C mismatches were found to stimulate human methyltransferase *in vitro* by a factor of six, hairpin formation could thus provide a structural basis for the specific hypermethylation observed for expanded $(CGG)_n$ arrays at the human *FMR1* gene. Gacy *et al.* (1995) determined that the threshold stability per hairpin (about 50 kcal) is also influenced by the DNA sequence flanking the repeat. Thus, only sequences of sufficient length, base composition and sequence environment can form hairpins of threshold energy, a notion which explains both the protective effects endowed by AGG interruptions at *FMR1* and the stability of a CCG run close to the CAG cluster expanded in Huntington's disease.

1.7 Conclusions

A variety of different types of mutational hotspot have now been recognized in the human genome. Spontaneous deamination of 5–methylcytosine within CpG dinucleotides is responsible for about a third of all single base-pair substitutions causing human genetic disease. This chemical mechanism of mutagenesis is also responsible for the progressive loss of CpG dinucleotides from human gene sequences over evolutionary time. Physical mechanisms of mutagenesis appear to be responsible for deletion hotspots and the expansion of unstable repeat sequences. However, the details of the exact mechanisms involved remain to be determined and their role in promoting DNA sequence change on an evolutionary timescale has yet to be studied formally. What is clear is that the types of mutational lesion evident in genome pathology are also to be found in genome evolution. Furthermore, the rate of change of a DNA sequence either during evolution or as a cause of pathology is a function both of the nature of the underlying DNA sequence and its chemical modification.

Our genes have evolved slowly, probably via a host of meandering and circuitous pathways, 'guided' through the myriad of erratic environmental influences by the moulding force of natural selection. Perhaps this hesitant evolutionary past accounts for present-day genes containing, encoded within their base sequences, the potential seeds of their own destruction. How apt in this context is Tennyson's description of nature: "so careful of the type she seems, so careless of the single life" ('In Memoriam A.H.H.', 1850).

References

Aaltonen LA, Peltomaki P, Mecklin JP. (1994) Replication errors in benign and malignant tumours from hereditary non-polyposis colorectal cancer patients. *Cancer Res.* **54**: 1645–1648.

Alber T. (1989) Mutational effects on protein stability. *Annu. Rev. Biochem.* **58**: 765–798.

Amor M, Parker KL, Globerman H, New MI, White PC. (1988) Mutation in the CYP21B gene (Ile→Asn) causes steroid 21-hydroxylase deficiency. *Proc. Natl Acad. Sci. USA* **85**: 1600–1604.

Antonarakis SE, Boehm CD, Giardina PJ, Kazazian HH. (1982) Non-random association of polymorphic restriction sites in the β-globin gene cluster. *Proc. Natl Acad. Sci. USA* **79**: 137–141.

Antonarakis SE, Boehm CD, Serjeant GR, Theisen CE, Dover GJ, Kazazian HH. (1984) Origin of the β^S-globin gene in Blacks: the contribution of recurrent mutation or gene conversion or both. *Proc. Natl Acad. Sci. USA* **81**: 853–856.

Armour JA, Patel I, Thein SL, Fey MF, Jeffreys AJ. (1989) Analysis of somatic mutations at human minisatellite loci in tumours and cell lines. *Genomics* **4**: 328–334.

Aslanidis C, Jansen G, Amemiya C *et al.* (1992) Cloning of the essential myotonic dystrophy region and mapping of the putative defect. *Nature* **355**: 548–551.

Bailey AD, Shen CKJ. (1993) Sequential insertion of *Alu* family repeats into specific sites of higher primates. *Proc. Natl Acad. Sci. USA* **90**: 7205–7209.

Barker D, Schäfer M, White R. (1984) Restriction sites containing CpG show a higher frequency of polymorphism in human DNA. *Cell* **36**: 131–138.

Bateson W. (1894) *Materials for the Study of Variation Treated with Especial Regard to Discontinuity in the Origin of Species.* MacMillan, London.

Bateson W. (1913) *Problems of Genetics.* Yale University Press, New Haven, CT.

Batzer MA, Deininger PL. (1991) A human-specific subfamily of *Alu* sequences. *Genomics* **9**: 481–487.

Behn-Krappa A, Hölker I, Sandaradura de Silva U, Doerfler W. (1991) Patterns of DNA methylation are indistinguishable in different individuals over a wide range of human DNA sequences. *Genomics* **11**: 1–7.

Benzer S. (1961) On the topography of the genetic fine structure. *Proc. Natl Acad. Sci. USA* **47**: 403–417.

Bird AP. (1980) DNA methylation and the frequency of CpG in animal DNA. *Nucleic Acids Res.* **8**: 1499–1504.

Bird AP. (1986) CpG-rich islands and the function of DNA methylation. *Nature* **321**: 209–213.

Bollag RJ, Waldman AS, Liskay RM. (1989) Homologous recombination in mammalian cells. *Annu. Rev. Genet.* **23**: 199–225.

Bottema CDK, Ketterling RP, Yoon H-S, Sommer SS. (1990) The pattern of factor IX germ-line mutation in Asians is similar to that of Caucasians. *Am. J. Hum. Genet.* **47**: 835–841.

Brook JD, McCurrach ME, Harley HG *et al.* (1992) Molecular basis of myotonic dystrophy: expansion of a trinucleotide (CTG) repeat at the 3′ end of a transcript encoding a protein kinase family member. *Cell* **68**: 799–808.

Brunner HG, Nillesen W, van Oost BA, Jansen G, Wieringa B, Ropers HH, Smeets HJM. (1992) Presymptomatic diagnosis of myotonic dystrophy. *J. Med. Genet.* **29**: 780–784.

Bullock P, Champoux JJ, Botchan M. (1985) Association of crossover points with topoisomerase I cleavage sites: a model for non-homologous recombination. *Science* **230**: 954–957.
Bulmer M, Wolfe KH, Sharp PM. (1991) Synonymous nucleotide substitution rates in mammalian genes: implications for the molecular clock and the relationship of mammalian orders. *Proc. Natl Acad. Sci. USA* **88**: 5974–5978.
Buxton J, Shelbourne P, Davies *et al.* (1992) Detection of an unstable fragment of DNA specific to individuals with myotonic dystophy. *Nature* **355**: 547–548.
Cambareri EB, Jensen BC, Schabtach E, Selker EU. (1989) Repeat-induced G-C to A-T mutations in *Neurospora*. *Science* **244**: 1571–1575.
Caskey CT, Pizzuti A, Fu Y-H, Fenwick RG, Nelson DL. (1992) Triplet repeat mutations in human disease. *Science* **256**: 784–789.
Casula L, Murru S, Pecorara M *et al.* (1990) Recurrent mutations and three novel rearrangements in the factor VIII gene of hemophilia A patients of Italian descent. *Blood* **75**: 662–670.
Chen X, Mariappan SVS, Catasti P, Ratliff R, Moyzis RK, Laayoun A, Smith SS, Bradbury EM, Gupta G. (1995) Hairpins are formed by the single DNA strands of the fragile X triplet repeats: structure and biological implications. *Proc. Natl Acad. Sci. USA* **92**: 5199–5203.
Chung MY, Ranum LP, Duvick LA, Servadio A, Zoghbi HY, Orr HT. (1993) Evidence for a mechanism predisposing to intergenerational CAG repeat instability in spinocerebellar ataxia type I. *Nature Genetics* **5**: 254–258.
Collins DW, Jukes TH. (1994) Rates of transition and transversion in coding sequences since the human–rodent divergence. *Genomics* **20**: 386–396.
Cooper DN, Krawczak M. (1989) Cytosine methylation and the fate of CpG dinucleotides in vertebrate genomes. *Hum. Genet.* **83**: 181–189.
Cooper DN, Krawczak M. (1990) The mutational spectrum of single base-pair substitutions causing human genetic disease: patterns and predictions. *Hum. Genet.* **85**: 55–74.
Cooper DN, Krawczak M. (1991) Mechanisms of insertional mutagenesis in human genes causing genetic disease. *Hum. Genet.* **87**: 409–415.
Cooper DN, Krawczak M. (1993) *Human Gene Mutation.* BIOS Scientific Publishers, Oxford.
Coulondre C, Miller JH, Farabaugh PJ, Gilbert W. (1978) Molecular basis of base substitution hotspots in *Escherichia coli*. *Nature* **274**: 775–780.
Craig JM, Bickmore WA. (1994) The distribution of CpG islands in mammalian chromosomes. *Nature Genetics* **7**: 376–382.
Creighton TE. (1993) *Proteins: Structures and Molecular Properties.* W.H. Freeman, New York, pp. 105–133.
Davies J, Yamagata H, Shelbourne P *et al.* (1992) Comparison of the myotonic dystrophy associated CTG repeat in European and Japanese populations. *J. Med. Genetics* **29**: 766–769.
De Filippis V, Sander C, Vriend G. (1994) Predicting local structural changes that result from point mutations. *Protein Eng.* **7**: 1203–1208.
Dobkin CS, Ding X-H, Jenkins EC *et al.* (1991) Prenatal diagnosis of fragile X syndrome. *Lancet* **338**: 957–958.
Drake JW. (1970) *The Molecular Basis of Mutation.* Holden Day, San Francisco.
Dutrillaux B. (1979) Chromosomal evolution in primates: tentative phylogeny from *Microcebus maximus* (prosimian) to man. *Hum. Genet.* **48**: 257–314.
Eichler EE, Holden JJA, Popovich BW, Reiss AL, Snow K, Thibodeau SN, Richards CS, Ward PA, Nelson DL. (1994) Length of uninterrupted CGG repeats determines instability in the FMR1 gene. *Nature Genetics* **8**: 88–94.
Ellegren H, Primmer CR, Sheldon BC. (1995) Microsatellite 'evolution': directionality or bias? *Nature Genetics* **11**: 360–362.
Eyre-Walker A. (1994) DNA mismatch repair and synonymous codon evolution in mammals. *Mol. Biol. Evol.* **11**: 88–98.
Fearon ER, Cho KR, Nigro JM *et al.* (1990) Identification of a chromosome 18q gene that is altered in colorectal cancers. *Science* **247**: 49–56.

Fishel R, Lescoe MK, Rao MR, Copeland NG, Jenkins NA, Garber J, Kane M, Kolodner R. (1993) The human mutator gene homolog *MSH2* and its association with hereditary nonpolyposis colon cancer. *Cell* 75: 1027–1038.

Fitch DH, Bailey WJ, Tagle DA, Goodman M, Sien L, Slightom JL. (1991) Duplication of the gamma-globin gene mediated by L1 long interspersed repetitive elements in an early ancestor of simian primates. *Proc. Natl Acad. Sci. USA* **88**: 7396–7400.

Fu Y-H, Kuhl DPA, Pizzuti A *et al.* (1991) Variation of the CGG repeat at the fragile X site results in genetic instability: resolution of Sherman paradox. *Cell* **67**: 1047–1058.

Fu Y-H, Pizzuti A, Fenwick RG *et al.* (1992) An unstable triplet repeat in a gene related to myotonic muscular dystrophy. *Science* **255**: 1256–1258.

Gacy AM, Goellner G, Juranic N, Macura S, McMurray CT. (1995) Trinucleotide repeats that expand in human disease form hairpin structures *in vitro*. *Cell* **81**: 533–540.

Goodchild NL, Wilkinson DA, Mager DL. (1993) Recent evolutionary expansion of a subfamily of RTVL-H human endogenous retrovirus-like elements. *Virology* **196**: 778–788.

Green PM, Montandon AJ, Bentley DR, Ljung R, Nilsson IM, Giannelli F. (1990) The incidence and distribution of CpG→TpG transitions in the coagulation factor IX gene. A fresh look at CpG mutational hotspots. *Nucleic Acids Res.* **18**: 3227–3231.

Gritzmacher CA. (1989) Molecular aspects of heavy-chain class switching. *Crit. Rev. Immunol.* **9**: 173–200.

Hanawalt PC. (1990) Selective DNA repair in active genes. *Acta Biol. Hung.* **41**: 77–91.

Harley HG, Brook JD, Floyd J, Rundle SA, Crow S, Walsh KV, Thibault MC, Harper PS, Shaw DJ. (1991) Detection of linkage disequilibrium between the myotonic dystrophy locus and a new polymorphic DNA marker. *Am. J. Hum. Genet.* **49**: 68–75.

Harley HG, Brook JD, Rundle SA, Crow S, Reardon W, Buckler AJ, Harper PS, Housman DE, Shaw DJ. (1992) Expansion of an unstable DNA region and phenotypic variation in myotonic dystrophy. *Nature* **355**: 545–546.

Harley HG, Rundle SA, MacMillan JC *et al.* (1993) Size of the unstable CTG repeat sequence in relation to phenotype and parental transmission in myotonic dystrophy. *Am. J. Hum. Genet.* **52**: 1164–1174.

Harwood J, Tachibana A, Meuth M. (1991) Multiple dispersed spontaneous mutations: A novel pathway of mutation in a malignant human cell line. *Mol. Cell. Biol.* **11**: 3163–3170.

Henthorn PS, Smithies O, Mager DL. (1990) Molecular analysis of deletions in the human β-globin gene cluster: deletion junctions and locations of breakpoints. *Genomics* **6**: 226–237.

Hess ST, Blake JD, Blake RD. (1994) Wide variations in neighbour-dependent substitution rates. *J. Mol. Biol.* **236**: 1022–1033.

Hirst M, Knight S, Davies K *et al.* (1991) Prenatal diagnosis of fragile X syndrome. *Lancet* **338**: 956–957.

Horiuchi Y, Kawaguchi H, Figueroa F, O'hUigin C, Klein J. (1993) Dating the primigenial *C4–CYP21* duplication in primates. *Genetics* **134**: 331–339.

Hu X, Worton RG. (1992) Partial gene duplication as a cause of human disease. *Hum. Mutat.* **1**: 3–12.

Hu X, Ray PN, Worton RG. (1991) Mechanisms of tandem duplication in the Duchenne muscular dystrophy gene include both homologous and nonhomologous intrachromosomal recombination. *EMBO J.* **10**: 2471–2477.

Hwu HR, Roberts JW, Davidson EH, Britten RJ. (1986) Insertion and/or deletion of many repeated DNA sequences in human and higher ape evolution. *Proc. Natl Acad. Sci. USA* **83**: 3875–3879.

Ikemura T, Wada K. (1991) Evident diversity of codon usage patterns of human genes with respect to chromosome banding patterns and chromosome numbers; relation between nucleotide sequence data and cytogenetic data. *Nucleic Acids Res.* **19**: 4333–4339.

Imbert G, Kretz C, Johnson K, Mandel JL. (1993) Origin of the expansion mutation in myotonic dystrophy. *Nature Genetics* **4**: 72–76.

Kariya Y, Kato K, Hayashizaki Y, Himeno S, Tarui S, Matsubara K. (1987) Revision of consensus sequence of human *Alu* repeat–a review. *Gene* **53**: 1–10.

Kawamura S, Saitou N, Ueda S. (1992) Concerted evolution of the primate immunoglobulin alpha-gene through gene conversion. *J. Biol. Chem.* **267**: 7359–7357.

Kazazian HH, Wong C, Youssoufian H, Scott AF, Phillips DG, Antonarakis SE. (1988) Haemophilia A resulting from *de novo* insertion of L1 sequences represents a novel mechanism for mutation in man. *Nature* **332**: 164–166.

Koeberl DD, Bottema CDK, Ketterling RP, Bridge PJ, Lillicrap DP, Sommer SS. (1990) Mutations causing hemophilia B: direct estimate of the underlying rates of spontaneous germ-line transitions, transversions and deletions in a human gene. *Am. J. Hum. Genet.* **47**: 202–217.

Kondrashov AS, Crow JF. (1993) A molecular approach to estimating the human deleterious mutation rate. *Hum. Mutat.* **2**: 229–234.

Kornreich R, Bishop DF, Desnick RJ. (1990) α-Galactosidase A gene rearrangements causing Fabry disease. *J. Biol. Chem.* **265**: 9319–9326.

Krawczak M, Cooper DN. (1991) Gene deletions causing human genetic disease: mechanisms of mutagenesis and the role of the local DNA sequence environment. *Hum. Genet.* **86**: 425–441.

Krawczak M, Cooper DN. (1996) Single base-pair substitutions in pathology and evolution: two sides to the same coin. *Hum. Mutat.*, in press.

Krawczak M, Reitsma PH, Cooper DN. (1995) The mutational demography of protein C deficiency. *Hum. Genet.* **96**: 142–146.

Kunkel TA, Soni A. (1988) Mutagenesis by transient misalignment. *J. Biol. Chem.* **263**: 14784–14789.

Kuryavyi VV, Jovin TM. (1995) Triad-DNA: a model for trinucleotide repeats. *Nature Genetics* **9**: 339–341.

Lakich D, Kazazian HH, Antonarakis SE, Gitschier J. (1993) Inversions disrupting the factor VIII gene are a common cause of severe haemophilia A. *Nature Genetics* **5**: 236–241.

Langlois S, Kastelein JJP, Hayden MR. (1988) Characterization of six partial deletions in the low-density lipoprotein (*LDL*) receptor gene causing familial hypercholesterolemia (FH). *Am. J. Hum. Genet.* **43**: 60–68.

Larsen F, Gundarsen G, Lopez X, Prydz H. (1992) CpG islands as gene markers in the human genome. *Genomics* **13**: 1095–1107.

Lavedan C, Hofmann-Radvanyi H, Shelbourne P *et al.* (1993) Myotonic dystrophy: size- and sex-dependent dynamics of CTG meiotic instability, and somatic mosaicism. *Am. J. Hum. Genet.* **52**: 875–883.

Lehrman MA, Russell DW, Goldstein JL, Brown MS. (1987) *Alu- Alu* recombination deletes splice acceptor sites and produces secreted low-density lipoprotein receptor in a subject with familial hypercholesterolemia. *J. Biol. Chem.* **262**: 3354–3361.

Leib-Mösch C, Brack-Werner R, Werner T, Bachmann M, Faff O, Erfle V, Hehlmann R. (1990) Endogenous retroviral elements in human DNA. *Cancer Res.* **50**: 5636S-5642S.

Liebhaber SA, Goossens M, Kan YW. (1981) Homology and concerted evolution at the α1 and α2 loci of human α-globin. *Nature* **290**: 26–29.

Loeb LA, Kunkel TA. (1982) Fidelity of DNA synthesis. *Annu. Rev. Biochem.* **52**: 429–457.

Loeb LA, Preston BD. (1986) Mutagenesis by apurinic/apyrimidinic sites. *Annu. Rev. Genet.* **20**: 201–230.

MacDonald ME, Novelletto A, Lin C *et al.* (1992) The Huntington's disease candidate region exhibits many different haplotypes. *Nature Genetics* **1**: 99–103.

Mager DL, Goodchild NL. (1989) Homologous recombination between the LTRs of a human retrovirus-like element causes a 5-kb deletion in two siblings. *Am. J. Hum. Genet.* **45**: 848–854.

Mahadevan M, Tsilfidis C, Sabourin L *et al.* (1992) Myotonic dystrophy mutation: an unstable CTG repeat in the 3′ untranslated region of the gene. *Science* **255**: 1253–1255.

Matsuno Y, Yamashiro Y, Yamamoto K, Hattori Y, Yamamoto K, Ohba Y, Miyaji T. (1992) A possible example of gene conversion with a common β-thalassaemia mutation and *Chi* sequence present in the β-globin gene. *Hum. Genet.* **88**: 357–358.

Meireles CM, Schneider MP, Sampaio MI *et al.* (1995) Fate of a redundant gamma-globin gene in the atelid clade of New World monkeys: implications concerning fetal globin gene expression. *Proc. Natl Acad. Sci. USA* **92**: 2607–2611.

Modrich P. (1987) DNA mismatch correction. *Annu. Rev. Biochem.* **56**: 435–466.

Monnat RJ, Hackmann AFM, Chiaverotti TA. (1992) Nucleotide sequence analysis of human hypoxanthine phosphoribosyltransferase (*HPRT*) gene deletions. *Genomics* **13**: 777–787.

Mullan M, Crawford F, Axelman K, Houlden H, Lilius L, Winblad B, Lannfelt L. (1992) A pathogenic mutation for probable Alzheimer's disease in the APP gene at the N-terminus of β-amyloid. *Nature Genetics* **1**: 345–347.

Naylor JA, Green PM, Rizza CR, Giannelli F. (1993) Analysis of factor VIII mRNA reveals defects in everyone of 28 haemophilia A patients. *Hum. Mol. Genet.* **2**: 11–17.

Nei M. (1987) *Molecular Evolutionary Genetics*. Columbia University Press, New York.

Nicholls RD, Fischel-Ghodsian N, Higgs DR. (1987) Recombination at the human α-globin gene cluster: sequence features and topological constraints. *Cell* **49**: 369–378.

Nürnberg P, Roewer L, Neitzel H *et al.* (1989) DNA fingerprinting with the oligonucleotide probe (CAC)5/(GTG)5: somatic stability and germline mutations. *Hum. Genet.* **84**: 75–78.

Oberlé I, Rousseau F, Heitz D, Kretz C, Devys D, Hanauer A, Boué J, Bertheas MF, Mandel JL. (1991) Instability of a 550 base pair DNA segment and abnormal methylation in fragile X syndrome. *Science* **252**: 1097–1102.

Ohta T. (1991) Role of the diversifying selection and gene conversion in evolution of major histocompatibility complex loci. *Proc. Natl Acad. Sci. USA* **88**: 6716–6720.

Pakula AA, Sauer RT. (1989) Genetic analysis of protein stability and function. *Annu. Rev. Genet.* **23**: 289–310.

Pergolizzi RG, Erster SH, Goonewardena P, Brown WT. (1992) Detection of full fragile X mutation. *Lancet* **339**: 271–272.

Pizzuti A, Pieretti M, Fenwick RG, Gibbs RA, Caskey CT. (1992) A transposon-like element in the deletion-prone region of the dystrophin gene. *Genomics* **13**: 594–600.

Powers PA, Smithies O. (1986) Short gene conversion in the human fetal globin gene region: a by-product of chromosome pairing during meiosis? *Genetics* **112**: 343–358.

Raisonnier A. (1991) Duplication of the apolipoprotein C-1 gene occurred about forty million years ago. *J. Mol. Evol.* **32**: 211–219.

Rearden A, Magnet A, Kudo S, Fukuda M. (1993) Glycophorin B and glycophorin E genes arose from the glycophorin A ancestral gene via two duplications during primate evolution. *J. Biol. Chem.* **268**: 2260–2267.

Richards RI, Sutherland GR. (1994) Simple repeat DNA is not replicated simply. *Nature Genetics* **6**: 114–116.

Richards RI, Holman K, Friend K *et al.* (1992) Evidence of founder chromosomes in fragile X syndrome. *Nature Genetics* **1**: 257–260.

Rideout WM, Coetzee GA, Olumi AF, Jones PA. (1990) 5-Methylcytosine as an endogenous mutagen in the human LDL receptor and p53 genes. *Science* **249**: 1288–1290.

Rousseau F, Heitz D, Biancalana V *et al.* (1991) Direct diagnosis by DNA analysis of the fragile X syndrome of mental retardation. *N. Engl. J. Med.* **325**: 1673–1681.

Rousseau F, Heitz D, Mandel J-L. (1992) The unstable and methylatable mutations causing the fragile X syndrome. *Hum. Mutat.* **1**: 91–96.

Rubinsztein DC, Amos W, Leggo J *et al.* (1994) Mutational bias provides a model for the evolution of Huntington's disease and predicts a general increase in disease prevalence. *Nature Genetics* 7: 525–530.

Saitou N, Ueda S. (1994) Evolutionary rates of insertion and deletion in noncoding nucleotide sequences of primates. *Mol. Biol. Evol.* **11**: 504–512.

Schaaper RM, Danforth BN, Glickman BW. (1986) Mechanisms of spontaneous mutagenesis: an analysis of the spectrum of spontaneous mutation in the *Escherichia coli* lacI gene. *J. Mol. Biol.* **189**: 273–284.

Schimenti JC. (1994) Gene conversion and the evolution of gene families in mammals. In: *Molecular Evolution of Physiological Processes* (ed. DM Fambrough). Rockefeller University Press, New York. pp. 85–91.

Schirmer RH. (1979) *Principles of Protein Structure*. Springer-Verlag, New York.

Schorderet DF, Gartler SM. (1992) Analysis of CpG suppression in methylated and nonmethylated species. *Proc. Natl Acad. Sci. USA* **89**: 957–961.

Setlow P. (1976) Nearest neighbour frequencies in deoxyribonucleic acids. In: *CRC Handbook of Biochemistry and Molecular Biology*, Vol. 2: *Nucleic Acids*, 3rd Edn (ed. GD Fasman). CRC Press, Cleveland, OH, pp. 312–318.

Shapiro LJ, Yen P, Pomerantz D, Martin E, Rolewic L, Mohandas T. (1989) Molecular studies of deletions at the human steroid sulphatase locus. *Proc. Natl Acad. Sci. USA* **86**: 8477–8481.

Shen S, Slightom JL, Smithies O. (1981) A history of the human fetal globin gene duplication. *Cell* **26**: 191–203.

Shen J-C, Rideout WM, Jones PA. (1992a) High frequency mutagenesis by a DNA methyltransferase. *Cell* **71**: 1073–1080.

Shen J-C, Creighton S, Jones PA, Goodman MF. (1992b) A comparison of the fidelity of copying 5-methylcytosine and cytosine at a defined DNA template site. *Nucleic Acids Res.* **20**: 5119–5125.

Shih C-C, Stoye JP, Coffin JM. (1988) Highly preferred targets for retrovirus integration. *Cell* **53**: 531–537.

Shih A, Coutavas EE, Rush MG. (1991) Evolutionary implications of primate endogenous retroviruses. *Virology* **182**: 495–502.

Skandalis A, Ford BN, Glickman BW. (1994) Strand bias in mutation involving 5-methylcytosine deamination in the human *hprt* gene. *Mutation Res.* **314**: 21–26.

Smit AFA, Tóth G, Riggs AD, Jurka J. (1995) Ancestral, mammalian-wide sub-families of LINE-1 repetitive sequences. *J. Mol. Biol.* **246**: 401–417.

Smith GR. (1983) Chi hotspots of generalized recombination. *Cell* **34**: 709–710.

Steinberg RA, Gorman KB. (1992) Linked spontaneous CG→TA mutations at CpG sites in the gene for protein kinase regulatory subunit. *Mol. Cell. Biol.* **12**: 767–772.

Steinhuber S, Brack M, Hunsmann G, Schwelberger H, Dierich MP, Vogetseder W. (1995) Distribution of human endogenous retrovirus HERV-K genomes in humans and different primates. *Hum. Genet.* **96**: 188–192.

Stoeckert CJ, Collins FS, Weissman S. (1984) Human fetal globin DNA sequences suggest novel conversion event. *Nucleic Acids Res.* **12**: 4469–4479.

Stoppa-Lyonnet D, Duponchel C, Meo T *et al.* (1991) Recombinational biases in the rearranged C1-inhibitor genes of hereditary angioedema patients. *Am. J. Hum. Genet.* **49**: 1055–1062.

Sutherland GR, Richards RI. (1995) Simple tandem DNA repeats and human genetic disease. *Proc. Natl Acad. Sci. USA* **92**: 3636–3641

Sutherland GR, Haan EA, Kremer E, Lynch M, Pritchard M, Yu S, Richards RI. (1991) Hereditary unstable DNA: a new explanation for some old genetic questions? *Lancet* **338**: 289–292.

Takagi Y, Omura T, Go M. (1991) Evolutionary origin of thyroglobulin by duplication of esterase gene. *FEBS Lett.* **282**: 17–22.

Thacker J. (1985) The molecular nature of mutations in cultured mammalian cells: a review. *Mutat. Res.* **150**: 431–442.

Ting CN, Rosenberg MP, Snow CM, Samuelson LC, Meisler MH. (1992) Endogenous retroviral sequences are required for tissue-specific expression of a human salivary amylase gene. *Genes Dev.* **6**: 1457–1465.

Valdez AM, Slatkin M, Freimer NB. (1993) Allele frequencies of microsatellite loci: the stepwise mutation model revisited. *Genetics* **133**: 737–749.

Verkerk AJMH, Pieretti M, Sutcliffe JS *et al.* (1991) Identification of a gene (*FMR1*) containing a CGG repeat coincident with a breakpoint cluster region exhibiting length variation in fragile X syndrome. *Cell* **65**: 905–914.

Vnencak-Jones CL, Phillips JA. (1990) Hot spots for growth hormone gene deletions in homologous regions outside of *Alu* repeats. *Science* **250**: 1745–1748.

Vnencak-Jones CL, Phillips JA, Chen EY, Seeburg PH. (1988) Molecular basis of human growth hormone deletions. *Proc. Natl Acad. Sci. USA* **85**: 5615–5619.

Vogel F, Kopun M. (1977) Higher frequencies of transitions among point mutations. *J. Mol. Evol.* **9**: 159–180.

Vogel F, Motulsky AG. (1986) *Human Genetics: Problems and Approaches*. Springer-Verlag, Berlin.

Vogel F, Röhrborn G. (1965) Mutationsvorgänge bei der Entstehung von Hämoglobinvarianten. *Humangenetik* **1**: 635–650.

Wacey AI, Krawczak M, Kakkar VV, Cooper DN. (1994) Determinants of the factor IX mutational spectrum in haemophilia B: an analysis of missense mutations using a multi-domain molecular model of the activated protein. *Hum. Genet.* **94**: 594–608.

Wada K, Wada Y, Doi H, Ishibashi F, Gojibori T, Ikemura T. (1991) Codon usage tabulated from the GenBank genetic sequence data. *Nucleic Acids Res.* **19** (Suppl.): 1981–1986.

Wapenaar MC, Kievits T, Hart KA *et al.* (1988) A deletion hot spot in the Duchenne muscular dystrophy gene. *Genomics* **2**: 101–108.

Weaver DT, DePamphilis ML. (1982) Specific sequences in native DNA that arrest synthesis by DNA polymerase α. *J. Biol. Chem.* **257**: 2075–2086.

Weber JL. (1990) Informativeness of human $(dC\text{-}dA)_n \cdot (dG\text{-}dT)_n$ polymorphisms. *Genomics* **7**: 524–530.

Wichman HA, Van den Bussche RA, Hamilton MJ, Baker RJ. (1992) Transposable elements and the evolution of genome organization in mammals. *Genetica* **86**: 287–293.

Wilbur JW. (1985) Codon equilibrium II: its use in estimating silent substitution rates. *J. Mol. Evol.* **21**: 182–191.

Willems PJ. (1994) Dynamic mutations hit double figures. *Nature Genetics.* **8**: 213–215.

Wu XW, Muzny DM, Lee CC, Caskey CT. (1992) Two independent mutational events in the loss of urate oxidase during hominoid evolution. *J. Mol. Evol.* **34**: 78–84.

Yeldandi AV, Yeldandi V, Kumar S, Murthy CV, Wang XD, Alvares K, Rao MS, Reddy JK. (1991) Molecular evolution of the urate oxidase-encoding gene in hominoid primates: nonsense mutations. *Gene* **109**: 281–284.

Yen PH, Li X-M, Tsai S-P, Johnson C, Mohandas T, Shapiro LJ. (1990) Frequent deletions of the human X chromosome distal short arm result from recombination between low copy repetitive elements. *Cell* **61**: 603–610.

Yu S, Pritchard M, Kremer E *et al.* (1991) Fragile X genotype characterized by an unstable region of DNA. *Science* **252**: 1179–1181.

Zhu ZB, Jian B, Volanakis JE. (1994) Ancestry of SINE-R.C2, a human-specific retroposon. *Hum. Genet.* **93**: 545–551.

2

Evolution of human proteins by exon-shuffling

L. Patthy

2.1 Introduction

The discovery of split genes and the suggestion that intronic recombination could facilitate the shuffling of exons (Gilbert, 1978) had a great impact on theories of genome evolution. The fascination with the idea of rapid construction of novel genes from parts of old ones led to the formulation of exon-shuffling hypotheses that assigned an all-pervading role to this evolutionary mechanism. The various 'introns early' theories suggested that introns and RNA splicing are the relics of the primordial RNA world, and that the 'genes in pieces' organization of the eukaryotic genome is in fact the primitive original form (Darnell, 1978; Darnell and Doolittle, 1986; Doolittle, 1978; Gilbert, 1986). It was further proposed that all extant genes were constructed from a limited number of exon types by exon-shuffling (Dorit *et al.*, 1990) and that all exons encode protein structural–functional units from which the proteins were assembled (Blake 1978, 1979, 1983). The fact that introns are usually missing from prokaryote genomes was explained by assuming that bacteria eliminated their introns, thereby gaining increased efficiency (Darnell, 1978; Darnell and Doolittle, 1986; Doolittle, 1978).

A different scenario was suggested by various 'introns late' theories. According to these hypotheses, the prokaryotic genes resemble the ancestral ones and the introns were inserted later in genes of eukaryotes (Cavalier-Smith, 1985; Cech, 1985; Crick, 1979; Hickey and Benkel, 1986; Orgel and Crick, 1980; Sharp, 1985). Consequently, intronic recombination and exon-shuffling could not play a major role in the assembly of the most ancient genes.

Human Genome Evolution, edited by M. Jackson, T. Strachan and G. Dover.

In the 1980s the introns early theories of exon-shuffling dominated the field of protein evolution. It has become part of the folklore of molecular biologists that all protein genes (including those that were formed before the prokaryote–eukaryote split) were assembled via intron-mediated recombination from exon modules that code for folding domains or structural elements. In support of this hypothesis, it was frequently claimed that there is a general correlation between protein structure and the exon–intron structure of the genes of the most ancient proteins, such as the various glycolytic enzymes and globins (Blake, 1981; Branden *et al.*, 1984; Craik *et al.*, 1980, 1982; Duester *et al.*, 1986; Go, 1981, 1983, 1987; Ireland *et al.*, 1986; Liaud *et al.*, 1992; Lonberg and Gilbert, 1985; Marchionni and Gilbert, 1986; McKnight *et al.*, 1986; Michelson *et al.*, 1985; Stone *et al.*, 1985a,b; Straus and Gilbert, 1985; Tittiger *et al.*, 1993). The general conclusion from these studies was that exons do indeed correspond to building blocks (such as α helices or β sheets) from which all the genes were assembled by intronic recombination (Gilbert and Glynias, 1993).

From the mid-1980s it has become increasingly difficult to reconcile the introns early hypotheses with emerging information about the origin and evolution of introns and their possible role in the construction of novel genes by exon-shuffling. The major inconsistencies between introns early hypotheses and empirical data are summarized below.

(i) In the case of many ancient protein genes, no obvious correspondence was observed between protein structure and the location of introns (Benyajati *et al.*, 1981; Lee *et al.*, 1987; Quinto *et al.*, 1982; Venta *et al.*, 1985; Weber and Kabsch, 1994), raising serious doubts about a general correlation between intron–exon structure and protein structure (Cornish-Bowden, 1985; Lewin, 1982; Patthy, 1987, 1991a; Rogers, 1990).

(ii) Contrary to the assumptions of 'introns early' hypotheses and in harmony with the 'introns late' hypotheses, it became clear that introns may be inserted into genes. It is incorrect, therefore, to view all introns as primordial assembly points. In the case of serpins, members of the trypsin family, the actin, myosin, tubulin and globin families, intron insertions had to be assumed to provide a plausible explanation for the marked differences in the exon–intron structures of closely related homologous genes (Cavalier-Smith, 1985; Dibb and Newman, 1989; Dixon and Pohajdak, 1992; Leicht *et al.*, 1982; Patthy, 1990; Prochownik *et al.*, 1985; Rogers, 1985; Wright, 1993). In fact, there is now compelling evidence that introns are mostly of recent origin (Palmer and Logsdon, 1991). The actual mechanisms of insertion and propagation of some self-splicing introns are understood in great detail (Belfort 1991, 1993; Dujon, 1989; Lambowitz, 1993; Lambowitz and Belfort, 1989; Morl and Schmelzer, 1990; Mueller *et al.*, 1993; Perlman and Butow, 1989; Schmidt *et al.*, 1994; Thompson and Herrin, 1994). The mechanisms responsible for the insertion of spliceosomal introns are also becoming clear (Belfort, 1993;

Grivell, 1994; Hickey *et al.*, 1989; Patthy, 1994; Rogers, 1989). An important consequence of these findings is that since the exon–intron pattern is the result of intron insertion and removal, the genomic organization of a gene does not necessarily reflect the structure that existed at the time of its formation.

(iii) It has also become clear that introns suitable for exon-shuffling by intronic recombination appeared at a relatively late stage of evolution. As a result, exon-shuffling could not play a major role in the construction of ancient proteins (Patthy, 1987, 1991a,b). The most ancient intron types, the self-splicing introns, are unlikely to contribute to exon-shuffling by intronic recombination since they encode an essential function that makes their sequence intolerant to intronic recombination (Patthy, 1987, 1991a,b, 1994).

Exon-shuffling could have become significant, however, when spliceosomal introns evolved. Only a tiny portion of these introns is essential for splicing, creating the potential for intronic recombination without any loss of splice function. The short conserved segments essential for splicing (the 5′ and 3′ splice sites and the branch site) are found in the vicinity of the exon–intron boundaries. These are usually separated by very long DNA tracts that account for most of the intron and that are dispensable for splicing. Thus, the majority of the intron is 'junk' and is highly tolerant to deletions and insertions. Middle repetitive sequences (such as *Alu* repeats) may be inserted in these regions of introns and these may increase the chances of misalignment and recombination of non-homologous introns.

Another important feature of pre-mRNA introns is that splice sites of different introns are equivalent and interchangeable (cf. alternative splicing). This is essential for exon-shuffling since recombination in spliceosomal introns leads to the formation of a chimeric intron with 5′ and 3′ splice sites derived from different, non-orthologous, introns. As might be evident from studies on abnormal low-density lipoprotein (LDL) receptor genes (Section 2.4), such chimeric introns are usually spliced as efficiently as their two progenitors.

Thus, the very large size of vertebrate spliceosomal introns, the presence of middle repetitive sequences in such introns, as well as their tolerance to the structural changes that accompany recombination, make them ideal for intronic recombination, and exon-shuffling. This is not true for self-splicing introns. Since these introns carry out their own excision, they have to conserve a unique, catalytically active three-dimensional structure that makes them sensitive to structural changes. For these introns, a recombination event involving a non-orthologous intron would be likely to yield a chimera which lacks some of the essential catalytic elements and which would, therefore, be deficient in splicing activity (Patthy, 1987, 1991a,b).

Since spliceosomal introns evolved relatively recently from group II self-splicing introns (Cavalier-Smith, 1991; Cech, 1986; Copertino and Hallick, 1993; Jacquier, 1990; Palmer and Logsdon, 1991; Saldanha *et al.*, 1993; Sharp, 1994), exon-shuffling could play a major role only in the construction of proteins which have evolved after the spliceosomal intron (Patthy, 1987, 1991a,b, 1994).

(iv) In view of the above, it is not surprising that the unquestionable cases of exon-shuffling are apparently restricted to proteins unique to higher eukaryotes and which are, therefore, assumed to have evolved relatively recently (Patthy, 1991a,b, 1994). Analysis of the evolutionary distribution of proteins assembled from modules by intronic recombination now suggests that exon-shuffling became significant at the time of the appearance of the first multicellular animals and could have contributed to the explosive nature of metazoan radiation (Patthy, 1994).

A search for modules and modular proteins has brought many examples from all major groups of metazoa (Patthy, 1994). The fact that proteins composed of modules familiar from vertebrate genes have been found in Hydrozoa, nematodes, molluscs, arthropods and echinoderms indicates that the machinery of exon-shuffling was available before their divergence. It is noteworthy that although proteins assembled from modules have already been found in all major phyla of the animal kingdom, there is practically no evidence for related modular proteins in plants or fungi. It must be emphasized that the spliceosomal introns of plants seem to be less suitable for intronic recombination than those of vertebrates. Unlike vertebrate introns, the spliceosomal introns of plants are recognized *in toto* rather than as the simple assembly of splice junctions and a branch point. As a result, they are not as tolerant of structural changes (Patthy, 1994).

In summary, studies on the evolution of introns, evolution of modules and evolutionary distribution of modular proteins all suggest that exon-shuffling is a relatively late development.

(v) Analysis of established cases of modular protein evolution has shown that only a special group of exons (exon-sets) are really valuable for exon-shuffling (Patthy, 1987, 1991a,b, 1994). As will be discussed in Section 2.2, the splice junctions of the shuffled exon have to be 'in phase' with those of its new neighbours, otherwise a shift in the reading frame would occur that would obliterate the protein information of the shuffled exon as well as that of the downstream exons of the recipient gene (Patthy, 1987). Consequently, the gene structures of proteins created by exon-shuffling are characterized not only by showing a correlation between the domain structure of the proteins and location of the introns but also by a characteristic intron-phase distribution (Section 2.4; Patthy, 1987, 1991a,b, 1994). If these criteria are applied to the exon–intron structures of the genes of ancient proteins, it is clear that their hypothetical 'modules' do not conform to these rules of exon-shuffling (Patthy 1987, 1991a).

As a result of the developments summarized above, the introns late hypothesis is gaining increasing acceptance; indeed, one of the chief proponents of the introns early theory has now also come to the conclusion that extension of the exon-shuffling hypothesis to the most evolutionarily ancient genes is untenable (Stoltzfus *et al.*, 1994).

In contrast with the case of evolutionarily ancient protein genes, there is now solid evidence for a major role of exon-shuffling in the modular assembly of many proteins unique to higher eukaryotes which are assumed to have evolved relatively recently. The first piece of strong evidence for exon-shuffling came from studies on proteases of blood coagulation and fibrinolysis (Bányai *et al.*, 1983; Ny *et al.*, 1984; Patthy, 1985) In the past decade, however, it has become clear that this evolutionary mechanism has been widely used in the creation of a variety of multi-domain proteins.

This chapter will survey only the unquestionable cases of exon-shuffling, the principles and mechanism of this process, and will assess its significance in the evolution of the human genome. The emphasis will be on human proteins but, as will be clear from the evolutionary history of these proteins, in most cases the exon-shuffling events leading to the construction of these modular proteins occurred long before the appearance of the human race. When discussing the evolution of human modular proteins, comparison with orthologues from other animal species will thus be inevitable.

2.2 Modular proteins produced by exon-shuffling

Exon-shuffling played a major role in the construction of various proteases of the fibrinolytic and blood coagulation cascades (*Table 2.1*). These enzymes possess a catalytic domain homologous with trypsin-like serine proteases. However, unlike the simple digestive enzymes, they have very large 'non-catalytic' segments. The non-protease parts of the different blood coagulation and fibrinolytic enzymes display a remarkable variation in function, size and structure and they are organized into distinct structural–functional and folding domains, such as kringles, vitamin-K-dependent calcium-binding domains and domains related to epidermal growth factor (EGF) (*Figure 2.1*). The structural–functional autonomy of these domains suggested that they may have been joined to the protease domain independently to build the non-catalytic regions of these regulatory proteases. The first evidence supporting this hypothesis was our finding that a region of tissue plasminogen activator (t-PA) is homologous with the finger-domains of fibronectin, a constituent of the extracellular matrix that is otherwise unrelated to serine proteases (Bányai *et al.*, 1983). Furthermore, another region of t-PA was shown to be homologous with EGF (Bányai *et al.*, 1983), a small molecule derived from a large precursor unrelated to proteases. On the basis of these observations, the non-protease parts of regulatory proteases of blood coagulation and fibrinolysis and a variety of other proteins were proposed to have been assembled from a

▶ p. 43

Table 2.1. Vertebrate proteins assembled from modules by exon-shuffling

Constituents of the extracellular matrix		
Laminin A[2]	Restrictin[14]	Type VI collagen α_1 and α_2 chains[20]
Merosin[3]	Nidogen/entactin[15]	Type VI collagen α_3 chain[21]
Epiligrin[4]	Cartilage link protein	Type VII collagen[22]
Laminin B1	Cartilage matrix protein	Type IX collagen[23]
Laminin B2[5]	Cartilage oligomeric matrix protein[16]	Type V, XI collagen[23]
Laminin B3[6]	Osteonectin	Type XII collagen[24]
Laminin B1k[7]	Undulin[17]	Type XIV collagen[25]
Perlecan[8]	Submaxillary mucins	Surfactant protein A and D, conglutinin, mannose binding protein[26]
Agrin[9]	Intestinal mucin MUC-2[18]	Fibulin[27]
Fibronectin	F-spondin[19]	Fibrillin[28]
Aggrecan[10]	Vitronectin	
Versican[11]	Type I procollagen[4]	
Neurocan[12]		
Tenascin/hexabrachion[13]		
Serine proteases involved in blood coagulation, fibrinolysis and complement activation		
Prothrombin[29]	Urokinase	Complement factors C2
Factor VII	Tissue plasminogen activator	Complement factor I[31]
Factor IX	Factor XII	Plasma prekallikrein[32]
Factor X	Complement factors C1r	Factor XI
Protein C	Complement factors C1s	
Plasminogen[30]	Complement factors B	
Diverse members of the trypsin family of serine proteases		
Apolipoprotein(a)[33]	Hepatocyte growth factor-related proteins[35]	Haptoglobin
Apolipoprotein(a)-related proteins[34]	Hepatocyte growth factor activator[36]	Enterokinase[37]
Proteins involved in the regulation of blood coagulation		
Factors V and VIII[38]	Thrombospondin 2[40]	Lipoprotein-associated coagulation inhibitor[43]
Factor XIII b chain	Thrombospondin 3[41]	
Protein S	Thrombospondin 4[42]	
Thrombospondin 1[39]	Von Willebrand factor	
Proteins involved in the regulation of the complement system		
Complement factor C6[44]	Complement factor C9	Decay-accelerating factor
Complement factor C7	Complement factor H and related proteins[45]	Membrane cofactor protein
Complement factors C8α and C8β	C4b-binding proteins[46]	Properdin[47]
Metalloproteases		
Type IV collagenases[48]	Bone morphogenetic protein 1	Cysteine-rich metalloproteases[50]
Meprin α and β[49]		
Diverse receptors, cell adhesion proteins, membrane-associated proteins		
Cytokine receptors[51]	Antigen-processing receptor DEC-205[60]	Lymph node homing receptor[67]
TGF-β type III receptor[52]	Insulin-like growth factor II receptor[61]	Thrombomodulin
Urokinase receptor[53]	Interleukin-2-receptor α chain	Contactin, N-CAM, DCC protein and related proteins[68]
LDL-receptor	Netrins[62]	Uromodulin[52]
LDL-receptor-related proteins[54]	Neurexins[63]	Zona pellucida proteins Zp2, Zp3[52,69]
VLDL-receptor[55]	Integrins Mac-1, LFA-1, p150,95, VLA-2[64]	Zymogen granule membrane protein GP-2[52,70]
Scavenger receptor[56]	Integrin β4	Kallman syndrome protein[71]
Scavenger-receptor-related proteins[57]	Endothelial leukocyte adhesion molecule 1[65]	A4 amyloid protein precursor[72]
Complement receptors CR1, CR2	Granule membrane protein of platelets[66]	ARIA/heregulin/Neu differentiation factor[73]
Mannose receptor[58]		Carcinoma-associated antigens GA733-1, GA733-2[74]
Phospholipase A2 receptor[59]		

Table 2.1. Vertebrate proteins assembled from modules by exon-shuffling – *continued*

Diverse receptors, cell adhesion proteins, membrane-associated proteins – continued		
P14 protein[75] TSG6 protein[76] A5 antigen[77] Epidermal growth factor precursor	*Notch* homologs, *Xotch*, *Tan*-1[78] Cell surface antigen 114/A10[79] Fetal antigen 1[80]	Pref-1 protein[81] Polycystic kidney disease protein PKD1[82]
Diverse receptor protein tyrosine phosphatases		
R-PTP-κ[83] R-PTP-P1 and P2[84] R-PTP-NE3[85]	R-PTP-σ[86] SAP1[87] LAR protein[88]	GLEPP1 protein[89]
Diverse receptor protein tyrosine kinases		
DDR, Tyro 10, Ptk-3 receptor tyrosine kinases[90] *ret* proto-oncogene[91] PDGF, FGF, KGF receptors[92] FLT, NYK/FLK-1 receptor protein tyrosine kinases[93] Neurotrophin receptors[94]	Ror receptor protein[95,96] tyrosine kinases[96] Insulin receptor[97] IGF-I receptor[98] Axl/Ufo, Sky, Tyro 3 receptor protein tyrosine kinases[99]	Eph/elk family of receptor protein tyrosine kinases[100] hyk, tek, tie family of receptor protein tyrosine kinases[101]
Miscellaneous		
Immunoglobulins Apolipoprotein H (β_2-glycoprotein) Thyroid peroxidase Prostaglandin endoperoxidase synthase α_2 subunit of DHP-sensitive calcium channel protein[102] Plasma cell membrane glycoprotein PC-1 Pregnancy-associated protein A[103] Anaphylatoxins C3a, C4a, C5a[104] Inter-α-trypsin inhibitor light chain	Inter-α-trypsin inhibitor heavy chain[102] Thyroglobulin Insulin-like growth factor binding proteins cyr61/fisp-12 family of growth factors[105] Leukocyte protease inhibitor Pancreatic spasmolytic polypeptide Stomach hSP protein Intestinal sucrase-isomaltase α-Glucosidase Placental protein 11[106] Follistatin Follistatin-related protein[107]	Testican[108] Whey proteins Seminal fluid proteins PDC-109, BSPA3[109] TGF-β1 binding protein Milk fat globule membrane protein[110] Ia antigen-associated invariant chain (Ii) Mac-2 binding protein[111] Muscle proteins titin, C-protein, M-protein, skelemin[112]

Only references not included in an earlier list (ref. 1; Patthy, 1991b) are indicated.
[2]Nissinen *et al.* (1991), [3]Vuolteenaho *et al.* (1994), [4]Ryan *et al.* (1994), [5]Kallunki *et al.* (1991), [6]Pulkkinen *et al.* (1995), [7]Gerecke *et al.* (1994), [8]Cohen *et al.* (1993), Kallunki and Tryggvason, (1992), Murdoch *et al.* (1992), Noonan *et al.* (1991), [9]Rupp *et al.* (1991, 1992), Smith *et al.* (1992), Tsim *et al.* (1992), [10]Doege *et al.* (1991, 1994), [11]Naso *et al.* (1994), [12]Rauch *et al.* (1992), [13]Bristow *et al.* (1993), Gulcher *et al.* (1991), Matsumoto *et al.* (1992a,b), Nies *et al.* (1991), [14]Norenberg *et al.* (1992), [15]Durkin *et al.* (1995), [16]Oldberg *et al.* (1992), Newton *et al.* (1994), [17]Just *et al.* (1991), [18]Gum *et al.* (1994), [19]Klar *et al.* (1992), [20]Hayman *et al.* (1991), Saitta *et al.* (1992), Walchli *et al.* (1992), [21]Stokes *et al.* (1991), Zanussi *et al.* (1992), [22]Christiano *et al.* (1994), [23]Bork (1992), van der Rest and Bruckner (1993), [24]Yamagata *et al.* (1991), [25]Gerecke *et al.* (1993), Trueb and Trueb, (1992), Walchli *et al.* (1993), [26]Crouch *et al.* (1993), Kawasaki *et al.* (1994), [27]Pan *et al.* (1993b), [28]Corson *et al.* (1993), Dietz *et al.* (1991), Lee *et al.* (1991), Maslin *et al.* (1991), Yin *et al.* (1995), Zhang *et al.* (1994), [29]Banfield *et al.* (1994), [30]Ichinose (1992), [31]Kunnath-Muglia *et al.* (1993), Vyse *et al.* (1994), [32]Beaubien *et al.* (1991), [33]Ichinose (1995), [34]Byrne *et al.* (1994), (1995), [35]Han *et al.* (1991), Miyazawa *et al.* (1991), [36]Miyazawa *et al.* (1993), [37]Kitamoto *et al.* (1994, 1995), Matsushima *et al.* (1994), [38]Cripe *et al.* (1992), [39]Lawler *et al.* (1991, 1995), [40]Shingu and Bornstein, (1993), [41]Bornstein *et al.* (1993), Vos *et al.* (1992), [42]Lawler *et al.* (1993, 1995), [43]van der Logt *et al.* (1991), [44]Hobart *et al.* (1993), [45]Cooper and Attie, (1992), Skerka *et al.* (1991), (1993), [46]Aso *et al.* (1991), Hillarp *et al.* (1993), Moffat *et al.* (1992), Sanchez-Corral *et al.* (1993), [47]Nolan *et al.* (1992), [48]Aimes *et al.* (1994), Collier *et al.* (1991), Fini *et al.* (1994), Huhtala *et al.* (1991), Marti *et al.* (1993), Masure *et al.* (1993), Tanaka *et al.* (1993), [49]Gorbea *et al.* (1993), Jiang *et al.* (1992), [50]Perry *et al.* (1994), Weskamp and Blobel, (1994), Wolfsberg *et al.* (1993), [51]Imamura *et al.* (1994), Kosugi *et al.* (1995), Nakagawa *et al.*

(continued overleaf)

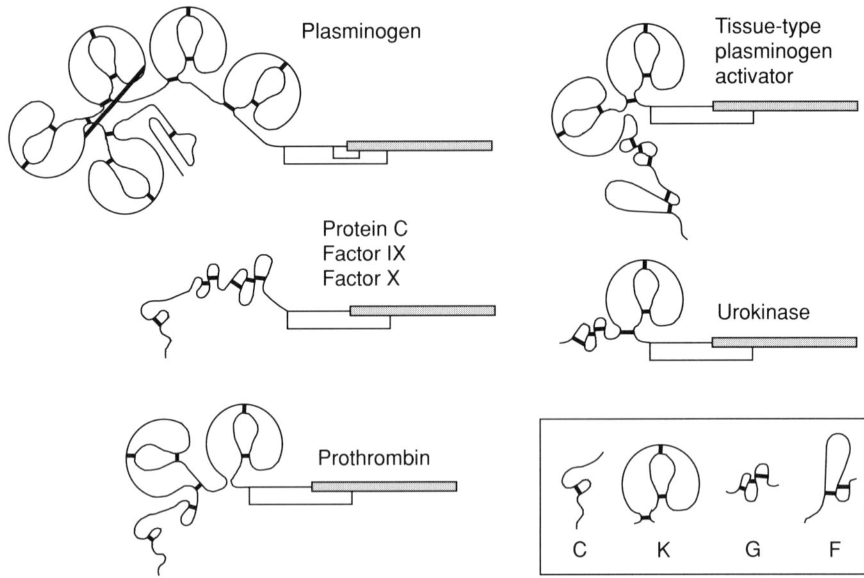

Figure 2.1. Modular architecture of the proteases of blood coagulation and fibrinolysis. The bars represent the trypsin homologue serine protease domains. Inset: a key showing representations of the vitamin-K-dependent calcium-binding module (C), kringle module (K), growth factor module (G) and the finger module (F), from which these proteases were assembled. Reproduced from Patthy (1985) with permission from Cell Press.

(1994), Pleiman *et al.* (1991), Tuypens *et al.* (1992), [52]Bork and Sander (1992), [53]Behrendt *et al.* (1991), Casey *et al.* (1994), Suh *et al.* (1994), Wang *et al.* (1995), [54]Mehta *et al.* (1991), Nimpf *et al.* (1994), Saito *et al.* (1994), [55]Sakai *et al.* (1994), [56]Emi *et al.* (1993), [57]Elomaa *et al.* (1995), [58]Kim *et al.* (1992), [59]Ancian *et al.* (1995), Ishizaki *et al.* (1994), Lambeau *et al.* (1994), [60]Jiang *et al.* (1995), [61]Szebenyi and Rotwein, (1994), [62]Serafini *et al.* (1994), [63]Ushkaryov *et al.* (1992), [64]Fleming *et al.* (1993), [65]Collins *et al.* (1991), Larigan *et al.* (1992), [66]Auchampach *et al.* (1994), Strubel *et al.* (1993), [67]Dowbenko *et al.* (1991), [68]Cho *et al.* (1994), Connelly *et al.* (1994), Davis *et al.* (1993), Grumet *et al.* (1991), Hasler *et al.* (1993), Kayyem *et al.* (1992), Volkmer *et al.* (1992), Yoshihara *et al.* (1994), Zuellig *et al.* (1992), [69]Bork (1993), [70]Hoops and Rindler (1991), [71]del Castillo *et al.* (1992), Franco *et al.* (1991), Legouis *et al.* (1991), [72]Okado and Okamoto, (1992), [73]Falls *et al.* (1993), Holmes *et al.* (1992), Wen *et al.* (1992), [74]Linnenbach *et al.* (1993), [75]Lecain *et al.* (1991), [76]Lee *et al.* (1992), [77]Beckman and Bork (1993), Takagi *et al.* (1991), [78]Ellisen *et al.* (1991), Weinmaster *et al.* (1991), [79]Dougherty *et al.* (1989), [80]Jensen *et al.* (1994), [81]Smas *et al.* (1994), [82]Glücksmann-Kuis *et al.* (1995), [83]Jiang *et al.* (1993), [84]Pan *et al.* (1993b), [85]Walton *et al.* (1993), [86]Yan *et al.* (1993), [87]Matozaki *et al.* (1994), [88]O'Grady *et al.* (1994), Sreuli *et al.* (1992), [89]Thomas *et al.* (1994), [90]Alves *et al.* (1995), Johnson *et al.* (1993), Lai and Lemke (1994), Sanchez *et al.* (1994), [91]Iwamoto *et al.* (1993), Kwok *et al.* (1993), [92]Miki *et al.* (1991), [93]Oelrichs *et al.* (1993), [94]Lamballe *et al.* (1991), Schneider and Schweiger (1991), [95]Masiakowski and Carroll (1992), [96]Jennings *et al.* (1993), [97]Schaefer *et al.* (1992), [98]Abbott *et al.* (1992), [99]Lai *et al.* (1994), Mark *et al.* (1994), Ohashi *et al.* (1994), Schulz *et al.* (1993), [100]Lindberg and Hunter (1990), Maisonpierre *et al.* (1993), [101]Dumont *et al.* (1993), Partanen *et al.* (1992), Ziegler *et al.* (1993), [102]Bork and Rohde, (1991), [103]Kristensen *et al.* (1994), [104]Pan *et al.* (1994), [105]Bork (1993b), [106]Jenne, (1991), [107]Zwijsen *et al.* (1994), [108]Alliel *et al.* (1993), [109]Brauer and Scheit (1991), [110]Larocca *et al.* (1991), [111]Koths *et al.* (1993), [112]Noguchi *et al.* (1992), Price and Gomer (1993), Tan *et al.* (1993).

common pool of modules (Patthy, 1985). This hypothesis assumed that the common ancestor of regulatory proteases was similar to the simple trypsin-like enzymes and that all regulatory modules were inserted between the signal-peptide and zymogen-activation domain of the ancestral protease (Patthy, 1985, 1990).

In agreement with the predictions of this hypothesis, the complex regulatory proteases of the trypsin family all turned out to be mosaics assembled from modules found in a variety of other proteins (*Table 2.1*). These proteases include not only the enzymes of the blood coagulation, fibrinolytic and complement cascades (Patthy, 1993) but also enterokinase, the initiator of the intestinal digestion cascade (Kitamoto *et al.*, 1994, 1995; Matsushima *et al.*, 1994).

Our modular assembly hypothesis implied that the gene pieces encoding the modules of plasma proteases display an unusual mobility: they undergo repeated insertions, tandem duplications, exchanges and deletions during evolution (Patthy, 1985). We have assumed that exon-shuffling is the mechanism that permits such unusual mobility of these protein domains (Bányai *et al.*, 1983; Patthy, 1985). Our prediction that introns should be found at the boundaries of these mobile modules was borne out by subsequent studies on the gene for t-PA (Ny *et al.*, 1984). The correlation between the mosaic structure of t-PA protein and the exon–intron organization of its gene thus provided the first solid evidence for a role of introns in exon-shuffling.

It soon became clear that modular protein evolution by exon-shuffling is a quite widely used evolutionary strategy: a rapidly growing number of proteins and genes turned out to have evolved by this mechanism. *Table 2.1* lists most of the modular proteins produced by exon-shuffling that have been identified in humans and other vertebrates. A survey of the list of modular proteins reveals that they are mostly extracellular (proteins of body fluids, constituents of the extracellular matrix, extracellular parts of transmembrane proteins, receptor proteins); their modules (*Table 2.2*) are usually rich in disulphide bonds (e.g. kringles, EGF modules, complement B-type modules). There are a few intracellular muscle proteins assembled from immunoglobulin and fibronectin type III. However, a common feature of all the proteins listed in *Table 2.1* is that they are apparently restricted to the animal kingdom with no counterparts in prokaryotes, plants or fungi (Patthy, 1994).

2.3 Principles of modular exchange by exon-shuffling

Analysis of the gene structures of the first modular proteins to be identified (such as prothrombin, factor IX, factor X, protein C, plasminogen, urokinase, t-PA, interleukin-2-receptor α chain, haptoglobin, fibronectin and the LDL receptor) revealed some unique features of the introns that participated in the shuffling of modules (Patthy, 1987). The introns demarcating both the 5′ and 3′ boundaries of the kringle, growth factor, finger, fibronectin type II,

Table 2.2. List of class 1-1 modules used most frequently in exon-shuffling

Abbreviation	Module names
B	Complement B-type module, SCR or short consensus repeat, sushi domain, CR or complement protein regulatory repeat
C	Vitamin-K-dependent calcium-binding module, GLA domain, γ-carboxy glutamate domain
CF	Contact factor module, apple domain, factor XI domain
CYT	Cytokine-receptor module
C1r	Complement C1r module, CUB module
C7	Complement factor 7 module
DISC	Discoidin lectin module, factor V/VIII type C domain
G	Epidermal growth factor module, EGF-like domain
F	Finger module, fibronectin type I repeat
FIB	Fibulin module, anaphylatoxin module
FN2	Fibronectin type II module
FN3	Fibronectin type III module
FS	Follistatin module
Ig	Immunoglobulin module
INH	Kunitz-type trypsin inhibitor module
K	Kringle module
LDL	LDL receptor module
LK	Link protein module, PTR or proteoglycan tandem repeat, hyaluronan-binding domain
LMA	Laminin A module, G domain of laminin A
LMB	Laminin B module, EGF-like domain of laminins
LMC	Laminin C module
LMD	Laminin D module, laminin VI domain
LN	C-type lectin module, CRD or carbohydrate-recognition domain
M	Meprin module, A5 module, MAM module
NT	Notch protein module
PP	Pancreatic spasmolytic polypeptide module, P-domain, trefoil domain
SC	Scavenger receptor module
SLT	Slit protein module
TBP	TGF-β1-binding protein module, fibrillin module
THY	Thyroglobulin module
TSP	Thrombospondin module, type 1 repeat of thrombospondin
UM	Uromodulin module, zona pellucida protein domain
UPAR	Urokinase receptor module, Ly6 domain
VN	Vitronectin module, somatomedin B domain
vWA	A-type module of von Willebrand factor
vWB	B-type module of von Willebrand factor
vWC	C-type module of von Willebrand factor
vWD	D-type module of von Willebrand factor
WH	Whey protein module, WAP domain, 4-disulphide core domain

fibronectin type III, complement B and LDL receptor modules always split the reading frame in phase 1; that is, between the first and second nucleotides of a codon (Patthy, 1987).

The significance of the observation that introns of identical phase are found at both the 5′ and 3′ boundaries of the various modules lies in the fact that each of the basic steps of module-shuffling (insertion, deletion, duplication of modules by intronic recombination) joins the 5′ splice junction of one intron to the 3′ splice junction of another intron. It is easy to see that the two intron partners involved in recombination must belong to the same phase class; otherwise, recombination would shift the reading frame and would lead to loss of protein information downstream of the recombination point (Patthy, 1987). Insertion, deletion and duplication of a module by intronic recombination could satisfy this phase-compatibility requirement only if the module is 'symmetrical' in the sense that the two introns flanking the module are of the same phase (*Figure 2.2a*). When 'non-symmetrical' modules (i.e. when the introns at the 5′ and 3′ boundaries of the module are of different phase) are shuffled by intronic recombination, this phase-compatibility rule will be violated, resulting in a shift of the reading frame (*Figure 2.2b*). In other words, only the three symmetrical module groups (class 1-1, class 0-0 and class 2-2,

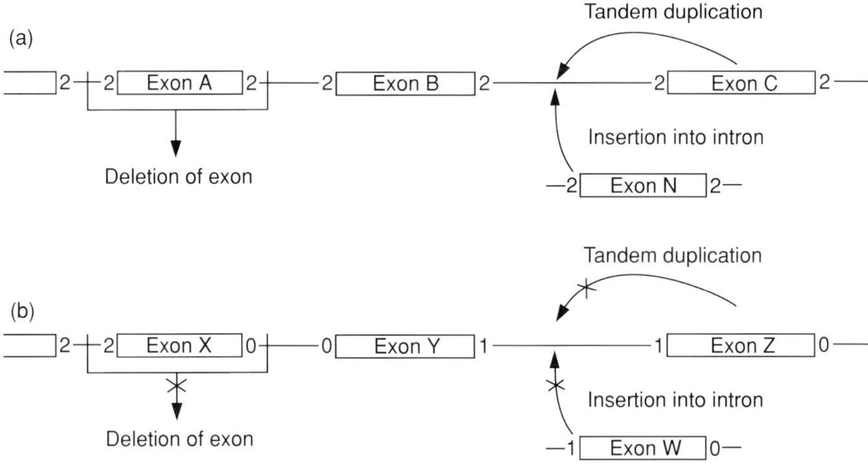

Figure 2.2. Intron phase and exon-shuffling. The boxes represent exons, the thin lines show the introns and the numbers indicate the phase class of the 5' and 3' splice junctions of the introns. It should be noted that deletion, insertion and duplication of modules by intronic recombination joins the 5' splice junction of one intron to the 3' splice junction of another intron. If the module is 'symmetrical' (i.e. when the introns at the 5' and 3' boundaries of the module are of identical phase), and the recipient also belongs to the same phase class, then the reading frame is not affected (a). However, deletion, duplication or insertion of 'non-symmetrical' exons (i.e. when the introns at the 5' and 3' boundaries of the exon are of different phase) will disrupt the reading frame (b). Reproduced from Patthy (1987) with permission from Cell Press.

depending on whether phase 1, phase 0 or phase 2 introns are found at both boundaries of the modules) are really useful in exon-shuffling; non-symmetrical exons or exon-sets (such as class 0-1, class 0-2, class 1-2 or class 1-0) are practically useless in modular protein evolution. [It must be mentioned that the vast majority of the proteins reviewed in this chapter have been constructed from class 1-1 modules; much fewer class 0-0 and class 2-2 modules are known. The reasons for the predominance of class 1-1 modules have been discussed by Patthy (1994)].

The various modules used in the construction of the proteases of blood coagulation, fibrinolysis and complement activation all turned out to belong to the class 1-1 group (Patthy, 1990, 1991b, 1993, 1994). On the basis of the phase-compatibility rule, this observation implies that these modules required a phase 1 target intron for their insertion into the ancestral protease gene. In our hypothesis for the evolution of plasma proteases (Patthy, 1985, 1987), we have assumed that this target intron was located at the boundary separating the signal-peptide domain from the catalytic domain. Analysis of the gene structures of trypsin-type proteases has added strength to our prediction that a phase 1 intron would have been present between the signal-peptide and catalytic domains of the ancestral protease (Patthy 1987, 1990). During the course of the assembly process, insertions and duplications of, for example, class 1-1 growth factor, kringle, finger and fibronectin type II modules, complement B and von Willebrand factor type A modules divided, duplicated and thus proliferated phase 1 introns. This eventually led to gene structures in which all intermodule introns belong to the phase 1 class (*Figure 2.3*). The dominance of a single intron phase class in the non-catalytic chains of plasma proteases and related proteins is thus a necessary consequence of their evolution by exon-shuffling. In other words, exon-shuffling is reflected not only in a correlation between the domain organization of the protein and exon–intron organization of its gene but also in a characteristic phase distribution of intermodule introns (Patthy, 1987). The fact that the structure of the gene of a modular protein conforms to these rules may actually be used as evidence that it has evolved by exon-shuffling (Patthy, 1988). The genes for factor XIII b chain, interleukin-2 receptor α chain, selectins, lipoprotein-associated coagulation inhibitor, cartilage link protein, fibronectin, factor H and tenascin provide typical examples of such a correlation: the C-type lectin, growth factor and complement B-type modules, finger modules, fibronectin type II and III modules and Kunitz inhibitor modules are usually encoded by discrete class 1-1 exons or class 1-1 exon-sets; that is, the intermodule introns are phase 1 (*Figure 2.4*).

The general validity of these rules was supported by studies on the genes of numerous modular proteins (Patthy, 1991b, 1994). Some more recent examples include the genes of complement factor I (Vyse *et al.*, 1994), apolipoprotein(a) (Ichinose, 1995), very low-density lipoprotein (VLDL) receptor (Sakai *et al.*, 1994), entactin (Durkin *et al.*, 1995), type VII collagen (Christiano *et al.*, 1994), aggrecan (Doege *et al.*, 1994; Shinomura *et al.*,

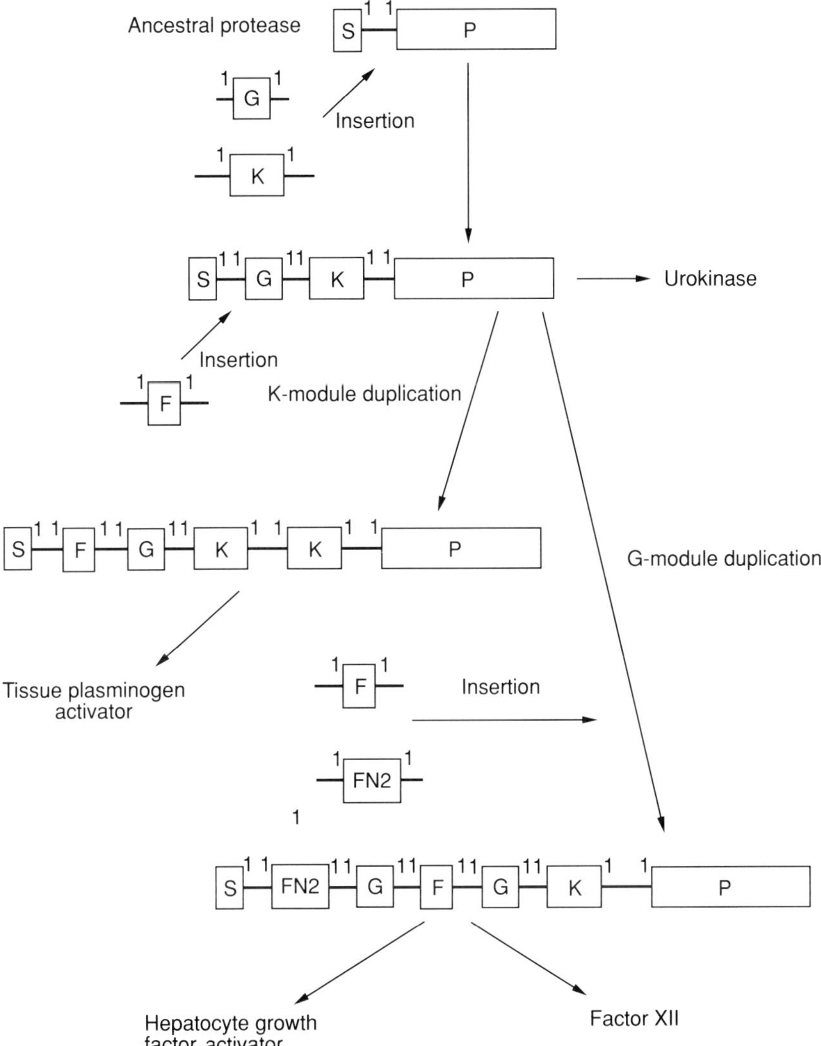

Figure 2.3. Formation of the ancestors of urokinase, tissue plasminogen activator, factor XII and hepatocyte growth factor activator by acquisition of class 1-1 modules. The numbers indicate the phase of the 5' and 3' splice junctions of the introns participating in the assembly process (for simplicity, the introns present in the protease domains and in the kringle modules are not shown.) It should be noted that all module insertions occurred in a phase 1 intron between the signal-peptide domain (S) and the trypsin-homologue domain (P) of the ancestral trypsin-type protease. Since insertions and duplications of class 1-1 growth factor (G), kringle (K), finger (F), and fibronectin type II (FN2) modules divide and duplicate phase 1 introns, modular assembly necessarily resulted in gene structures in which all intermodule introns are phase 1.

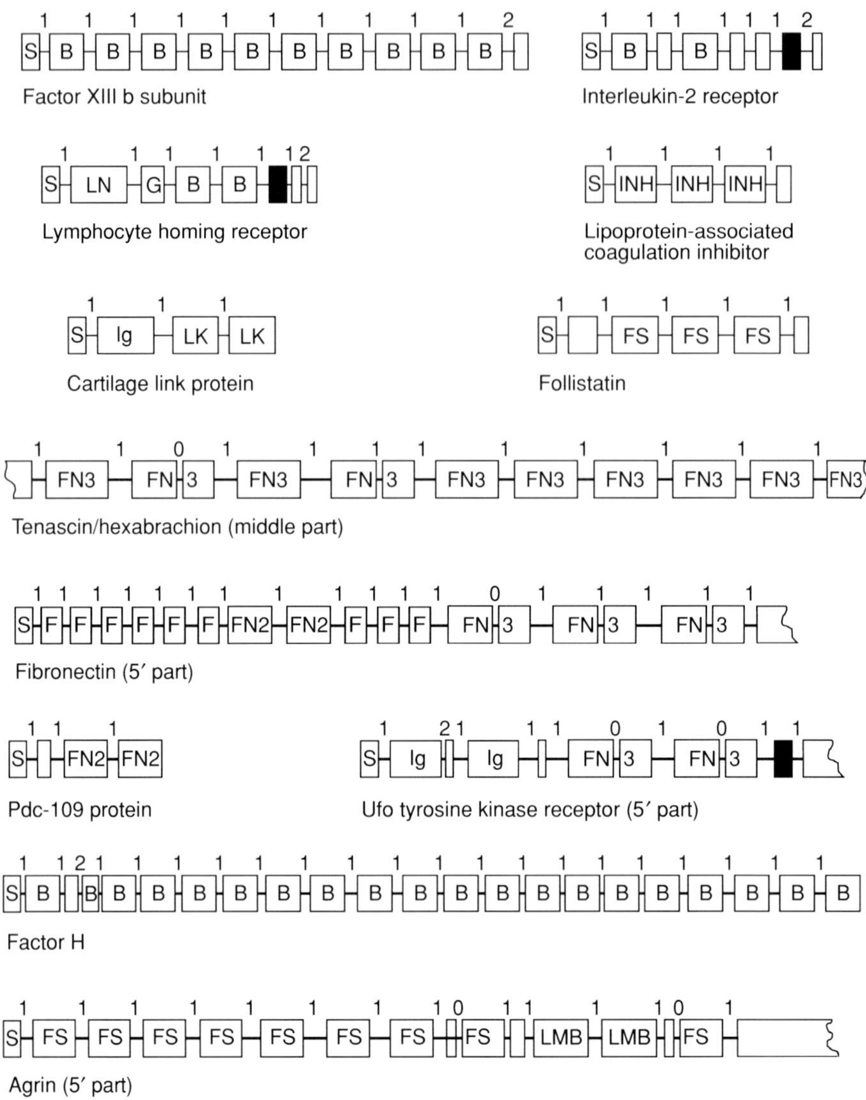

Figure 2.4. Some gene structures still reflect their original assembly from class 1-1 protein modules. Modules shown are: complement B-type (B), C-type lectin (LN), fibronectin type II (FN2), fibronectin type III (FN3), finger (F), follistatin (FS), growth factor (G), immunoglobulin (Ig), Kunitz-type inhibitor modules (INH), laminin B (LMB) and link protein (LK) modules. Note that introns separating these modules from each other or from the signal-peptide (S) domain are phase 1. Also note that some modules are split by internal introns (e.g. FN3 modules and one of the B modules of factor H). The black boxes represent exons encoding transmembrane domains of receptor proteins.

1995) and urokinase receptor (Casey *et al.*, 1994; Suh *et al.*, 1994; Wang *et al.*, 1995).

It must be pointed out that there are cases where the gene structure does not correlate with the modular organization of a protein but where the 'expected' introns are missing from the module boundaries. For example, in the gene for thrombomodulin, the C-type lectin module and the six EGF-like modules are not flanked by introns (Jackman *et al.*, 1987). Similarly, there is practically no correlation between the module organization and exon–intron structure in the case of laminins (Kallunki *et al.*, 1991; Vuolteenaho *et al.*, 1990). Introns are absent from the boundaries separating the numerous class 1-1 EGF modules of the *Notch* protein gene of *Drosophila* and the *lin-12* gene of *Caernorhabditis elegans* (Greenwald, 1985; Kidd *et al.*, 1986; Wharton *et al.*, 1985; Yochem *et al.*, 1988). In the case of the complement C6 gene, introns are missing from the boundaries of its class 1-1 thrombospondin, LDL receptor, EGF and C7-related modules (Hobart *et al.*, 1993). Since these proteins are composed of modules that have been clearly shown in many other cases to be duplicated and inserted into new locations by recombination in phase 1 introns, it may be assumed that they also arose by exon-shuffling, even though the 'original' introns are already missing from their genes.

The most plausible explanation for such a lack of correlation is that the original exon–intron organization of these modular proteins has been obscured by removal and insertion of introns (Patthy, 1994). Since erosion of the original genomic structure progresses with time, it is not surprising that evolutionarily ancient genes usually show less perfect correlation with protein structure than recently assembled ones. This point may be illustrated by the exon–intron structures of laminin genes. Laminins are believed to be among the oldest modular proteins of Metazoa since they are already present in Hydrozoa (Sarras *et al.*, 1994) and their domain organization has remained practically unchanged during subsequent evolution. Consistent with the great age of these genes, most of the original phase 1 introns are already missing from the boundaries of the laminin B-type modules of *Drosophila* and vertebrate laminin B1 and B2 chain genes (Chi *et al.*, 1991; Gow *et al.*, 1993; Kallunki *et al.*, 1991; Vuolteenaho *et al.*, 1990). However, in the case of the *Drosophila* laminin A chain, phase 1 introns are still present at the boundaries of six of its laminin B-type modules (Kusche-Gullberg *et al.*, 1992; MacKrell *et al.*, 1993). Significantly, these laminin B modules are present only in the *Drosophila* protein but are absent from the vertebrate homologue, suggesting that they are the products of more recent duplications; that is, these 'younger' introns had a better chance of survival.

Another interesting example comes from studies on the evolutionary history of genes related to that encoding the LDL receptor. Vertebrate LDL-receptor-related proteins (such as LRP and megalin) and the LDL-receptor-related protein of *C. elegans* have practically identical arrangements of EGF and LDL receptor modules. The only major difference is in the most amino-terminal parts of the molecules where the *C. elegans* protein has six

LDL modules (Yochem and Greenwald, 1993) while there are two copies in human LRP and seven copies in rat megalin (Herz *et al.*, 1988; Raychowdhury *et al.*, 1989; Saito *et al.*, 1994), suggesting that these parts have been constructed more recently than the rest of the molecule. Interestingly, these youngest LDL receptor modules are the only ones that are still encoded by separate class 1-1 exons in the *C. elegans* gene.

2.4 Mechanism of modular evolution by exon-shuffling

Investigations of abnormal LDL receptor genes causing familial hypercholesterolaemia have revealed many cases where intronic recombination has resulted in deletion or duplication of EGF and LDL receptor modules of this protein. Comparison of the structures of normal and abnormal LDL receptor genes has shown that misalignment and recombination of introns flanking the modules, involving middle repetitious sequences (frequently *Alu* repeats), were responsible for the deletion or duplication of individual modules or entire sets of modules of this mosaic protein (Hobbs *et al.*, 1986; Kigawa *et al.*, 1993; Lehrman *et al.*, 1987; Miyake *et al.*, 1989). It must be emphasized that the mechanism that causes deletion or duplication of modules is the same as the *Alu*-mediated intronic recombination affecting a variety of non-modular proteins (such as apolipoprotein B, β-hexosaminidase, C1 inhibitor, antiplasmin, antithrombin and lysyl hydroxylase) in some human inherited disorders (Heikkinen *et al.*, 1994; Huang *et al.*, 1989; Miura *et al.*, 1989; Myerovitz and Hogikyan, 1987; Olds *et al.*, 1993; Stoppa-Lyonnet *et al.*, 1990). The only difference is that in the case of modular proteins a 'viable' multi-domain protein with an altered number of modules is produced, whereas in the case of non-modular proteins an integral part of a folding domain is deleted.

It seems plausible that the same mechanism of intronic recombination that operates during tandem duplications and deletions of modules is also respon-

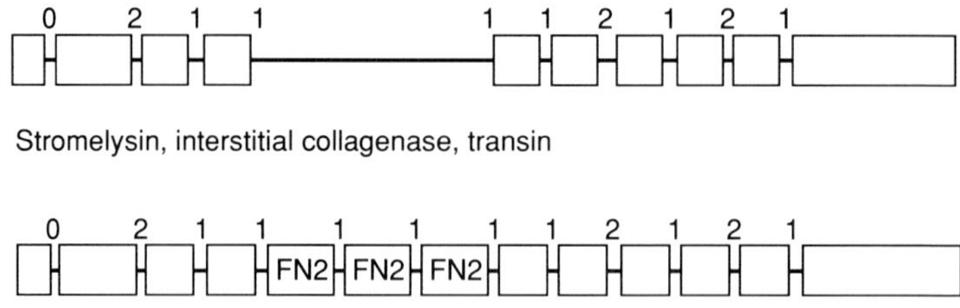

Figure 2.5. Comparison of the exon–intron structures of the more ancestral forms of the matrixin family (stromelysin, interstitial collagenase, transin) with the structure of the gene for the 72-kDa type IV collagenase. Note that a phase 1 intron was present in the common ancestor of these metalloproteases exactly where the class 1-1 fibronectin type II module was inserted to give rise (via internal triplication of the FN2 module) to the type IV collagenase.

sible for their transfer from one gene to another. This intronic recombination model of module-shuffling requires that a 'target' intron of the proper phase must be present at the location of the insertion site. There are indeed some cases of exon-shuffling where the exon–intron structure of the recipient gene (before module insertion) is known and where it is clear that module insertion has indeed occurred into a pre-existing intron.

The case of complex regulatory proteases has been discussed above (and illustrated in *Figure 2.3*): the various class 1-1 modules of the proteases of the fibrinolytic, blood coagulation and complement activation cascades were all inserted into a phase 1 intron that was present in the common ancestor of trypsin-like proteases. Subsequent class 1-1 module insertions have also occurred in phase 1 introns (Patthy, 1987, 1990). A similar conclusion can be reached from an analysis of the evolutionary history of the zinc metalloproteases, the matrixins (Patthy, 1991b). Comparison of the gene structures of interstitial collagenases, stromelysins, transins and type IV collagenases revealed a perfect conservation of the location and phase class of introns. Significantly, a phase 1 intron is present in the genes of all members of this family exactly where the class 1-1 fibronectin type II module was inserted later to give rise to the more complex type IV collagenases (*Figure 2.5*).

Analysis of the evolution of the α chains of integrins also argues for the 'intronic recombination' scenario (Fleming *et al.*, 1993; Patthy, 1991b). The α chains of the CD11b, LFA-1 and p150,95 integrins differ from simpler, more ancestral forms of α chains (such as those of PS2 of *Drosophila* and platelet glycoprotein IIb) in that they contain an insert related to the A-type modules of von Willebrand factor. Significantly, in the *PS2* gene of *Drosophila* and in the gene for platelet glycoprotein IIb, a phase 1 intron is present exactly where the class 1-1 von Willebrand factor A-type module was inserted into an ancestral gene for an integrin α chain to give rise to the α chains of the CD11b, LFA-1 and p150,95 group of integrins (Brown *et al.*, 1989; Heidenreich *et al.*, 1990).

These findings support the intronic-recombination model where modules are inserted into pre-existing introns of the target gene. Furthermore, it seems likely that the same genomic features of modules that facilitate their duplication and/or deletion by intronic recombination are also responsible for their transfer from the donor to the recipient gene. This is supported by the fact that recombination in *Alu* repeats is known to predispose *Alu*-rich regions to excision and tranposition (Calabretta *et al.*, 1982). We assume that recombination of middle repetitious sequences (such as *Alu* repeats) present in the introns that flank the modules facilitate module transposition by a similar mechanism. It is well known that symmetrical modules frequently undergo repeated tandem duplications resulting in proteins that contain numerous tandem copies of modules (Patthy, 1991b). Intermodule introns greatly facilitate contraction and expansion of such modular genes; during this process looping out and excision of modules may occur and the excised modules may be inserted in introns of other genes on distant chromosomal locations.

2.5 Evolutionary significance of modular protein evolution

The fact that the majority of the proteins regulating blood coagulation, fibrinolysis and complement activation, most constituents of the extracellular matrix, cell adhesion proteins, receptor tyrosine kinases and receptor tyrosine phosphatases were constructed from modules underlines the extreme value of exon-shuffling. There are several unique features of this evolutionary mechanism that have made it so important. Proteins assembled from modules may carry a large collection of binding specificities that make such proteins valuable in regulatory or structural networks where multiple interactions are essential. For example, in the case of the multi-domain proteins of the extracellular matrix (e.g. laminins, proteoglycans, fibronectin, modular collagens), their numerous interactions with each other and with cell surface proteins are indispensable for the proper organization of the extracellular matrix. The various cell surface molecules that mediate the interaction of cells with other cells via homophilic and heterophilic interactions, or bind cells to matrix constituents, are also involved in a variety of protein–protein interactions. Similarly, the different modules of plasma proteases recognize their substrates and bind, for example, inhibitors, cofactors and phospholipid membranes, and through these interactions control the activation and activity of these enzymes.

Modular protein evolution by exon-shuffling is an extremely powerful mechanism, since acquisition of a new domain can bring about a sudden change in the specificity of the recipient. One of the most telling examples illustrating the advantage of gaining a module *de novo* is the case of gelatinases (type IV collagenases). As mentioned in Section 2.4, these enzymes arose as a result of the insertion of a fibronectin type II module into an intron of an ancestral metalloprotease of the matrixin family. Significantly, the type II modules of fibronectin are involved in collagen binding and the fibronectin type II modules of gelatinases also bind tightly to denatured collagens (Bányai and Patthy, 1991; Bányai *et al.*, 1990). The case of gelatinases also illustrates the role of selection in modular protein evolution. Although modules are shuffled at random, the chances that an inserted module is accepted depends on whether its insertion is deleterious, neutral or carries a function potentially useful for its new host protein. It seems obvious that the insertion of a collagen-binding type II module into an ancestral member of the collagenase family was accepted because its binding specificity for denatured collagens was biologically relevant for a collagenase.

Shuffling of modules is thus an extremely efficient mechanism for the creation of proteins with novel functions. Therefore, it is plausible to assume that the appearance of this mechanism had a great impact on evolution. It is interesting in this respect that the explosion of exon-shuffling does indeed coincide with a burst of evolutionary creativity: the 'Big Bang' of Metazoan radiation (Kerr, 1993, 1994; Levinton, 1992; Lipps and Signor, 1992; Morris, 1993). As a result of an enigmatic acceleration of evolution, different phyla of

Metazoa with different body plans appeared in a very short interval of the Cambrian period. It seems likely that exon-shuffling has contributed significantly to the rapid establishment of all major phyla of multicellular animals. It must be emphasized that many modular proteins produced by exon-shuffling are associated with, and are absolutely essential for, multicellularity. For example, the appearance of the constituents of the extracellular matrix is inextricably linked to the appearance of the first multicellular organisms (Har-el and Tanzer, 1993). Clearly, proteins of the extracellular matrix, membrane-associated proteins mediating cell–cell and cell–matrix interactions, and receptor proteins regulating cell–cell communications (such as receptor tyrosine kinases, receptor tyrosine phosphatases and growth factor receptors) are vital in permitting the cells of Metazoa to function in an integrated way; they are involved in controlling morphogenetic and differentiation processes, cell fate decisions and thus determine the basic body plans of Metazoa.

In summary, it seems that the evolution and spread of spliceosomal introns reached a critical point sometime before the Cambrian period and this led to a spectacular increase in the efficiency of modular protein evolution. The appearance of this efficient evolutionary mechanism permitted the rapid assembly of novel multi-domain proteins.

2.6 Evolutionary changes in the domain composition of modular proteins: defining the time of the formation of the basic architecture of human modular proteins

Accumulation of structural and evolutionary information on modular proteins of various vertebrate and invertebrate species is now such that in some cases it is possible to define the approximate time of their formation.

Not surprisingly, those human proteins that are involved in the most profound biological functions common to multicellular organisms usually have a domain organization that was established in the earliest stages of Metazoan evolution. Although sequence data are still sporadic, it is clear that many components of the extracellular matrix were invented at the time of Metazoan radiation (Har-el and Tanzer, 1993). It appears that the basic architecture of laminins has not changed since the divergence of arthropods and vertebrates (Chi and Hui, 1989; Garrison *et al.*, 1991). As discussed in Section 2.3, it is noteworthy that introns are missing from most module boundaries of laminin genes. It is possible that there was selective pressure to eliminate these introns, since the loss of intermodule introns could hinder module rearrangements and could help 'freeze' the biologically crucial module structure of these indispensable extracellular matrix proteins.

The modular organization of netrins, laminin-related proteins that guide axon migrations and promote axon outgrowth, is also conserved from nematodes to mammals (Ishii *et al.*, 1992; Serafini *et al.*, 1994). Osteonectin, another matrix constituent, also has the same domain organization in *C. ele-*

gans and mammals (McVey *et al.*, 1988; Schwarzbauer and Spencer, 1993). Conservation of this protein probably reflects the fact that it plays an essential role in cell–matrix interactions and control of developmental processes (Schwartzbauer and Spencer, 1993).

On the other hand, perlecan/heparan sulfate proteoglycan seems to be a more recent 'invention'. Although the UNC-52 protein of nematodes (Rogalski *et al.*, 1993) has many similarities to mammalian perlecans (Cohen *et al.*, 1993; Kallunki and Tryggvason, 1992; Murdoch *et al.*, 1992; Noonan *et al.*, 1991), it still lacks the amino-terminal heparan sulphate domain as well as the carboxyl-terminal laminin-A and EGF-like modules that are present in vertebrate perlecans.

Members of the *Notch* family are cell-surface proteins that play crucial roles in regulating development. *Notch* of *Drosophila* (Kidd *et al.*, 1986; Wharton *et al.*, 1985), *Xotch* of *Xenopus laevis* (Coffman *et al.*, 1990), *TAN-1* of humans (Ellisen *et al.*, 1991) and *Notch* homologues of other mammals (Weinmaster *et al.*, 1991) have an identical domain organization: their most striking feature is the presence of 36 tandem EGF-like repeats in their extracellular domain. The remarkable conservation of *Notch*-related proteins indicates that their modular assembly was completed and their basic biological function was established before the divergence of the ancestors of vertebrates and arthropods. The fact that the *Notch* homologues of *C. elegans, lin-12* and *glp-1* (Yochem and Greenwald, 1989; Yochem *et al.*, 1988) contain fewer EGF-like repeats suggests that the domain organization of the ancestral *Notch*-related gene was frozen in the arthropod–vertebrate line only after its divergence from the line of nematodes.

It appears that the architecture of nidogen/entactin, a component of the basement membrane of vertebrates, has been established relatively recently: the module organization of the entactin of a prochordate differs significantly from that of mammalian species (Nakae *et al.*, 1993). Similarly, comparison of the structures of Ror-type receptor tyrosine kinases from *C. elegans* (Wilson *et al.*, 1994), *Drosophila* (Wilson *et al.*, 1993), *Torpedo* (Jennings *et al.*, 1993) and humans (Masiakowski and Carroll, 1992) indicates that significant changes in domain organization occurred in the vertebrate lineage: whereas invertebrate homologues of Ror lack immunoglobulin modules, there are three of these module types in the *Torpedo* homologue, but there is only one in the human Ror receptors.

Unlike the examples described above, there are many modular proteins that are more restricted in their evolutionary distribution and have probably only been constructed in certain evolutionary lineages. [Some modular protein types found so far only in invertebrates have been discussed by Patthy (1994)]. The various regulatory proteases typical of mammalian blood coagulation and fibrinolysis were probably formed only in the vertebrate line since the blood-clotting enzymes of arthropods have different domain organization and were assembled from modules independently (Muta *et al.*, 1991).

It appears that there has been little change in the architecture of the majority of proteins listed in *Table 2.1* during vertebrate evolution. Thus, the domain organization of chicken and hagfish prothrombin (Banfield *et al.*, 1994), avian urokinase (Leslie *et al.*, 1990), *Xenopus* complement factor I (Kunnath-Muglia *et al.*, 1993), chicken agrin (Tsim *et al.*, 1992), various amphibian and avian thrombospondins (Newton *et al.*, 1994), chick aggrecan and versican (Li *et al.*, 1993; Shinomura *et al.*, 1993), chick cartilage link protein, cartilage matrix protein and type VI and type XIV collagens (Hayman *et al.*, 1991; Walchli *et al.*, 1992, 1993) and chicken type IV collagenase (Aimes *et al.*, 1994) are identical with those of their mammalian counterparts.

There are some proteins that are apparently younger and which have undergone major rearrangements during mammalian evolution. For example, the fact that human, bovine and rat orthologues of P selectins (GM-140-related proteins) contain a different number of complement B-type modules (Auchampach *et al.*, 1994; Johnston *et al.*, 1990; Strubel *et al.*, 1993) indicates that duplication and deletion of these modules by intronic recombination occurred following the divergence of these mammalian species. The fact that human ELAM-1 contains six complement B-type modules but the rabbit homologue has only five of these units also shows that exon duplications and exon deletions are quite common in the selectin family (Collins *et al.*, 1991; Larigan *et al.*, 1992). This conclusion is also supported by the observation that the murine lymphocyte homing receptor has two complement B modules that show complete identity at the nucleotide sequence level, including intron regions upstream and downstream of the splice sites (Dowbenko *et al.*, 1991), indicating that duplication of a complement B-type module occurred recently by intronic recombination. In the case of C4b-binding proteins it has been noted that homologues from different mammalian species may have different domain organization; for example, mouse C4b-binding protein has only six complement B-type repeats whereas the human homologue contains eight repeats (Kristensen *et al.*, 1987). Changes in the domain organization of some modular proteins may occur at an even more striking rate; for example, the complement C1 receptor homologues may undergo frequent rearrangements, expansion and contraction, leading to polymorphic forms with different numbers of modules (Hourcade *et al.*, 1990).

The most striking example illustrating the contraction and expansion of a modular protein gene is the human apolipoprotein(a) gene. In the human population this molecule shows a remarkable variation in structure with the number of kringle units ranging from 12 to 51 copies, suggesting a high rate of module duplications and deletions (Koschinsky *et al.*, 1990; Lackner *et al.*, 1991, 1993). In one apolipoprotein(a) variant, 24 of the 37 kringle-4-like units were found to have identical nucleotide sequences, indicating that they are the products of very recent duplications (McLean *et al.*, 1987). Duplication and deletion of kringle modules occurs quite frequently, as suggested by the observation that the isoforms containing different numbers of kringles do not follow a simple Mendelian pattern of inheritance: offspring often have apolipoprotein(a) isoforms that differ from those of their parents (Gaubatz *et al.*, 1990).

It must be emphasized that the modular proteins that still show significant variation in the number of modules [e.g. apolipoprotein(a), selectins, the group of C4b-binding proteins, factor-H-related proteins and complement receptors] have retained all their intermodule phase 1 introns. It seems likely that the presence of intermodule introns ensures that the plasticity of these genes is still preserved. On the other hand, the fact that the LDL receptor gene is prone to deleterious module rearrangements illustrates the dangers inherent in intronic recombinations.

References

Abbott AM, Bueno R, Pedrini MT, Murray JM, Smith RJ. (1992) Insulin-like growth factor I receptor gene structure. *J. Biol. Chem.* **267**: 10759–10763.

Aimes RT, French DL, Quigley JP. (1994) Cloning of a 72 kDa matrix metalloproteinase (gelatinase) from chicken embryo fibroblasts using gene family PCR: expression of the gelatinase increases upon malignant transformation. *Biochem. J.* **300**: 729–736.

Alliel PM, Perin JP, Jolles P, Bonnet FJ. (1993) Testican, a multidomain testicular proteoglycan resembling modulators of cell social behaviour. *Eur. J. Biochem.* **214**: 347–350.

Alves F, Vogel W, Mossie K, Millauer B, Höfler H, Ullrich A. (1995) Distinct structural characteristics of discoidin I subfamily receptor tyrosine kinases and complementary expression in human cancer. *Oncogene* **10**: 609–618.

Ancian P, Lambeau G, Mattei MG, Lazdunski M. (1995) The human 180-kDa receptor for secretory phospholipases A2. Molecular cloning, identification of a secreted soluble form, expression, and chromosomal localization. *J. Biol. Chem.* **270**: 8963–8970.

Aso T, Okamura S, Matsuguchi T, Sakamot N, Sata T, Niho Y. (1991) Genomic organization of the α chain of the human C4b-binding protein gene. *Biochem. Biophys. Res. Commun.* **174**: 222–227.

Auchampach JA, Oliver MG, Anderson DC, Manning AM. (1994) Cloning, sequence comparison and *in vivo* expression of the gene encoding rat P selectin. *Gene* **145**: 251–255.

Banfield DK, Irwin DM, Walz DA, MacGillivray RTA. (1994) Evolution of prothrombin: isolation and characterization of the cDNAs encoding chicken and hagfish prothrombin. *J. Mol. Evol.* **38**: 177–187.

Bányai L, Patthy L. (1991) Evidence for the involvement of type II domains in collagen binding by 72 kDa type IV procollagenase. *FEBS Lett.* **282**: 23–25.

Bányai L, Váradi A, Patthy L. (1983) Common evolutionary origin of the fibrin-binding structures of fibronectin and tissue-type plasminogen activator. *FEBS Lett.* **163**: 37–41.

Bányai L, Trexler M, Koncz S, Gyenes M, Sipos G, Patthy L. (1990) The collagen-binding site of type-II units of bovine seminal fluid protein PDC-109 and fibronectin. *Eur. J. Biochem.* **193**: 801–806 .

Beaubien G, Rosinski-Chupin I, Mattei MG, Mbikaz M, Chretien M, Seidah NG. (1991) Gene structure and chromosomal localization of plasma kallikrein. *Biochemistry* **30**: 1628–1635.

Beckman G, Bork P. (1993) An adhesive domain detected in functionally diverse receptors. *Trends Biochem. Sci.* **18**: 40–41.

Behrendt N, Ploug M, Patthy L, Houen G, Blasi F, Dano K. (1991) The ligand-binding domain of the cell-surface receptor for urokinase-type plasminogen activator. *J. Biol. Chem.* **266**: 7842–7847.

Belfort M. (1991) Self-splicing introns in prokaryotes: migrant fossils? *Cell* **11**: 9–11.

Belfort M. (1993) An expanding universe of introns. *Science* **262**: 1009–1010.

Benyajati C, Place AR, Powers DA, Sofer W. (1981) Alcohol dehydrogenase gene of *Drosophila melanogaster*: relationship of intervening sequences to functional domains in the protein. *Proc. Natl Acad. Sci. USA* **78**: 2717–2721.

Blake CCF. (1978) Do genes-in-pieces imply proteins-in-pieces? *Nature* **273**: 267.

Blake CCF. (1979) Exons encode protein functional units. *Nature* **277**: 598.

Blake CCF. (1981) Exons and the structure, function and evolution of haemoglobin. *Nature* **291**: 616.

Blake CCF. (1983) Exons and the evolution of proteins. *Trends Biochem. Sci.* **8**: 11–13.

Bork P. (1992) The modular architecture of vertebrate collagens. *FEBS Lett.* **307**: 49–54.

Bork P. (1993a) A trefoil domain in the major rabbit zona pellucida protein. *Protein Sci.* **2**: 669–670.

Bork P. (1993b) The modular architecture of a new family of growth regulators related to connective tissue growth factor. *FEBS Lett.* **327**: 125–130.

Bork P, Rohde K. (1991) More von Willebrand factor type A domains? Sequence similarities with malaria thrombospondin-related anonymous protein, dihydropyridine-sensitive calcium channel and inter-α-trypsin inhibitor. *Biochem. J.* **279**: 908–910.

Bork P, Sander C. (1992) A large domain common to sperm receptors (Zp2 and Zp3) and TGF-β type III receptor. *FEBS Letters* **300**: 237–240.

Bornstein P, Devarayalu S, Edelhoff S, Disteche CM. (1993) Isolation and characterization of the mouse thrombospondin 3 (Thbs3) gene. *Genomics* **15**: 607–613.

Branden CI, Eklund H, Cambillau C, Pryor AJ. (1984) Correlation of exons with structural domains in alcohol dehydrogenase. *EMBO. J.* **3**: 1307–1310.

Brauer C, Scheit KH. (1991) Characterization of the gene for the bovine seminal vesicle secretory protein SVSP109. *Biochim. Biophys. Acta* **1090**: 259–260.

Bristow J, Tee MK, Gitelman SE, Mellon SH, Miller WL. (1993) Tenascin-X: a novel extracellular matrix protein encoded by the human XB gene overlapping P450c21B. *J. Cell Biol.* **122**: 265–278.

Brown NH, King DL, Wilcox M, Kafatos FC. (1989) Developmentally regulated alternative splicing of *Drosophila* integrin PS2 α transcripts. *Cell* **59**: 185–195.

Byrne CD, Schwartz K, Meer K, Cheng JF, Lawn RM. (1994) The human apolipoprotein(a)/plasminogen gene cluster contains a novel homologue transcribed in liver. *Arterioscler. Thromb. Vasc. Biol.* **14**: 534–541.

Byrne CD, Schwartz K, Lawn RM. (1995) Loss of a splice donor site at a ‘skipped exon’ in a gene homologous to apolipoprotein(a) leads to an mRNA encoding a protein consisting of a single kringle domain. *Arterioscler. Thromb. Vasc. Biol.* **15**: 65–69.

Calabretta B, Robberson DL, Barrera-Saldana HA, Lambrou TP, Saunders GF. (1982) Genome instability in a region of human DNA enriched in *Alu* repeat sequences. *Nature* **296**: 219–225.

Casey JR, Petranka JG, Kottra J, Fleenor DE, Rosse WF. (1994) The structure of the urokinase-type plasminogen activator receptor gene. *Blood* **84**: 1151–1156.

Cavalier-Smith T. (1985) Selfish DNA and the origin of introns. *Nature* **315**: 283–284..

Cavalier-Smith T. (1991) Intron-phylogeny: a new hypothesis. *Trends Genet.* 7: 145–148.

Cech TR. (1985) Self-splicing RNA: implications for evolution. *Int. Rev. Cytol.* **93**: 3–22.

Cech TR. (1986) The generality of self-splicing RNA: relationship to nuclear mRNA splicing. *Cell* **44**: 207–210.

Chi HC, Hui CF. (1989) Primary structure of the *Drosophila* laminin B2 chain and comparison with human, mouse and *Drosophila* laminin B1 and B2 chains. *J. Biol. Chem.* **264**: 1543–1550.

Chi HC, Juminaga D, Wang SY, Wang SY, Hui CF. (1991) Structure of the *Drosophila* gene for the Laminin B2 chain. *DNA Cell Biol.* **10**: 451–466.

Cho KR, Oliner JD, Simons JW, Hedrick L, Fearon ER, Preisinger AC, Hedge P, Silverman GA, Vogelstein B. (1994) The DCC gene: structural analysis and mutations in colorectal carcinomas. *Genomics* **19**: 525–531.

Christiano AM, Hoffman GG, Chung-Honet LC, Lee S, Cheng W, Uitto J, Greenspan DS. (1994) Structural organization of the human type VII collagen gene (COL7A1),

composed of more exons than any previously characterized gene. *Genomics* **21**: 169–179.
Coffman C, Harris W, Kintner C. (1990) Xotch, the *Xenopus* homolog of *Drosophila Notch*. *Science* **249**: 1438–1441 .
Cohen IR, Grassel S, Murdoch AD, Iozzo RV. (1993) Structural characterization of the complete human perlecan gene and its promoter. *Proc. Natl Acad. Sci. USA* **90**: 10404–10408.
Collier IE, Bruns GAP, Goldberg GI, Gerhard DS. (1991) On the structure and chromosome location of the 72- and 92-kDa human type IV collagenase genes. *Genomics* **9**: 429–434.
Collins T, Williams A, Johnston GI, Kim J, Eddy R, Shows T, Gimbrone MA, Jr, Bevilacqua MP. (1991) Structure and chromosomal location of the gene for endothelial leukocyte adhesion molecule 1. *J. Biol. Chem.* **266**: 2466–2473.
Connelly MA, Grady RC, Mushinski JF, Marcu KB. (1994) PANG, a gene encoding a neuronal glycoprotein, is ectopically activated by intracisternal A-type particle long terminal repeats in murine plasmacytomas. *Proc. Natl Acad. Sci. USA* **91**: 1337–1341.
Cooper ST, Attie AD. (1992) Pig apolipoprotein R: a new member of the short consensus repeat family of proteins. *Biochemistry* **31**: 12328–12336.
Copertino DW, Hallick RB. (1993) Group II and group III introns of twintrons: potential relationships with nuclear pre-mRNA introns. *Trends Biochem. Sci.* **18**: 467–471.
Cornish-Bowden A. (1985) Are introns structural elements or evolutionary debris? *Nature* **313**: 434–435.
Corson GM, Chalberg SC, Dietz HC, Charbonneau NL, Sakai LY. (1993) Fibrillin binds calcium and is coded by cDNAs that reveal a multidomain structure and alternatively spliced exons at the 5′ end. *Genomics* **17**: 476–484.
Craik CS, Buchman SR, Beychok S. (1980) Characterization of globin domains: heme binding to the central exon product. *Proc. Natl Acad. Sci. USA* **77**: 1384–1388.
Craik CS, Sprang S, Fletterick R, Rutter WJ. (1982) Intron–exon splice junctions map at protein surfaces. *Nature* **299**: 180–182.
Crick F. (1979) Split genes and RNA splicing. *Science* **204**: 264–271.
Cripe LD, Moore KD, Kane WH. (1992) Structure of the gene for human coagulation factor V. *Biochemistry* **31**: 3777–3785.
Crouch, Rust K, Veile R, Doris-Keller H, Grosso L. (1993) Genomic organization of human surfactant protein D (SP-D). SP-D is encoded on chromosome 10q22.2–23.1. *J. Biol. Chem.* **268**: 2976–2983.
Darnell JE. (1978) Implications of RNA:RNA splicing in evolution of eukaryotic cells. *Science* **202**: 1257–1260.
Darnell JE, Doolittle WF. (1986) Speculations on the early course of evolution. *Proc. Natl Acad. Sci. USA* **83**: 1271–1275.
Davis JQ, McLaughlin T, Bennett V. (1993) Ankyrin-binding proteins related to nervous system cell adhesion molecules: candidates to provide transmembrane and intercellular connections in adult brain. *J. Cell. Biol.* **121**: 121–133.
del Castillo I, Cohen-Salmon M, Blanchard S, Lutfalla G, Petit C. (1992) Structure of the X-linked Kallmann syndrome gene and its homologous pseudogene on the Y chromosome. *Nature Genetics* **2**: 305–310.
Dibb NJ, Newman AJ. (1989) Evidence that introns arose at proto-splice sites. *EMBO J.* **8**: 2015–2021.
Dietz HC, Cutting CR, Pyeritz RE *et al.* (1991) Marfan syndrome caused by a recurrent *de novo* missense mutation in the fibrillin gene. *EMBO J.* **8**: 2015–2021 .
Dixon B, Pohajdak B. (1992) Did the ancestral globin gene of plants and animals contain only two introns? *Trends Biochem. Sci.* **17**: 486–488.
Doege JK, Sasaki M, Kimura T, Yamada Y. (1991) Complete coding sequence and deduced primary structure of the human cartilage large aggregating proteoglycan, aggrecan. Human-specific repeats, and additional alternatively spliced forms. *J. Biol. Chem.* **266**: 894–902.
Doege JK, Garrison K, Coulter SN, Yamada Y. (1994) The structure of the rat aggrecan gene and preliminary characterization of its promoter. *J. Biol. Chem.* **269**: 29232–29240.

Doolittle WF. (1978) Genes in pieces: were they ever together? *Nature* **272**: 581–582.

Dorit RL, Schoenbacher L, Gilbert W. (1990) How big is the universe of exons? *Science* **250**: 1342–1382.

Dougherty GJ, Kay RJ, Humphries RK. (1989) Molecular cloning of 114/A10, a cell surface antigen containing highly conserved repeated elements, which is expressed by murine hemopoietic progenitor cells and interleukin-3 dependent cell-lines. *J. Biol. Chem.* **264**: 6509–6514.

Dowbenko DJ, Diep A, Taylor BA, Lusis AJ, Lasky LA. (1991) Characterization of the murine homing receptor gene reveals correspondence between protein domains and coding exons. *Genomics* **9**: 270–277.

Duester G, Jornvall H, Hatfield GW. (1986) Intron-dependent evolution of the nucleotide-binding domains within alcohol dehydrogenase and related enzymes. *Nucleic Acids Res.* **14**: 1931–1941.

Dujon B. (1989) Group I introns as mobile genetic elements: facts and mechanistic speculations. *Gene* **82**: 91–114.

Dumont DJ, Gradwohl GJ, Fong GH, Auerbach R, Breitman ML. (1993) The endothelial-specific receptor tyrosine kinase, *tek*, is a member of a new subfamily of receptors. *Oncogene* **8**: 1293–1301.

Durkin ME, Wewer UM, Chung AE. (1995) Exon organization of the mouse entactin gene corresponds to the structural domains of the polypeptide and has regional homology to the low-density lipoprotein receptor gene. *Genomics* **26**: 219–228.

Ellisen LW, Bird J, West DC, Soreng AL, Reynolds TC, Smith SD, Sklar J. (1991) TAN-1, the human homolog of the *Drosophila Notch* gene, is broken by chromosomal translocations in T lymphoblastic neoplasms. *Cell* **66**: 649–661.

Elomaa O, Kangas M, Sahlberg C, Tuukkanen J, Sormunen R, Liakka J, Thesless I, Kraal G, Tryggvason K. (1995) Cloning of a novel bacteria-binding receptor structurally related to scavenger receptors and expressed in a subset of macrophages. *Cell* **80**: 603–609.

Emi M, Asaoka H, Matsumoto A *et al.* (1993) Structure, organization and chromosomal mapping of the human macrophage scavenger receptor gene. *J. Biol. Chem.* **268**: 2120–2125.

Falls DL, Rosen KM, Corfas G, Lane WS, Fischbach GD. (1993) ARIA, a protein that stimulates acetylcholine receptor synthesis is a member of the Neu ligand family. *Cell* **72**: 801–815.

Fini ME, Bartlett JD, Matsubara M, Rinehart WB, Mody MK, Girard MT, Rainville M. (1994) The rabbit gene for 92-kDa matrix metalloproteinase. Role of AP1 and AP2 in cell type-specific transcription. *J. Biol. Chem.* **269**: 28620–28628.

Fleming JC, Pahl HL, Gonzalez DA, Smithe TF, Tenen DG. (1993) Structural analysis of the CD11b gene and phylogenetic analysis of the α-integrin gene family demonstrate remarkable conservation of genomic organization and suggest early diversification during evolution. *J. Immunol.* **150**: 480–490.

Franco B, Guioli S, Praglioli A *et al.* (1991) A gene deleted in Kallmann's syndrome shares homology with neural cell adhesion and axonal path-finding molecules. *Nature* **353**: 529–536.

Garrison K, MacKrell AJ, Fessler JH. (1991) *Drosophila* laminin A chain sequence, interspecies comparison, and domain structure of a major carboxyl portion. *J. Biol. Chem.* **266**: 22899–22904.

Gaubatz JW, Ghanem KI, Guevara J, Jr, Nava ML, Patsch W, Morrisett JD. (1990) Polymorphic forms of human apoliporotein(a): inheritance and relationship of their molecular weights to plasma levels of lipoprotein(a). *J. Lipid. Res.* **31**: 603–613.

Gerecke DR, Foley JW, Castagnola P *et al.* (1993) Type XIV collagen is encoded by alternative transcripts with distinct 5′ regions and is a multidomain protein with homologies to von Willebrand's factor, fibronectin, and other matrix proteins. *J. Biol. Chem.* **268**: 12177–12184.

Gerecke DR, Wagman DW, Champliaud MF, Burgeson RE. (1994) The complete primary structure for a novel laminin chain, the laminin B1k chain. *J. Biol. Chem.* **269**: 11073–11080.

Gilbert W. (1978) Why genes in pieces? *Nature* **271**: 501.

Gilbert W. (1986) The RNA world. *Nature* **319**: 618.

Gilbert W, Glynias M. (1993) On the ancient nature of introns. *Gene* **135**: 137–144.

Glücksmann-Kuis MA, Tayber O, Woolf EA *et al.* (1995) Polycystic kidney disease: the complete structure of the PKD1 gene and its protein. *Cell* **81**: 289–298.

Go M. (1981) Correlation of DNA exonic regions with protein structural units in haemoglobin. *Nature* **291**: 90–92.

Go M. (1983) Modular structural units, exons and function in chicken lysozyme. *Proc. Natl Acad. Sci. USA* **80**: 1964–1968.

Go M. (1987) Protein structure and the origin of introns. *Cold Spring Harbor Symp. Quant. Biol.* **52**: 915–924.

Gorbea CM, Marchand P, Jiang W, Copeland NG, Gilbert DJ, Jenkins NA, Bond JS. (1993) Cloning, expression and chromosomal localization of the mouse meprin β subunit. *J. Biol. Chem.* **268**: 21035–21043.

Gow CH, Chang HY, Lih CJ, Chang TW, Hui CF. (1993) Analysis of the *Drosophila* gene for the laminin B1 chain. *DNA Cell. Biol.* **12**: 573–587.

Greenwald I. (1985) *lin-12*, a nematode homeotic gene, is homologous to a set of mammalian proteins that includes epidermal growth factor. *Cell* **43**: 583–590.

Grivell LA. (1994) Invasive introns. *Curr. Biol.* **4**: 161–164.

Grumet M, Mauro V, Burgoon MP, Edelman GM, Cunningham BA. (1991) Structure of a new nervous system glycoprotein, Nr-CAM, and its relationship to subgroups of neural cell adhesion molecules. *J. Cell. Biol.* **113**: 1399–1412.

Gulcher JR, Nies DE, Alexakos MJ, Ravikant NA, Sturgill ME, Marton LS, Stefansson K. (1991) Structure of the human hexabrachion (tenascin) gene. *Proc. Natl Acad. Sci. USA* **88**: 9438–9442.

Gum JR, Hicks JW, Toribara NW, Siddiki B, Kim YS. (1994) Molecular cloning of human intestinal mucin (MUC2) cDNA. *J. Biol. Chem.* **269**: 2440–2446.

Han S, Stuart LA, Friezner Degen SJ. (1991) Characterization of the DNF15S2 locus on human chromosome 3: identification of a gene coding for four kringle domains with homology to hepatocyte growth factor. *Biochemistry* **30**: 9768–9780.

Har-el R, Tanzer ML. (1993) Extracellular matrix 3: evolution of the extracellular matrix in invertebrates. *FASEB J.* **7**: 1115–1123.

Hasler TH, Rader C, Stoeckli ET, Zuellig RA, Sonderegger P. (1993) cDNA cloning, structural features and eukaryotic expression of human TAG-1/axonin-1. *Eur. J. Biochem.* **211**: 329–339.

Hayman AR, Koppel J, Trueb B. (1991) Complete structure of the chicken α2(VI) collagen gene. *Eur. J. Biochem.* **197**: 177–184.

Heidenreich R, Eisman R, Surrey S, Delgrosso K, Bennett JS, Schwartz E, Poncz M. (1990) Organization of the gene for platelet glycoprotein IIb. *Biochemistry* **29**: 1232–1244.

Heikkinen J, Hautala T, Kivirikko KI, Myllyla R. (1994) Structure and expression of the human lysyl hydroxylase gene (PLOD): introns 9 and 16 contain *Alu* sequences at the sites of recombination in Ehlers-Danlos syndrome type VI patients. *Genomics* **24**: 464–471.

Herz J, Hamann U, Rogne S, Myklebost O, Gausepohl H, Stanley KK. (1988) Surface location and high affinity for calcium of a 500-kd liver membrane protein closely related to the LDL-receptor suggest a physiological role as lipoprotein receptor. *EMBO J.* **7**: 4119–4127.

Hickey DA, Benkel B. (1986) Introns as relict retrotransposons: implications for the evolutionary origin of eukaryotic mRNA splicing mechanisms. *J. Theor. Biol.* **121**: 283–291.

Hickey DA, Benkel BF, Abukashawa SM. (1989) A general model for the evolution of nuclear pre-mRNA introns. *J. Theor. Biol.* **137**: 41–53.

Hillarp A, Pardo-Manuel F, Ruiz RR, de Cordoba SR, Dahlback B. (1993) The human C4b-binding protein β-chain gene. *J. Theor. Biol.* **268**: 15017–15023.

Hobart MJ, Fernie B, DiScipio RG. (1993) Structure of the human C6 gene. *Biochemistry* **32**: 6198–6205.

Hobbs HH, Brown MS, Goldstein JL, Russell DW. (1986) Deletion of exon encoding cysteine-rich repeat of low density lipoprotein receptor alters its binding specificity in a subject with familial hypercholesterolemia. *J. Biol. Chem.* **261**: 13114–13120.

Holmes WE, Sliwkowski MX, Akita RW *et al.* (1992) Identification of heregulin, a specific activator of p185^{erbB2}. *Science* **256**: 1205–1210.

Huang LS, Ripps ME, Korman SH, Deckelbaum RJ, Breslow JL. (1989) Hypobetalipoproteinemia due to an apolipoprotein B gene exon 21 deletion derived by *Alu–Alu* recombination. *J. Biol. Chem.* **264**: 11394–11400.

Huhtala P, Tuuttila A, Chow LT, Lohi J, Keski-Oja J, Tryggvason K. (1991) Complete structure of the human gene for 92-kDa type IV collagenase. Divergent regulation of expression for the 92- and 72-kilodalton enzyme genes in HT-1080 cells. *J. Biol. Chem.* **266**: 16485–16490.

Hoops TC, Rindler MJ. (1991) Isolation of the cDNA encoding glycoprotein-2 (GP-2), the major zymogen granule membrane protein. Homology to uromodulin/Tamm-Horsfall protein. *J. Biol. Chem.* **266**: 4257–4263.

Hourcade D, Miesner DR, Bee C, Zeldes W, Atkinson JP. (1990) Duplication and divergence of the amino-terminal coding region of the complement receptor 1 (CR1) gene. An example of concerted (horizontal) evolution within a gene. *J. Biol. Chem.* **265**: 974–980.

Ichinose A. (1992) Multiple members of the plasminogen-apolipoprotein(a) gene family associated with thrombosis. *Biochemistry* **31**: 3113–3118.

Ichinose A. (1995) Characterization of the apolipoprotein (a) gene. *Biochem. Biophys. Res. Commun.* **209**: 365–371.

Imamura F, Takaki S, Akagi K, Ando M, Yamamura KI, Takatsu K, Tominaga A. (1994) The murine interleukin-5 receptor α-subunit gene: characterization of the gene structure and chromosome mapping. *DNA Cell Biol.* **13**: 283–292.

Ireland RC, Kotarski MA, Johnston LA, Stadler U, Birkenmeier E, Kozak LP. (1986) Primary structure of the mouse glycerol-3-phosphate dehydrogenase gene. *J. Biol. Chem.* **261**: 11779–11785.

Ishii N, Wadsworth WG, Stern BD, Culotti JG, Hedgecock EM. (1992) *UNC-6*, a laminin-related protein, guides cell and pioneer axon migrations in *C. elegans*. *Neuron* **9**: 873–881.

Ishizaki J, Hanasaki K, Higashino KI, Kishino J, Kikuchi N, Ohara O, Arita H. (1994) Molecular cloning of pancreatic group I phospholipase A2 receptor. *J. Biol. Chem.* **269**: 5897–5904.

Iwamoto T, Taniguchi M, Asai N, Ohkusu K, Nakashima I, Takahashi M. (1993) cDNA cloning of mouse *ret* proto-oncogene and its sequence similarity to the cadherin superfamily. *Oncogene* **8**: 1087–1091.

Jackman RW, Beeler DL, Fritze L, Soff G, Rosenberg RD. (1987) Human thrombomodulin gene is intron depleted: nucleic acid sequences of the cDNA and gene predict protein structure and suggest sites of regulatory control. *Proc. Natl Acad. Sci. USA* **84**: 6425–6429.

Jacquier A. (1990) Self-splicing group II and nuclear pre-mRNA introns: how similar are they? *Trends Biochem. Sci.* **15**: 351–354.

Jenne D. (1991) Homology of placental protein 11 and pea seed albumin 2 with vitronectin. *Biochem. Biophys. Res. Commun.* **176**: 1000–1006.

Jennings CGB, Dyer SM, Burden SJ. (1993) Muscle-specific *trk*-related receptor with a kringle domain defines a distinct class of receptor tyrosine kinases. *Proc. Natl Acad. Sci. USA* **90**: 2895–2899.

Jensen CH, Krogh TN, Hojrup P, Clausen PP, Skjodt K, Larsson LI, Enghild JJ, Teisner B. (1994) Protein structure of fetal antigen 1 (FA1). A novel circulating human epidermal-growth-factor-like protein expressed in neuroendocrine tumors and its relation to the gene products of dlk and pG2. *Eur. J. Biochem.* **225**: 83–92.

Jiang W, Gorbea CM, Flannery AV, Beynon RJ, Grant GA, Bond JS. (1992) The α subunit of meprin A. Molecular cloning and sequencing, differential expression in inbred mouse strains and evidence for divergent evolution of the α and β subunits. *J. Biol. Chem.* **267**: 9185–9193.

Jiang YP, Wang H, D'Eustachio P, Musacchio JM, Schlessinger J, Sap J. (1993) Cloning and characterization of R-PTP-κ, a new member of the receptor protein tyrosine phosphatase family with a proteolytically cleaved cellular adhesion molecule-like extracellular region. *Mol. Cell. Biol.* **13**: 2942–2951.

Jiang W, Swiggard WJ, Heufler WJ, Peng M, Mirza A, Steinman RM, Nussenzweig MC. (1995) The receptor DEC-205 expressed by dendritic cells and thymic epithelial cells is involved in antigen processing. *Nature* **375**: 151–155.

Johnson JD, Edman JC, Rutter WJ. (1993) A receptor tyrosine kinase found in breast carcinoma cells has an extracellular discoidin I-like domain. *Proc. Natl Acad. Sci. USA* **90**: 5677–5681.

Johnston GI, Bliss GA, Newman PJ, McEver RP. (1990) Structure of the human gene encoding granule membrane protein 140, a member of the selectin family of adhesion receptors for leukocytes. *J. Biol. Chem.* **265**: 21381–21385..

Just M, Herbst H, Hummel M, Dürkop H, Tripier D, Stein H, Schuppan D. (1991) Undulin is a novel member of the fibronectin-tenascin family of extracellular matrix glycoproteins. *J. Biol. Chem.* **266**: 17326–17332.

Kallunki P, Tryggvason K. (1992) Human basement membrane heparan sulfate proteoglycan core protein: a 467-kD protein containing multiple domains resembling elements of the low density lipoprotein receptor, laminin, neural cell adhesion molecules, and epidermal growth factor. *J. Cell Biol.* **116**: 559–571.

Kallunki T, Ikonen J, Chow LT, Kallunki P, Tryggvason K. (1991) Structure of the human laminin B2 chain gene reveals extensive divergence from the laminin B1 chain gene. *J. Biol. Chem.* **266**: 221–228.

Kawasaki N, Itoh N, Kawasaki T. (1994) Gene organization and the 5′-flanking region sequence of conglutinin: a C-type lectin containing a collagen-like domain. *Biochem. Biophys. Res. Commun.* **198**: 597–604.

Kayyem JF, Roman JM, de la Rosa EJ, Schwarz U, Dreyer WJ. (1992) Bravo/Nr-CAM is closely related to the cell adhesion molecules L1 and Ng-CAM and has a similar heterodimer structure. *J. Cell Biol.* **118**: 1259–1270.

Kerr RA. (1993) Evolution's Big Bang gets even more explosive. *Science* **261**: 1274–1275.

Kerr RA. (1994) Timing evolution's early bursts. *Science* **267**: 33–34.

Kidd S, Kelley MR, Young MW. (1986) Sequence of the *Notch* locus of *Drosophila melanogaster*: relationship of the encoded protein to mammalian clotting and growth factors. *Mol. Cell. Biol.* **6**: 3094–3108.

Kigawa K, Kihara K, Miyake Y, Tajima S, Funahashi T, Yamamura T, Yamamoto A. (1993) Low-density lipoprotein receptor mutation that deletes exon 2 and 3 by *Alu–Alu* recombination. *J. Biochem.* **113**: 372–376.

Kim SJ, Ruiz N, Bezouska K, Drickamer K. (1992) Organization of the gene encoding the human macrophage mannose receptor (MRC1). *Genomics* **14**: 721–727.

Kitamoto Y, Yuan X, Wu Q, McCourt DW, Sadler JE. (1994) Enterokinase, the initiator of intestinal digestion, is a mosaic protease composed of a distinctive assortment of domains. *Proc. Natl Acad. Sci. USA* **91**: 7588–7592.

Kitamoto Y, Veile RA, Donis-Keller H, Sadler JE. (1995) cDNA sequence and chromosomal localization of human enterokinase, the proteolytic activator of trypsinogen. *Biochemistry* **34**: 4562–4568.

Klar A, Baldassare M, Jessell TM. (1992) F-spondin: a gene expressed at high levels in the floor plate encodes a secreted protein that promotes neural cell adhesion and neurite extension. *Cell* **69**: 95–110.

Koschinsky ML, Beiseiegel U, Henne-Bruns D, Eaton DL, Lawn RM. (1990) Apolipoprotein(a) size heterogeneity is related to variable number of repeat sequences in its mRNA. *Biochemistry* **29**: 640–644.

Kosugi H, Nakagawa Y, Hotta T, Saito H, Miyajima A, Arai KI, Yokota T. (1995) Structure of the gene encoding the α subunit of the human interleukin 3 receptor. *Biochem. Biophys. Res. Commun.* **208**: 360–367.

Koths K, Taylor E, Halenback R, Casipit C, Wang A. (1993) Cloning and characterization of a human Mac-2–binding protein, a new member of the superfamily defined by the macrophage scavenger receptor cysteine-rich domain. *J. Biol. Chem.* **268**: 14245–14249.

Kristensen T, Ogata RT, Chung LP, Reid KBM, Tack BF. (1987) cDNA structure of murine C4b-binding protein, a regulatory component of the serum complement system. *Biochemistry* **26**: 4668–4674.

Kristensen T, Oxvig C, Sand O, Hundahl Moller NP, Sottrup-Jensen L. (1994) Amino acid sequence of human pregnancy-associated plasma protein-A derived from cloned cDNA. *Biochemistry* **33**: 1592–1598.

Kunnath-Muglia LM, Chang GH, Sim RB, Day AJ, Ezekowitz RAB. (1993) Characterization of *Xenopus laevis* complement factor I structure - conservation of modular structure except for an unusual insert not present in human factor I. *Mol. Immunol.* **30**: 1249–1256.

Kusche-Gullberg M, Garrison K, MacKrell AJ, Fessler LI, Fessler JH. (1992) Laminin A chain: expression during *Drosophila* development and genomic sequence. *EMBO J.* **11**: 4519–4527.

Kwok JBJ, Gardner E, Warner JP, Ponder BAJ, Mulligan LM. (1993) Structural analysis of the human *ret* protooncogene using exon trapping. *Oncogene* **8**: 2575–2582.

Lackner C, Boerwinkle E, Leffert CC, Rahmig T, Hobbs HH. (1991) Molecular basis of apolipoprotein(a) isoform size heterogeneity as revealed by pulse-field gel electrophoresis. *J. Clin. Invest.* **87**: 2153–2161.

Lackner C, Cohen JC, Hobbs HH. (1993) Molecular definition of the extreme size polymorphism in apolipoprotein(a). *Hum. Mol. Genet.* **2**: 933–940.

Lai C, Lemke G. (1994) Structure and expression of the *Tyro 10* receptor tyrosine kinase. *Oncogene* **9**: 877–883.

Lai C, Gore M, Lemke G. (1994) Structure, expression and activity of *Tyro 3*, a neural adhesion-related receptor tyrosine kinase. *Oncogene* **9**: 2567–2578.

Lamballe F, Klein R, Barbacid M. (1991) *trkC*, a new member of the *trk* family of tyrosine protein kinase, is a receptor for neurotrophin-3. *Cell* **66**: 967–979.

Lambeau G, Ancian P, Barhanin J, Lazdunski M. (1994) Cloning and expression of a membrane receptor for secretory phospholipases A2. *J. Biol. Chem.* **269**: 1575–1578.

Lambowitz AM. (1993) Introns as mobile genetic elements. *Annu. Rev. Biochem.* **62**: 587–622.

Lambowitz AM, Belfort M. (1989) Infectious introns. *Cell* **56**: 323–326.

Larigan JD, Tsang TC, Rumberger JM, Burns DK. (1992) Characterization of cDNA and genomic sequences encoding rabbit ELAM-1: conservation of structure and functional interactions with leukocytes. *DNA Cell Biol.* **11**: 149–162.

Larocca D, Peterson JA, Urrea R, Kuniyoshi J, Bistrain AM, Cerainin RL. (1991) A M_r 46 000 human milk fat globule protein that is highly expressed in human breast tumors contains factor VIII-like domains. *Cancer Res.* **51**: 4994–4998.

Lawler J, Duquette M, Ferro P, Copeland NG, Gilbert DJ, Jenkins NA. (1991) Characterization of the murine thrombospondin gene. *Genomics* **11**: 587–600.

Lawler J, Duquette M, Whittaker CA, Adams JC, McHenry K, DeSimone DW. (1993) Identification and characterization of thrombospondin-4, a new member of the thrombospondin gene family. *J. Cell Biol.* **120**: 1059–1067.

Lawler J, McHenry K, Duquette M, Derick L. (1995) Characterization of human thrombospondin 4. *J. Biol. Chem.* **270**: 323–326.

Lecain E, Zelenika D, Laine MC, Rhyner T, Pessac B. (1991) Isolation of a novel cDNA corresponding to a transcript expressed in the choroid plexus and leptomeninges. *J. Neurochem.* **56**: 2133–2138.

Lee CP, Kao MC, French BA, Putney SD, Chang SH. (1987) The rabbit muscle phosphofructokinase gene. Implications for protein structure, function and tissue specificity. *J. Biol. Chem.* **262**: 4195–4199.

Lee B, Godfrey M, Vitale E, Hori H, Mattei MG, Sarfarazi M, Tsipouras P, Ramirez F, Hollister DW. (1991) Linkage of Marfan syndrome and a phenotypically related disorder to two different fibrillin genes. *Nature* **352**: 330–339.

Lee TH, Wisniewski HG, Vilcek J. (1992) A novel secretory tumor necrosis factor-inducible protein (TSG6) is a member of the family of hyaluronate binding proteins, closely related to the adhesion receptor CD44. *J. Cell. Biol.* **116**: 545–557.

Legouis R, Hardelin JP, Levilliers J *et al.* (1991) The candidate gene for the X-linked Kallmann syndrome encodes a protein related to adhesion molecules. *Cell* **67**: 423–425.

Lehrman MA, Goldstein JL, Russell DW, Brown MS. (1987) Duplication of seven exons in LDL receptor gene caused by *Alu–Alu* recombination in a subject with familial hypercholesterolemia. *Cell* **48**: 827–835.

Leicht M, Long GL, Chandra T, Kurachi K, Kidd VJ, Mace M, Jr, Davie EW, Woo SLC. (1982) Sequence homology and structural comparison between the chromosomal human α_1-antitrypsin and chicken ovalbumin genes. *Nature* **297**: 655–659.

Leslie ND, Kessler CA, Bell SM, Degen JL. (1990) The chicken urokinase-type plasminogen activator gene. *J. Biol. Chem.* **265**: 1339–1344.

Levinton JS. (1992) The Big Bang of animal evolution. *Sci. Am.* **266/11**: 52–59.

Lewin R. (1982) On the origin of introns. *Science* **217**: 921–922.

Li H, Schwartz NB, Vertel BM. (1993) cDNA cloning of chick cartilage chondroitin sulfate (aggrecan) core protein and identification of a stop codon in the aggrecan gene associated with the chondrodystrophy, nanomelia. *J. Biol. Chem.* **268**: 23504–23511.

Liaud MF, Brinkmann H, Cerff R. (1992) The β-tubulin gene family of pea: primary structures, genomic organization and intron-dependent evolution of genes. *Plant. Mol. Biol.* **18**: 639–651.

Lindberg RA, Hunter T. (1990) cDNA cloning and characterization of *eck*, an epithelial cell receptor protein-tyrosine kinase in the *eph/elk* family of protein kinases. *Mol. Cell. Biol.* **10**: 6316–6324.

Linnenbach AJ, Seng BA, Wu S, Robbins S, Scollon M, Pyrc JJ, Druck T, Huebner K. (1993) Retroposition in a family of carcinoma-associated antigen genes. *Mol. Cell. Biol.* **13**: 1507–1515.

Lipps JH, Signor PW. (1992) (eds) *Origin and Early Evolution of Metazoa.* Plenum Press, New York..

Lonberg N, Gilbert W. (1985) Intron/exon structure of the chicken pyruvate kinase gene. *Cell* **40**: 81–90.

MacKrell AJ, Kusche-Gullberg M, Garrison K, Fessler JH. (1993) Novel *Drosophila* laminin A chain reveals structural relationships between laminin subunits. *FASEB J.* **376**: 375–381.

Maisonpierre PC, Barruzueta NX, Yancopoulos GD. (1993) *Ehk-1* and *Ehk-2*: two novel members of the *Eph* receptor-like tyrosine kinase family with distinctive structures and neuronal expression. *Oncogene* **8**: 3277–3288.

Marchionni M, Gilbert W. (1986) The triosephosphate isomerase gene from maize: introns antedate the plant–animal divergence. *Cell* **46**: 133–141.

Mark MR, Scadden DT, Wang Z, Gu Q, Goddard A, Godowski PJ. (1994) *rse*, a novel receptor-type tyrosine kinase with homology to *Axl/Ufo*, is expressed at high levels in the brain. *J. Biol. Chem.* **269**: 10720–10728.

Marti HP, McNeil L, Davies M, Martin J, Lovett DH. (1993) Homology cloning of rat 72 kDa type IV collagenase: cytokine and second-messenger inducibility in glomerular mesangial cells. *Biochem. J.* **291**: 441–446.

Masiakowski P, Carroll RD. (1992) A novel family of cell surface receptors with tyrosine kinase-like domain. *J. Biol. Chem.* **267**: 26181–26190.

Maslin CL, Corson GM, Maddox BK, Glanville RW, Sakai LY. (1991) Partial sequence of a candidate gene for the Marfan syndrome. *Nature* **352**: 334–337.

Masure S, Nys G, Fiten P, van Damme J, Opdenakker G. (1993) Mouse gelatinase B. cDNA cloning, regulation of expression and glycosylation in WEHI-3 macrophages and gene organisation. *Eur. J. Biochem.* **218**: 129–141.

Matozaki T, Suzuki T, Uchida T *et al.* (1994) Molecular cloning of a human transmembrane-type protein tyrosine phosphatase and its expression in gastrointestinal cancers. *J. Biol. Chem.* **269**: 2075–2081.

Matsumoto KI, Ishihara N, Ando A, Inoko H, Ikemura T. (1992a) Extracellular matrix protein tenascin-like gene found in human MHC class III region. *Immunogenetics* **36**: 400–403.

Matsumoto KI, Arai M,, Ishihara N, Ando A, Inoko H, Ikemura T. (1992b) Cluster of fibronectin type III repeats found in the human major histocompatibility complex class III region shows the highest homology with the repeats in an extracellular matrix protein, tenascin. *Genomics* **12**: 485–491.

Matsushima M, Ichinose M, Yahagi N *et al.* (1994) Structural characterization of porcine enteropeptidase. *J. Biol. Chem.* **269**: 19976–19982.

McKnight GL, O'Hara PJ, Parker ML. (1986) Nucleotide sequence of the triosephosphate isomerase gene from *Aspergillus nidulans*: implications for a differential loss of introns. *Cell* **46**: 143–147.

McLean JW, Tomlinson JE, Kuang WJ, Eaton DL, Chen EY, Fless GM, Scanu AM, Lawn RM. (1987) cDNA sequence of human apolipoprotein(a) is homologous to plasminogen. *Nature* **300**: 132–137.

McVey JH, Nomura S, Kelly P, Mason IJ, Hogan BLM. (1988) Characterization of the mouse SPARC/osteonectin gene. Intron/exon organization and unusual promoter region. *J. Biol. Chem.* **263**: 11111–11116.

Mehta KD, Chen WJ, Goldstein JL *et al.* (1991) The low density lipoprotein receptor in *Xenopus laevis*. I. Five domains that resemble the human receptor. *J. Biol. Chem.* **266**: 10406–10414.

Michelson AM, Blake CCF, Evans ST, Orkin SH. (1985) Structure of the human phosphoglycerate kinase gene and the intron-mediated evolution and dispersal of the nucleotide-binding domain. *Proc. Natl Acad. Sci. USA* **82**: 6965–6969.

Miki T, Fleming TP, Bottaro DP, Rubin JS, Ron D, Aaronson SA. (1991) Expression cDNA cloning of the KGF receptor by creation of a transforming autocrine loop. *Science* **251**: 72–75.

Miura O, Sugahara Y, Nakamura Y, Hirosawa S, Aoki N. (1989) Restriction fragment length polymorphism caused by a deletion involving *Alu* sequences within the human α_2-plasmin inhibitor gene. *Biochemistry* **28**: 4934–4938.

Miyake Y, Tajima S, Funahashi T, Yamamoto A. (1989) Analysis of a recycling-impaired mutant of low density lipoprotein receptor in familial hypercholesterolemia. *J. Biol. Chem.* **264**: 16584–16590.

Miyazawa K, Kitamura A, Kitamura N. (1991) Structural organization and the transcription initiation site of the human hepatocyte growth factor gene. *Biochemistry* **30**: 9170–9176 .

Miyazawa K, Shimomura T, Kitamura A, Kondo K, Morimoto Y, Kitamura N. (1993) Molecular cloning and sequence analysis of the cDNA for a human serine protease responsible for activation of hepatocyte growth factor. Structural similarity of the protease precursor to blood coagulation factor XII. *J. Biol. Chem.* **268**: 10024–10028.

Moffat GJ, Vik DP, Noack D, Tack BF. (1992) Complete structure of the murine C4b-binding protein gene and regulation of its expression by dexamethasone. *J. Biol. Chem.* **267**: 20400–20406.

Morl M, Schmelzer C. (1990) Integration of group II intron bl1 into a foreign RNA by reversal of the self-splicing reaction *in vitro*. *Cell* **60**: 629–636..

Morris SC. (1993) The fossil record and the early evolution of the metazoa. *Nature* **361**: 219–225.

Mueller MW, Allmaier M, Eskes R, Schweyen RJ. (1993) Transposition of group II intron aI1 in yeast and invasion of mitochondrial genes at new locations. *Nature* **366**: 174–176.

Murdoch AD, Dodge GR, Cohen I, Tuan RS, Iozzo RV. (1992) Primary structure of the human heparan sulfate proteoglycan from basement membrane (HSPG2/perlecan). A chimeric molecule with multiple domains homologous to the low density lipoprotein receptor, laminin, neural cell adhesion molecules, and epidermal growth factor. *J. Biol. Chem.* **267**: 8544–8557.

Muta T, Miyata T, Misumi Y, Tokunaga F, Nakamura T, Toh Y, Ikehara Y, Iwanaga S. (1991) Limulus factor C. An endotoxin-sensitive serine protease zymogen with a mosaic structure of complement-like, epidermal growth factor-like and lectin-like domains. *J. Biol. Chem.* **266**: 6554–6561.

Myerovitz R, Hogikyan ND. (1987) A deletion involving *Alu* sequences in the β-hexosaminidase α-chain gene of French Canadians with Tay-Sachs disease. *J. Biol. Chem.* **262**: 15396–15399.

Nakae H, Sugano M, Ishimori Y, Endo T, Obinata T. (1993) Ascidian entactin/nidogen. Implication of evolution by shuffling two kinds of cysteine-rich motifs. *Eur. J. Biochem.* **213**: 11–19.

Nakagawa Y, Kosugi H, Miyajima A, Arai KI, Yokota T. (1994) Structure of the gene encoding the α subunit of the human granulocyte-macrophage colony stimulating factor receptor. Implications for the evolution of the cytokine receptor superfamily. *J. Biol. Chem.* **269**: 10905–10912.

Naso MF, Zimmermann DR, Iozzo R. (1994) Characterization of the human versican gene and functional analysis of its promoter. *J. Biol. Chem.* **269**: 32999–33008.

Newton G, Weremowicz S, Morton CC, Copeland NG, Gilbert DJ, Jenkins NA, Lawler J. (1994) Characterization of human and mouse cartilage oligomeric matrix protein. *Genomics* **24**: 435–439.

Nies DE, Hemesath TJ, Kim JH, Gulcher JR, Stefansson K. (1991) The complete cDNA sequence of human hexabrachion (tenascin). A multidomain protein containing unique epidermal growth factor repeats. *J. Biol. Chem.* **266**: 2818–2823.

Nimpf J, Stifani S, Bilous PT *et al.* (1994) The somatic cell-specific low density lipoprotein receptor-related protein of the chicken. Close kinship to mammalian low density lipoprotein receptor gene family members. *J. Biol. Chem.* **269**: 212–219.

Nissinen M, Vuolteenaho R, Boot-Handford R, Kallunki P, Tryggvason K. (1991) Primary structure of the human laminin A chain. Limited expression in human tissues. *Biochem. J.* **276**: 369–379.

Noguchi J, Yanagisawa M, Imamura M *et al.* (1992) Complete primary structure and tissue expression of chicken pectoralis M-protein. *J. Biol. Chem.* **267**: 20302–20310.

Nolan KF, Kaluz S, Higgins JMG, Goundis D, Reid KBM. (1992) Characterization of the human properdin gene. *Biochem. J.* **287**: 291–297 .

Noonan DM, Fulle A, Valente P, Cai S, Horigan E, Sasaki M, Yamada Y, Hassell JR. (1991) The complete sequence of perlecan, a basement membrane heparan sulfate proteoglycan, reveals extensive similarity with laminin A chain, low density lipoprotein receptor and the neural cell adhesion molecule. *J. Biol. Chem.* **266**: 22939–22947.

Norenberg U, Wille H, Wolff JM, Frank R, Rathjen FG. (1992) The chicken neural extracellular matrix molecule restrictin: similarity with EGF-, fibronectin type III-, and fibrinogen-like motifs. *Neuron* **8**: 849–863.

Ny T, Elgh F, Lund B. (1984) The structure of human tissue-type plasminogen activator gene: correlation of intron and exon structures to functional and structural domains. *Proc. Natl Acad. Sci. USA* **81**: 5355–5359.

Oelrichs RB, Reid HH, Berrnard O, Ziemiecki A, Wilks AF. (1993) *NYK/FLK*-**1**: a putative receptor protein tyrosine kinase isolated from E19 embryonic neuroepithelium is expressed in endothelial cells of the developing embryo. *Oncogene* **8**: 11–18.

O'Grady P, Krueger NX, Streuli M, Saito H. (1994) Genomic organization of the human LAR protein tyrosine phosphatase gene and alternative splicing in the extracellular fibronectin type III domains. *J. Biol. Chem.* **269**: 25193–25199.

Ohashi K, Mizuno K, Kuma KI, Miyata T, Nakamura T. (1994) Cloning of the cDNA for a novel receptor tyrosine kinase, *Sky*, predominantly expressed in brain. *Oncogene* **9**: 699–705.

Okado H, Okamoto H. (1992) A *Xenopus* homologue of the human β-amyloid precursor protein: developmental regulation of its gene expression. *Biochem. Biophys. Res. Commun.* **189**: 1561–1568.

Oldberg A, Antonsson P, Lindblom K, Heinegard D. (1992) COMP (cartilage oligomeric matrix protein) is structurally related to the thrombospondins. *J. Biol. Chem.* **267**: 22346–22350.

Olds RJ, Lane DA, Chowdhury V, de Stefano V, Leone G, Thein SL. (1993) Complete nucleotide sequence of the antithrombin gene: evidence for homologous recombination causing thrombophilia. *Biochemistry* **32**: 4216–4224.

Orgel LE, Crick FHC.(1980) Selfish DNA: the ultimate parasite. *Nature* **284**: 604–607.

Palmer JD, Logsdon JM, Jr. (1991) The recent origins of introns. *Curr. Opin. Gen. Dev.* **1**: 470–477.

Pan MG, Rim C, Lu KP, Florio T, Stork PJS. (1993a) Cloning and expression of two structurally distinct receptor-linked protein-tyrosine phosphatases generated by RNA processing from a single gene. *J. Biol. Chem.* **268**: 19284–19291.

Pan TC, Sasaki T, Zhang RZ, Fassler R, Timpl R, Chu ML. (1993b) Structure and expression of fibulin-2, a novel extracellular matrix protein with multiple EGF-like repeats and consensus motifs for calcium binding. *J. Cell. Biol.* **123**: 1269–1277.

Partanen J, Armstrong E, Makela TP, Korhonen J, Sandberg M, Renkonen R, Knuutila S, Huebner K, Alitalo K. (1992) A novel endothelial cell surface receptor tyrosine kinase with extracellular epidermal growth factor homology domains. *Mol. Cell. Biol.* **12**: 1698–1707.

Patthy L. (1985) Evolution of the proteases of blood coagulation and fibrinolysis by assembly from modules. *Cell* **41**: 657–663.

Patthy L. (1987) Intron-dependent evolution: preferred types of exons and introns. *FEBS Lett.* **214**: 1–7.

Patthy L. (1988) Detecting distant homologies of mosaic proteins. Analysis of the sequences of thrombomodulin, thrombospondin, complement components C9, C8α and C8β, vitronectin and plasma cell membrane glycoprotein PC-1. *J. Mol. Biol.* **202**: 689–696.

Patthy L. (1990) Evolutionary assembly of blood coagulation proteins. *Semin. Thromb. Haemost.* **16**: 245–259.

Patthy L. (1991a) Exons - original building blocks of proteins? *BioEssay* **13**: 187–192.

Patthy L. (1991b) Modular exchange principles in proteins. *Curr. Opin. Struct. Biol.* **1**: 351–361.

Patthy L. (1993) Modular design of proteases of coagulation, fibrinolysis, and complement activation: implications for protein engineering and structure–function studies. *Meth. Enzymol.* 222: 10–21.

Patthy L. (1994) Exons and introns. *Curr. Opin. Struct. Biol.* **4**: 383–392.

Perlman PS, Butow RA. (1989) Mobile introns and intron-encoded proteins. *Science* **246**: 1106–1109..

Perry ACF, Barker HL, Jones R, Hall L. (1994) Genetic evidence for an additional member of the metalloproteinase-like, disintegrin-like, cysteine-rich (MDC) family of mammalian proteins and its abundant expression in the testis. *Biochim. Biophys. Acta* **1207**: 134–137.

Pleiman CM, Gimpel SD, Park LS, Harada H, Taniguchi T, Ziegler SF. (1991) Organization of the murine and human interleukin-7 receptor genes: two mRNAs generated by differential splicing and presence of a type I-interferon-inducible promoter. *Mol. Cell. Biol.* **11**: 3052–3059.

Price MG, Gomer RH. (1993) Skelemin, a cytoskeletal M-disc periphery protein, contains motifs of adhesion/recognition and intermediate filament proteins. *J. Biol. Chem.* **268**: 21800–21810.

Prochownik EV, Bock SC, Orkin SH. (1985) Intron structure of the human antithrombin III gene differs from that of other members of the serine protease inhibitor superfamily. *J. Biol. Chem.* **260**: 9608–9612.

Pulkkinen L, Gerecke DR, Christiano AM, Wagman DW, Burgeson RE, Uitto J. (1995) Cloning of the β3 chain gene (LAMB3) of human laminin 5, a candidate gene in junctional epidermolysis bullosa. *Genomics* **25**: 192–198.

Quinto C, Quiroga M, Swain WF, Nikovits WC, Jr, Standring DN, Pictete RL, Valenzuela P, Rutter WJ. (1982) Rat carboxypeptidase A: cDNA sequence and preliminary characterization of the gene. *Proc. Natl Acad. Sci. USA* **79**: 31–35.

Rauch U, Karthikeyan L, Maurel P, Margolis RU, Margolis RK. (1992) Cloning and primary structure of neurocan, a developmentally regulated, aggregating chondroitin sulfate proteoglycan of brain. *J. Biol. Chem.* **267**: 19536–19547.

Raychowdhury R, Niles JL, McCluskey RT, Smith JA. (1989) Autoimmune target in Heymann nephritis is a glycoprotein with homology to the LDL receptor. *Science* **244**:1163–1165.

Rogalski TM, Williams BD, Mullen GP, Moerman DG. (1993) Products of the *unc*-52 gene in *Caenorhabditis elegans* are homologous to the core protein of the mammalian basement membrane heparan sulfate proteoglycan. *Genes Dev.* 7: 1471–1484.

Rogers JH. (1985) Exon-shuffling and intron insertion in serine protease genes. *Nature* **315**: 458–459.

Rogers JH. (1989) How were introns inserted into nuclear genes? *Trends Genet.* **5**: 213–216.

Rogers JH. (1990) The role of introns in evolution. *FEBS Lett.* **268**: 339–343.

Rupp F, Payan DG, Magill-Solc, Cowan DM, Scheller RH. (1991) Structure and expression of a rat agrin. *Neuron* **6**: 811–823.

Rupp F, Ozcelik T, Linial M, Peterson K, Francke U, Scheller R. (1992) Structure and chromosomal localization of the mammalian agrin gene. *J. Neurosci.* **12**: 3535–3544.

Ryan MC, Tizard R, VanDevanter DR, Carter WG. (1994) Cloning of the LamA3 gene encoding the α3 chain of the adhesive ligand epiligrin. Expression in wound repair. *J. Biol. Chem.* **269**: 22779–22787.

Saito A, Pietromonaco S, Loo AKC, Farquhar MG. (1994) Complete cloning and sequencing of rat gp330/'megalin', a distinctive member of the low density lipoprotein receptor gene family. *Proc. Natl Acad. Sci. USA* **91**: 9725–9729.

Saitta B, Timpl R, Chu ML. (1992) Human α2(VI) collagen gene. Heterogeneity at the 5′ untranslated region generated by an alternative exon. *J. Biol. Chem.* **267**: 6188–6196.

Sakai J, Hoshino A, Takahashi S, Miura Y, Ishii H, Suzuki H, Kawarabayashi Y, Yamamoto T. (1994) Structure, chromosome location, and expression of the human very low density lipoprotein receptor gene. *J. Biol. Chem.* **269**: 2173–2182.

Saldanha R, Mohr G, Belfort M, Lambowitz AM. (1993) Group I and group II introns. *FASEB J.* 7: 15–24.

Sanchez MP, Tapley P, Saini S, He B, Pulido D, Barbacid M. (1994) Multiple tyrosine kinases in rat hippocampal neurons: isolation of *Ptk-3*, a receptor expressed in proliferative zones of the developing brain. *Proc. Natl Acad. Sci. USA* **91**:1819–1823.

Sanchez-Corral P, de Villena FPM, Rey-Campos J, de Cordoba SR. (1993) C4BPAL1, a member of the human regulator of complement activation (RCA) gene cluster that resulted from the duplication of the gene coding for the α-chain of C4b-binding protein. *Genomics* **17**: 185–193.

Sarras MP, Yan L, Grens A, Zhang X, Agbas A, Huff JK, St John PL, Abrahamson DR. (1994) Cloning and biological function of laminin in *Hydra vulgaris*. *Devel. Biol.* **164**: 312–324.

Schaefer EM, Erickson HP, Federwisch M, Wollmer A, Ellis L. (1992) Structural organization of the human insulin receptor ectodomain. *J. Biol. Chem.* **267**: 23393–23402.

Schmidt WM, Schweyen RJ, Wolf K, Mueller MW. (1994) Transposable group II introns in fission and budding yeast. Site-specific genomic instabilities and formation of group II IVS plDNAs. *J. Mol. Biol.* **243**: 157–166.

Schneider R, Schweiger M. (1991) A novel modular mosaic of cell adhesion motifs in the extracellular domains of the neurogenic *trk* and *trkB* tyrosine kinase receptors. *Oncogene* **6**: 1807–1811.

Schulz AS, Schleithof L, Faust M, Bartram CR, Janssen JWG. (1993) The genomic structure of the human *UFO* receptor. *Oncogene* **8**: 509–513.

Schwarzbauer JE, Spencer CS. (1993) The *Caenorhabditis elegans* homologue of the extracellular calcium binding protein SPARC/osteonectin affects nematode body morphology and mobility. *Mol. Biol. Cell.* **4**: 941–952.

Serafini T, Kennedy TE, Galko MJ, Mirzayan, Jessell TM, Tessier-Lavigne M. (1994) The netrins define a family of axon outgrowth-promoting proteins homologous to *C. elegans UNC-6*. *Cell* **78**: 409–424.

Sharp PA. (1985) On the origin of RNA splicing and introns. *Cell* **42**: 397–400.

Sharp PA. (1994) Split genes and RNA splicing. *Cell* **77**: 805–815.

Shingu T, Bornstein P. (1993) Characterization of the mouse thrombospondin 2 gene. *Genomics* **16**: 78–84.

Shinomura T, Nishida Y, Ito K, Kimata K. (1993) cDNA cloning of PG-M, a large chondroitin sulfate proteoglycan expressed during chondrogenesis in chick limb buds. Alternative spliced multiforms of PG-M and their relationship to versican. *J. Biol. Chem.* **268**: 14461–14469.

Shinomura T, Zako M, Ito K, Ujita M, Kimata K. (1995) The gene structure and organization of mouse PG-M, a large chondroitin sulfate proteoglycan. Genomic background for the generation of multiple PG.M transcripts. *J. Biol. Chem.* **270**: 10328–10333.

Skerka C, Horstman RD, Zipfel PF. (1991) Molecular cloning of a human serum protein structurally related to complement factor H. *J. Biol. Chem.* **266**: 12015–12020.

Skerka C, Kuhn S, Gunther K, Lingelbach, Zipfel PF. (1993) A novel short consensus repeat-containing molecule is related to human complement factor H. *J. Biol. Chem.* **268**: 2904–2908.

Smas CM, Green D, Sul HS. (1994) Structural characterization and alternate splicing of the gene encoding the preadipocyte EGF-like protein pref-1. *Biochemistry* **33**: 9257–9265.

Smith MA, Magill-Solc C, Rupp F, Yao YMM, Schilling JW, Snow P, McMahan UJ. (1992) Isolation and characterization of a cDNA that encodes an agrin homolog in the marine ray. *Mol. Cell. Neurosci.* **3**: 406–417.

Stokes DG, Saitta B, Timpl R, Chu ML. (1991) Human α3(VI) collagen gene. Characterization of exons coding for the amino-terminal globular domain and alternative splicing in normal and tumor cells. *J. Biol. Chem.* **266**: 8626–8633.

Stoltzfus A, Spences DF, Zuker M, Logsdon JM, Doolittle WF, Jr. (1994) Testing the exon theory of genes: the evidence from protein structure. *Science* **265**: 202–207.

Stone EM, Rothblum KN, Alevy MC, Kuo TM, Schwartz RJ. (1985a) Complete sequence of the chicken glyceraldehyde-3-phosphate dehydrogenase gene. *Proc. Natl Acad. Sci. USA* **82**: 1628–1632.

Stone EM, Rothblum KN, Schwartz RJ. (1985b) Intron-dependent evolution of chicken glyceraldehyde phosphate dehydrogenase gene. *Nature* **313**: 498–500.

Stoppa-Lyonnet D, Careter PE, Meo T. (1990) Clusters of intragenic Alu repeats predispose the human C1 inhibitor locus to deleterious rearrangements. *Proc. Natl Acad. Sci. USA* **87**: 1551–1555.

Straus D, Gilbert W. (1985) Genetic engineering in the Precambrian: structure of the chicken triosephosphate isomerase gene. *Mol. Cell. Biol.* **5**: 3497–3506.

Streuli M, Krueger NX, Ariniello PD, Tang M, Munro JM, Blattler WA, Adler DA, Dietsche CM, Saito H. (1992) Expression of the receptor-linked protein tyrosine phosphatase LAR: proteolytic cleavage and shedding of the CAM-like extracellular region. *EMBO J.* **11**: 897–907.

Strubel NA, Nguyen M, Kansas GS, Tedder TF, Bischoff J. (1993) Isolation and characterization of a bovine cDNA encoding a functional homolog of human P-selectin. *Biochem. Biophys. Res. Commun.* **192**: 338–344.

Suh TT, Nerlov C, Dano K, Degen JL. (1994) The murine urokinase-type plasminogen activator receptor gene. *J. Biol. Chem.* **269**: 25992–25998.

Szebenyi G, Rotwein P. (1994) The mouse insulin-like growth factor II/cation-independent mannose 6-phosphate (IGF-II/MPR) receptor gene: molecular cloning and genomic organization. *Genomics* **19**: 120–129.

Takagi S, Hirata T, Agata K, Mochii M, Eguchi G, Fujisawa H. (1991) The A5 antigen, a candidate for the neuronal recognition molecule, has homologies to complement components and coagulation factors. *Neuron* **7**: 295–307.

Tan KO, Sater GR, Myers AM *et al.* (1993) Molecular characterization of avian muscle titin. *J. Biol. Chem.* **268**: 22900–22907.

Tanaka H, Hojo K, Yoshida H, Yoshioka T, Sugita K. (1993) Molecular cloning and expression of the mouse 105-kDa gelatinase cDNA. *Biochem. Biophys. Res. Commun.* **190**: 732–740.

Thomas PE, Wharram BL, Goyal M, Wiggins JE, Holzman LB, Wiggins RC. (1994) GLEPP1, a renal glomerular epithelial cell (podocyte) membrane protein-tyrosine phosphatase. *J. Biol. Chem.* **269**: 19953–19962.

Thompson AJ, Herrin DL. (1994) A chloroplast group I intron undergoes the first step of reverse splicing into host cytoplasmic 5.8S rRNA. Implications for intron-mediated RNA recombination, intron transposition and 5.8S rRNA structure. *J. Mol. Biol.* **236**: 455–468.

Tittiger C, Whyard S, Walker VK. (1993) A novel intron site in the triosephosphate isomerase gene from the mosquito *Culex tarsalis*. *Nature* **361**: 470–472.

Trueb J, Trueb B. (1992) Type XIV collagen is a variant of undulin. *Eur. J. Biochem.* **207**: 549–557.

Tsim KWK, Ruegg MA, Escher G, Kroger S, McMahan UJ. (1992) cDNA that encodes active agrin. *Neuron* **8**: 677–689.

Tuypens T, Plaetinck G, Baker E, Sutherland G, Brusselle G, Fiers W, Devos R, Tavernier J. (1992) Organization and chromosomal localization of the human interleukin 5 receptor α-chain gene. *Eur. Cytokine Net.* **3**: 451–459.

Ushkaryov YA, Petrenko AG, Geppert M, Südhof TC. (1992) Neurexins: Synaptic cell surface proteins related to the α-latrotoxin receptor and laminin. *Science* **257**: 50–56.

van der Logt CPE, Reitsma PH, Bertina RM. (1991) Intron–exon organization of the human gene coding for the lipoprotein-associated coagulation inhibitor: the factor Xa dependent inhibitor of the extrinsic pathway of coagulation. *Biochemistry* **30**: 1571–1577.

van der Rest M, Bruckner P. (1993) Collagens: diversity at the molecular and supramolecular levels. *Curr. Opin. Struct. Biol.* **3**: 430–436.

Venta PJ, Montgomery JC, Hewett-Emmett, Wiebauer K, Tashian RE. (1985) Structure and exon to protein domain relationship of the mouse carbonic anhydrase II gene. *J. Biol. Chem.* **260**: 12130–12135.

Volkmer H, Hassel B, Wolff JM, Frank R, Rathjen FG. (1992) Structure of the axonal surface recognition molecule neurofascin and its relationship to a neural subgroup of the immunoglobulin superfamily. *J. Cell Biol.* **118**: 149–161.

Vos HL, Devarayalu S, de Vries Y, Bornstein P. (1992) Thrombospondin 3 (Thbs3), a new member of the thrombospondin gene family. *J. Biol. Chem.* **267**: 12192–12196.

Vuolteenaho R, Chow LT, Tryggvason K. (1990) Structure of the human laminin B1 chain gene. *J. Biol. Chem.* **265**: 15611–15616.

Vuolteenaho R, Nissinen M, Sainio K *et al.* (1994) Human laminin M chain (merosin): complete primary structure, chromosomal assignment, and expression of the M and A chain in human fetal tissues. *J. Cell Biol.* **124**: 381–394.

Vyse TJ, Bates GP, Walport MJ, Morley BJ. (1994) The organization of the human complement factor I gene (IF): a member of the serine protease gene family. *Genomics* **24**: 90–98.

Walchli C, Koller E, Trueb J, Trueb B. (1992) Structural comparison of the chicken genes for α1(VI) and α2(VI) collagen. *Eur. J. Biochem.* **205**: 583–589.

Walchli C, Trueb J, Kessler B, Winterhalter KH, Trueb B. (1993) Complete primary structure of chicken collagen XIV. *Eur. J. Biochem.* **212**: 483–490.

Walton KM, Martell KJ, Kwak SP, Dixon JE, Largent BL. (1993) A novel receptor-type protein tyrosine phosphatase is expressed during neurogenesis in the olfactory neuroepithelium. *Neuron* **11**: 387–400.

Wang Y, Dang J, Johnson LK, Selhamer JJ, Doe WF. (1995) Structure of the human urokinase receptor gene and its similarity to CD59 and the Ly-6 family. *Eur. J. Biochem.* **227**: 116–122.

Weber K, Kabsch W. (1994) Intron positions in actin genes seem unrelated to the secondary structure of the protein. *EMBO J.* **13**: 1280–1286.

Weinmaster G, Roberts VJ, Lemke G. (1991) A homolog of *Drosophila Notch* expressed during mammalian development. *Development* **113**: 199–205.

Wen D, Peles E, Cupples R *et al.* (1992) Neu differentiation factor: a transmembrane glycoprotein containing an EGF domain and an immunoglobulin homology unit. *Cell* **69**: 559–572.

Weskamp G, Blobel CP. (1994) A family of cellular proteins related to snake venom disintegrins. *Proc. Natl Acad. Sci. USA* **91**: 2748–2751.

Wharton KA, Johansen KM, Xu T, Artavanis-Tsakonas S. (1985) Nucleotide sequence from the neurogenic locus *Notch* implies a gene product that shares homology with proteins containing EGF-like repeats. *Cell* **43**: 567–581.

Wilson C, Goberdhan DCI, Steller H. (1993) *Dror*, a potential neurotrophic receptor gene, encodes a *Drosophila* homolog of the vertebrate *Ror* family of *Trk*-related receptor tyrosine kinases. *Proc. Natl Acad. Sci. USA* **90**: 7109–7113.

Wilson R, Ainscough R, Anderson K *et al.* (1994) 2.2 Mb of contiguous nucleotide sequence from chromosome III of *C. elegans*. *Nature* **368**: 32–38. Genbank Accession no. Z35595.

Wolfsberg TG, Bazan JF, Blobel CP, Myles DG, Primakoff P, White JM. (1993) The precursor region of a protein active in sperm–egg fusion contains a metalloprotease and disintegrin domains: structural, functional and evolutionary implications. *Proc. Natl Acad. Sci. USA* **90**: 10783–10787.

Wright HT. (1993) Introns and higher-order structure in the evolution of serpins. *J. Mol. Evol.* **36**: 136–143.

Yamagata M, Yamada KM, Yamada SS, Shinomura T, Tanaka H, Nishida Y, Obara M, Kimata K. (1991) The complete primary structure of type XII collagen shows a chimeric molecule with reiterated fibronectin type III motifs, von Willebrand factor A motifs, a domain homologous to a noncollagenous region of type IX collagen and short collagenous domains with an Arg-Gly-Asp site. *J. Cell Biol.* **115**: 209–221.

Yan H, Grossman A, Wang H, D'Eustachio P, Mossie K, Musacchio JM, Silvennoinen O, Schlessinger J. (1993) A novel receptor tyrosine phosphatase-σ that is highly expressed in the nervous system. *J. Biol. Chem.* **268**: 24880–24886.

Yin W, Smiley E, Germiller J, Sanguineti C, Lawton T, Pereira L, Ramirez F, Bonadio J. (1995) Primary structure and developmental expression of Fbn-1, the mouse fibrillin gene. *J. Biol. Chem.* **270**: 1798–1806.

Yochem J, Greenwald I. (1989) *glp-1* and *lin-12*, genes implicated in distinct cell–cell interactions in *C. elegans*, encode similar transmembrane proteins. *Cell* **58**: 553–563.

Yochem J, Greenwald I. (1993) A gene for a low density lipoprotein-related protein in the nematode *Caenorhabditis elegans*. *Proc. Natl Acad. Sci. USA* **90**: 4572–4576.

Yochem J, Weston K, Greenwald I. (1988) The *Caenorhabditis elegans lin-12* gene encodes a transmembrane protein with overall similarity to *Drosophila Notch*. *Nature* **335**: 547–550.

Yoshihara Y, Kawasaki M, Tani A, Tamada A, Nagata S, Kagamiyama H, Mori K. (1994) BIG-1: A new TAG-1/F3–related member of the immunoglobulin superfamily with neurite outgrowth-promoting activity. *Neuron* **13**: 415–426.

Zanussi S, Doliana R, Segat D, Bonaldo P, Colombatti A. (1992) The human type VI collagen gene. mRNA and protein variants of the α3 chain generated by alternative splicing of an additional 5′ -end exon. *J. Biol. Chem.* **267**: 24082–24089.

Zhang H, Apfelroth SD, Hu W, Davis EC, Sangunieti C, Bonadio J, Mecham RP, Ramirez F. (1994) Structure and expression of fibrillin-2, a novel microfibrillar component preferentially located in elastic matrices. *J. Cell Biol.* **124**: 855–863.

Ziegler SF, Bird TA, Scheringer JA, Schooley KA, Baum PR. (1993) Molecular cloning and characterization of a novel receptor protein tyrosine kinase from human placenta. *Oncogene* **8**: 663–670.

Zuellig RA, Rader C, Schroeder A *et al.* (1992) The axonally secreted cell adhesion molecule, axonin-1. Primary structure, immunoglobulin-like and fibronectin-type-III-like domains and glycosyl-phosphatidylinositol anchorage. *Eur. J. Biochem.* **204**: 453–463.

Zwijsen A, Blockx H, Van Arnhem W, Willems J, Fransen L, Devos K, Raymackers J, Van de Voorde A, Slegers H. (1994) Characterization of rat C6 glioma-secreted follistatin-related protein (FRP). Cloning and sequence of the human homologue. *Eur. J. Biochem.* **225**: 937–946.

3

Evolution of the HLA complex

Austin L. Hughes

3.1 Introduction

The HLA complex is a region on human chromosome 6p21.3 which includes genes encoding the HLA (human leukocyte antigen) cell-surface glycoproteins. These molecules function to present peptides to T cells (Klein, 1986) and are clearly evolutionarily related, albeit distantly, to immunoglobulins. As a result, HLA genes are considered to be part of the immunoglobulin gene superfamily. Counterparts of the HLA genes are also clustered in other vertebrate species and the general term of major histocompatability complex (MHC) is used to describe any such vertebrate gene cluster. In humans and other placental mammals, the region of the genome containing the MHC genes also includes a number of other genes that are not related evolutionarily to the MHC genes. They include genes that encode proteins essential for the function of MHC proteins in cells, genes whose products have little or no functional interaction with MHC molecules, and genes of unknown function.

The MHC genes are remarkable for several reasons. Their most striking characteristic is the extremely high level of polymorphism found at certain MHC loci, which are among the most polymorphic known in any organism. Certain of these loci are characterized by 'trans-species' polymorphism (Lawlor *et al.*, 1988; Mayer *et al.*, 1988); that is, allelic lineages may be shared by related species, such as human and chimpanzee, having apparently been present in their common ancestor. By contrast, some other MHC loci are monomorphic or nearly so. The MHC gene family is divided into two subfamilies called class I and class II; the products of these two families differ in structure, level of expression and function. Class I and class II MHC genes also differ strikingly with respect to the patterns of homology among loci in different species of placental mammals. Comparison of the class II MHC

Human Genome Evolution, edited by M. Jackson, T. Strachan and G. Dover.

genes in different mammals enables the identification of orthologues (i.e. loci homologous by descent from a common ancestral locus without gene duplication; Hughes and Nei, 1990; Klein and Figueroa, 1986). In the case of the class I MHC, however, no such orthologous relationships have been observed between mammals of different orders (Hughes and Nei, 1989a).

Because these aspects of MHC biology can only be understood from an evolutionary perspective, many comparative and evolutionary studies of the MHC have been undertaken in recent years. Here I review the results of some of these studies. First, I give a brief introduction to the structure and function of MHC molecules. Then I consider the mechanism of maintenance of MHC polymorphism and the role of gene duplication in the history of the human class I and class II MHC gene families.

3.2 Structure and function of MHC molecules

Class I MHC molecules, which are expressed on almost all kinds of nucleated cells, present peptides to cytotoxic ($CD8^+$) T cells (Bjorkman and Parham, 1990). The peptides, which are typically nonamers (Rammensee *et al.*, 1995), are derived from breakdown of intracellular proteins (Monaco, 1992). The peptide is transported across the membrane of the endoplasmic reticulum by a transporter known as TAP (transporter associated with antigen processing and bound to the class I molecule, which is then transported to the cell surface. Ordinarily, in a healthy cell, the class I molecules bind 'self' peptides; that is, peptides derived from the cell's own proteins. The complex of self MHC and self peptide is not attacked by T cells, since self-reactive T cells are eliminated in development. However, when a cell is infected by a virus or other intracellular parasite, peptides derived from the parasite's proteins are presented on the surface of the infected cell by class I MHC molecules. When a cytotoxic T cell binds the complex of foreign peptide and self MHC, it kills the infected cell.

The recognition of MHC plus peptide involves a multimolecular complex which includes the T-cell receptor, the CD8 molecule on the T cell, and the class I molecule. The CD8 molecule is a heterodimer consisting of an α chain and a β chain (O'Rourke and Mescher, 1993). Both of these are members of the immunoglobulin superfamily, as is CD4, which plays an analogous role in the case of helper ($CD4^+$) T cells.

Class II molecules have a much more restricted expression than class I molecules, being expressed primarily on antigen-presenting cells of the immune system. The peptides presented by class II molecules, which are generally 13–17 amino acids in length (Rammensee *et al.*, 1995), are derived from proteins endocytosed by the antigen-presenting cell and are believed to be bound to the class II molecule in a lysosome-like compartment of the cell (Tulp *et al.*, 1994). The complex of self class II MHC molecule and foreign peptide on an antigen-presenting cell is recognized by a helper ($CD4^+$) T-cell. Recognition involves formation of a complex including the T-cell receptor and CD4 on the T cell and the class II molecule on the antigen-presenting cell.

The helper T cell then releases lymphokines that stimulate B cells (which produce antibodies) and macrophages to respond to the infection.

Both class I and class II molecules are heterodimers with four extracellular domains but they achieve a similar molecular structure in different ways (*Figure 3.1*). The class I α chain includes three extracellular domains (α_1, α_2, α_3). The α_3 domain associates non-covalently with β_2-microglobulin (β_2m). In mammals, β_2m is encoded by a gene outside the MHC region, but the gene has a distant evolutionary relationship to MHC genes. The α_3 domain contains a conserved seven amino acid loop which serves as a binding site for CD8 (O'Rourke and Mescher, 1993). The class II heterodimer consists of an α chain with two extracellular domains (α_1 and α_2) and a β chain with two extracellular domains (β_1 and β_2), both encoded by genes within the MHC region.

In a typical mammal, it appears that there are one to three polymorphic class I α chain loci, as well as a number of non-polymorphic class I loci. The polymorphic class I loci are often referred to as classical class I loci or class Ia loci, and their products are sometimes referred to as the 'major transplantation antigens' because of their role in transplant rejection. In humans, the classical class I loci are designated *HLA-A*, *HLA-B* and *HLA-C* (*Figure 3.2*). In addition to the classical class I loci, mammalian species so far studied have a variable number of non-polymorphic class I loci which are called non-classical class I loci or class Ib loci. Class Ib have reduced levels of expression compared with class Ia loci and their function is not understood at present (Howard, 1987). Examples of class Ib loci in the human genome are *HLA-G*, which is expressed in the trophoblast, and *HLA-E* (*Figure 3.2*).

Bjorkman and colleagues (1987a,b) determined the crystal structure of the human class I MHC molecule HLA-A2.1, the product of an allele at the *HLA-A* locus. The structure revealed that portions of the α_1 and α_2 domains

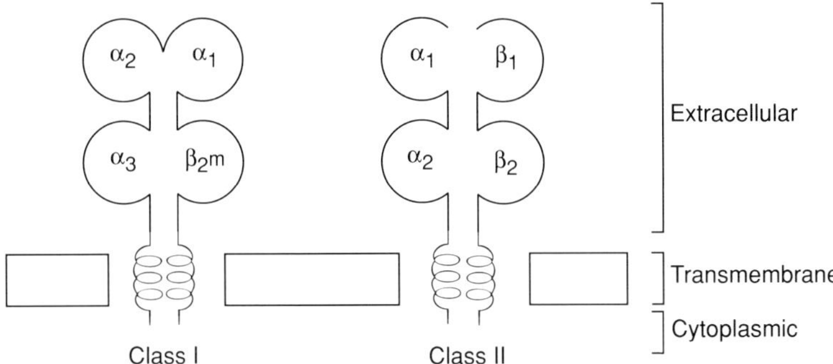

Figure 3.1. Structure of the class I and class II MHC molecules. The class I α chain consists of three extracellular domains: α_1, α_2, and α_3. The α_3 domain associates non-covalently with β_2-microglobulin (β_2m). The peptide-binding region is contained in the α_1 and α_2 domains. The class II heterodimer consists of α and β chains which have two extracellular domains each. The peptide-binding region is made up of portions of α_1 and β_1.

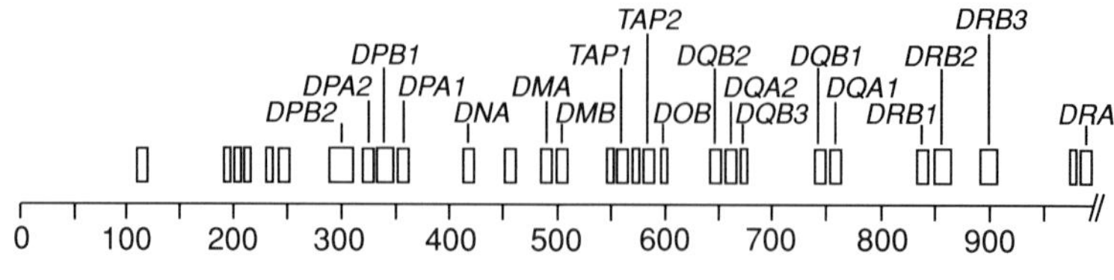

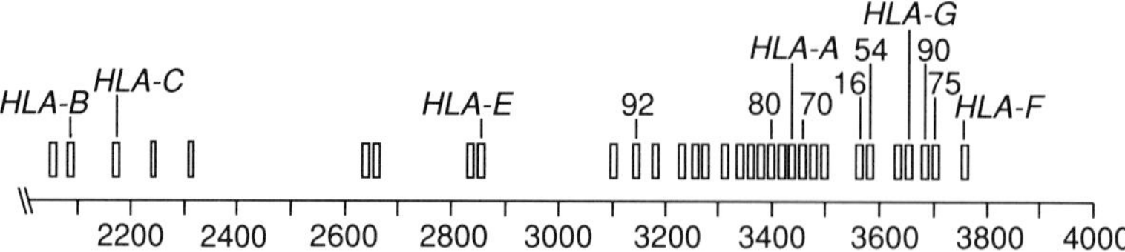

Figure 3.2. Map of class I and class II MHC genes on human chromosome 6p21.3. The group of unrelated genes sometimes called class III MHC, which lies between class I and class II clusters, is not shown. Numbers on *x* axes indicate distance in base pairs. Derived from Trowsdale (1995).

together form a groove at the top of the molecule (*Figure 3.3*). This groove is formed by a β-pleated sheet bound by two α helices. Because the groove is at the top of the molecule and thus able to interact with the T-cell receptor, it was deduced to be the site which binds peptides and presents them to T cells. This peptide-binding region (PBR) was called the antigen recognition site by Bjorkman *et al.* (1987a,b), who also noted that polymorphic amino acid residues are found mainly in this region.

The class II MHC genes of mammals are arranged in a number of separate regions, each of which typically contains at least one α chain gene and one or more β chain genes. In humans, there are three regions including polymorphic class II loci, designated *HLA-DR*, *HLA-DQ* and *HLA-DP* (*Figure 3.2*). Each of these regions contains a polymorphic β chain locus; in the *DR* region, there are between one and three β chain loci, depending on the haplotype. Only in the *DQ* region is the α chain gene polymorphic as well. In humans, the genes encoding α chains are designated *A* (e.g. *DRA*), and those encoding β chains are designated *B* (e.g. *DRB1*).

There are a number of class II loci that are non-polymorphic. Although some of these genes have parallels with class I non-classical loci, they have not been given any special designation in the literature. One particularly interesting example is the pair of loci *DMB* and *DMA* (*Figure 3.2*) (Kelly *et al.*, 1991; see Section 3.5).

A hypothetical structure for the class II heterodimer was initially proposed on the basis of analogy with the class I structure (Brown *et al.*, 1988). Recently, the structure of one class II molecule has been established (Brown *et al.*, 1993), and shown to be similar to the previous hypothetical model. At the top of the class II molecule, two α helices form a groove flanking a β-pleated

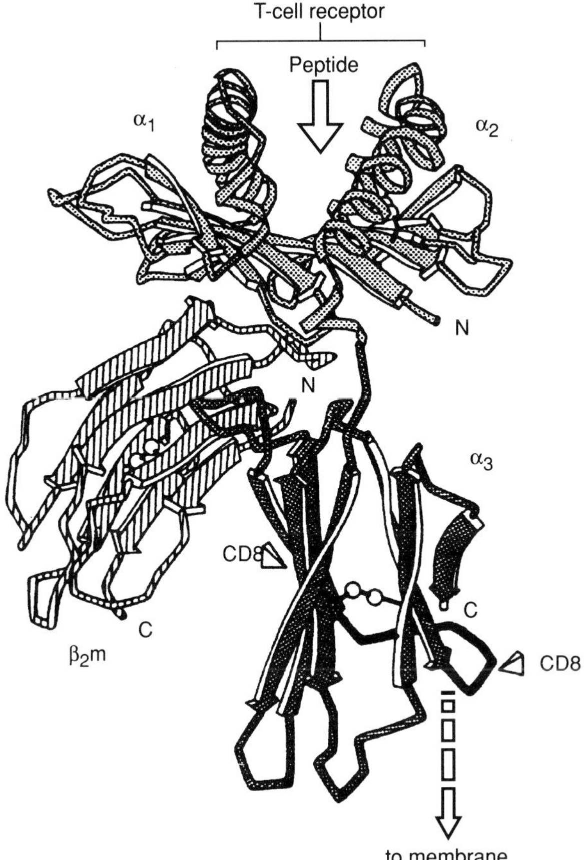

Figure 3.3. Molecular structure of class I MHC. The β strands are indicated by arrows in the amino to carboxyl direction; the α helices are shown as coiled ribbons. In the groove at the top of the molecule, peptides are bound and recognized by T cells (Lawlor *et al.*, 1990). Reproduced, with permission, from the *Annual Review of Immunology*, volume 8, © 1990, by Annual Reviews Inc.

sheet. One α helix and about half of the β-pleated sheet is contributed by the α_1 domain of the α chain, while the β_1 domain of the β chain contributes the other α helix and the remainder of the β-pleated sheet.

3.3 Maintenance of MHC polymorphism

The fact that MHC loci are polymorphic was known before the function of MHC molecules was understood. Thus, it was difficult to imagine an evolutionary mechanism that would explain MHC polymorphism. The polymorphism of class I MHC loci was discovered through their role in mediating transplant rejection, yet it seemed very unlikely that this role could be directly related to the selective pressures acting on the MHC in nature.

Later, when the function of the class I MHC in immune recognition was worked out, its role in antigen presentation suggested to Doherty and Zinkernagel (1975) a mechanism whereby overdominant selection (heterozygote advantage) might operate at MHC loci. They had evidence that products of different alleles at MHC loci differ in their ability to bind and present specific foreign peptides. Given such differences, in a population exposed to a variety of pathogens, an individual heterozygous at all or most MHC loci would presumably be able to present more kinds of foreign peptides than a homozygote. Thus, a heterozygote would be resistant to a wider array of pathogens than a homozygote.

This argument was initially not very popular with immunologists because it was difficult to obtain evidence of a correlation between MHC heterozygosity and survival. However, an analysis of allelic frequencies at human MHC loci provided strong support for the hypothesis that MHC polymorphisms are maintained by balancing selection (Hedrick and Thompson, 1983). In the case of neutral polymorphism, we expect to find one common allele and one or a few rare alleles. At polymorphic MHC loci in humans, by contrast, there are a large number of alleles of intermediate frequency, a pattern suggesting balancing selection (Hedrick and Thompson, 1983).

Hughes and Nei (1988, 1989b) approached the question of selection at MHC loci by comparing rates of nucleotide substitution. In protein-coding regions of DNA, point mutations (mutations changing a single base pair) can be classified as synonymous or non-synonymous. A synonymous mutation is one that, because of the redundancy built into the genetic code, does not change the amino acid sequence of the polypeptide encoded. A non-synonymous mutation is one that changes the amino acid sequence. To understand the evolutionary forces that have acted on two related genes since they diverged from a common ancestor, it is useful to estimate the number of synonymous nucleotide substitutions per synonymous site (d_S) and the number of non-synonymous nucleotide substitutions per non-synonymous site (d_N).

In estimating these quantities, the number of synonymous sites and the number of non-synonymous sites are first counted in the two sequences (Nei and Gojobori, 1986). A synonymous site is one (like the third codon position of glycine codons) at which every possible mutation would be synonymous. A non-synonymous site is one (like all second codon positions) at which every possible mutation would be non-synonymous. At certain sites some mutations are synonymous and some non-synonymous. Such a site is counted as some fraction synonymous and some fraction non-synonymous. For example, at the third position of the lysine codon AAA, one possible mutation (to AAG) is synonymous, whereas the other two (to AAC and AAT) are non-synonymous. This site would thus be counted as one-third of a synonymous site and two-thirds of a non-synonymous site.

In the case of functional genes, d_S is expected to exceed d_N. Because most amino acid changes will disrupt protein structure and thus are harmful to the organism, most non-synonymous mutations will be eliminated by conservative or 'purifying' natural selection. Synonymous mutations, being selectively neutral, or nearly so, will be fixed or lost by genetic drift. Thus, in the case of functional genes, d_S is expected to exceed d_N. In the case of a non-functional pseudogene, however, we expect d_S and d_N to be about equal, since in this case mutations at all codon positions are neutral. On the other hand, when natural selection is acting to favour changes at the amino acid level, it is predicted that d_N will exceed d_S.

Hughes and Nei (1988) reasoned that, on Doherty and Zinkernagel's (1975) hypothesis that MHC polymorphisms are maintained by overdominant selection relating to peptide binding and thus to pathogen resistance, such selection would act to favour amino acid differences in the PBR. Therefore d_N should exceed d_S in the codons of the PBR. This is precisely what they found. *Figure 3.4* shows the means of d_S and d_N for pairwise comparisons among alleles at the human class I loci *A*, *B* and *C*. These quantities were computed separately for the 57 codons that encode the residues in the PBR, and for the remainder of the α_1 and α_2 domains. In the PBR, mean d_N is about three times as high as mean d_S, whereas in the remainder of the gene d_S exceeds d_N. Similar results have been found for the class I MHC alleles from

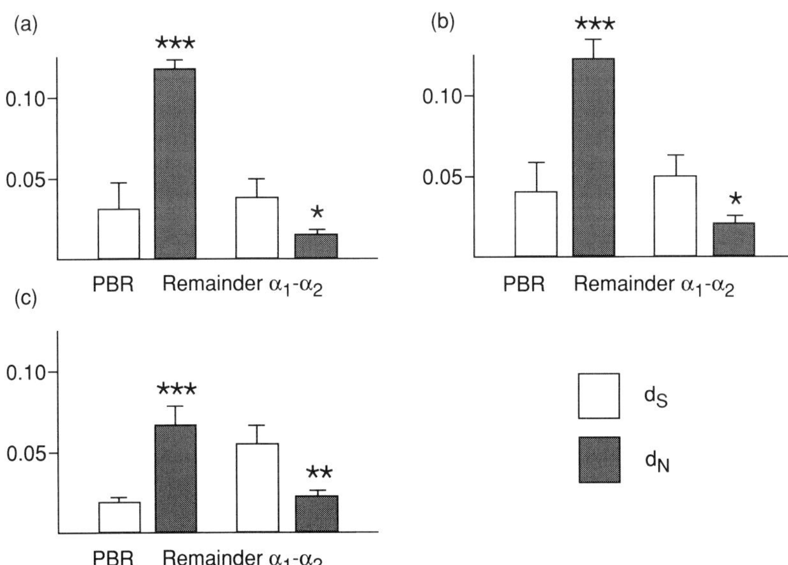

Figure 3.4. Mean numbers of synonymous (d_S) and non-synonymous (d_N) nucleotide substitutions per site in pairwise comparisons among human class I alleles: (a) *HLA-A*, (b) *HLA-B*, and (c) *HLA-C*. PBR, codons of the peptide-binding region; Remainder α_1-α_2, remainder of the α_1 and α_2 domains. Tests of the hypothesis that $d_S = d_N$: $*p < 0.05$; $***p < 0.001$.

the mouse, bovine and a number of other mammalian species (Garber *et al.*, 1993; Hughes and Nei, 1988; Watkins *et al.*, 1991).

Tables 3.1 and *3.2* show similar calculations for polymorphic class II MHC loci from human and mouse. At most polymorphic loci in both species, d_N is enhanced in the PBR (*Table 3.1*). The one exception is *DQA1* of humans. This locus has a relatively limited number of alleles although in some comparisons the alleles are quite divergent from each other and may be very ancient (Gyllensten and Erlich, 1989). In the *DR* region of humans, we find d_N greater than d_S in the PBR both in comparisons among alleles at one locus and in comparisons among loci (*Table 3.2*). Thus, in the case of the multiple *DR* β chains of humans, natural selection appears to have acted both to enhance polymorphism and to enhance diversity among loci.

Clearly, MHC polymorphism is maintained by some form of balancing selection that favours diversity in the PBR. This selection presumably relates to the advantage such diversity confers in enhancing the individual's ability to bind a wide array of foreign peptides and thus to resist a wide array of pathogens. The discovery of a human class I MHC allele and a class II haplotype associated with resistance to *Plasmodium falciparum* malaria in West Africa (Hill *et al.*, 1991) provides further support for the hypothesis that selection on MHC loci is driven by pathogens. Whether the selection acting on MHC loci is overdominant or dependent on frequency is difficult to determine at present. However, certain models of frequency-dependent selection can be ruled out since they make predictions incompatible with what we know of MHC polymorphism (Takahata and Nei, 1990). In addition, Doherty and Zinkernagel's (1975) model of overdominant selection at MHC loci has a strong biological appeal. This model is consistent with all available data and its biological basis (disease resistance) bears a straightforward relationship to MHC function.

Analysis of MHC DNA sequences has revealed still another interesting fact. Certain MHC alleles (or, more accurately, allelic lineages) are very ancient, having persisted since before the speciation events that have given rise to extant animal species. This pattern of long-term persistence of families of MHC alleles has been called 'trans-species polymorphism' (Mayer *et al.*, 1988), because the polymorphism predates speciation and persists through the speciation process. Actually, such long-lasting polymorphism is further evidence that some form of balancing selection is operating at MHC loci. Neutral polymorphisms, free to drift in frequency within a population, are transient when compared with polymorphisms maintained by overdominant selection with other forms of balancing selection (Takahata and Nei, 1990).

3.4 Class Ia and Ib genes

In all mammals studied in detail, the class I MHC region includes the highly expressed polymorphic class I classical or Ia loci, the class I non-classical or

Table 3.1. Numbers of synonymous (d_S) and non-synonymous (d_N) nucleotide substitutions per 100 sites between class II MHC alleles in mice and humans. Numbers of sequences compared are in parentheses

Comparisons (no. of sequences)	Peptide-binding region		Remainder, D1		D2	
	d_S	d_N	d_S	d_N	d_S	d_N
β-Chain loci						
Mouse						
Ab vs. *Ab* (9)	0.6 ± 0.8	21.7 ± 4.0***	4.3 ± 1.9	6.0 ± 1.3	9.2 ± 2.5	1.2 ± 0.5**
Eb vs. *Eb* (6)	5.2 ± 3.8	22.9 ± 4.4**	1.3 ± 1.0	3.1 ± 0.9	1.0 ± 0.8	0.5 ± 0.3
Human						
DPB vs. *DPB* (6)	1.7 ± 2.4	9.8 ± 3.0*	1.8 ± 1.4	1.9 ± 0.7	4.3 ± 1.9	0.7 ± 0.4
DQB vs. *DQB* (9)	7.3 ± 4.2	19.3 ± 3.6*	11.1 ± 2.9	5.7 ± 1.2	5.4 ± 2.0	1.7 ± 0.6
α-Chain loci						
Mouse						
Aa vs. *Aa* (6)	3.6 ± 3.3	15.8 ± 3.7*	2.7 ± 1.6	4.1 ± 1.0	7.6 ± 2.2	0.7 ± 0.4***
Human						
DPA vs. *DPA* (4)	0.0 ± 0.0	2.1 ± 1.5	5.1 ± 2.6	0.7 ± 0.5	2.4 ± 1.4	0.7 ± 0.4
DQA vs. *DQA* (6)	23.2 ± 12.0	7.2 ± 2.7	9.7 ± 3.0	8.4 ± 1.7	4.1 ± 1.7	1.4 ± 0.5

Tests of the hypothesis that $d_S = d_N$: * $p < 0.05$; ** $p < 0.01$; *** $p < 0.001$.
D1 = α_1 or β_1 domain; D2 = α_2 or β_2 domain.

Table 3. 2. Numbers of synonymous (d_S) and non-synonymous (d_N) nucleotide substitutions per 100 sites between class II MHC alleles in humans

Comparisons (no. of sequences)	Peptide-binding region		Remainder, D1		D2	
	d_S	d_N	d_S	d_N	d_S	d_N
DRB1 vs. *DRB1* (23)	6.1 ± 3.9	24.7 ± 3.5***	9.5 ± 2.5	3.4 ± 0.8*	6.1 ± 1.7	2.3 ± 0.6*
vs. *DRB3* (27)	6.3 ± 4.1	27.3 ± 4.3***	9.8 ± 3.4	4.3 ± 1.2	9.4 ± 2.7	2.4 ± 0.7*
vs. *DRB4* (24)	6.7 ± 4.6	31.2 ± 5.3***	12.5 ± 4.2	8.5 ± 2.2	7.1 ± 2.2	3.8 ± 1.0
vs.. *DRB5* (27)	4.2 ± 3.9	31.0 ± 5.5***	10.0 ± 3.5	3.3 ± 0.8	8.7 ± 2.6	3.9 ± 1.1
DRB3 vs. *DRB3* (4)	4.5 ± 4.5	10.3 ± 3.3	1.3 ± 1.3	1.2 ± 0.7	4.3 ± 1.9	0.9 ± 0.5
vs. *DRB4* (5)	3.8 ± 3.8	30.3 ± 7.9**	11.8 ± 5.0	9.1 ± 2.4	8.8 ± 3.4	4.6 ± 1.4
vs. *DRB5* (8)	3.7 ± 3.4	38.6 ± 8.8***	11.8 ± 4.7	4.4 ± 1.5	11.8 ± 3.8	3.9 ± 1.3*
DRB4 vs. *DRB5* (5)	4.6 ± 3.4	33.5 ± 8.6**	8.6 ± 4.0	6.9 ± 2.1	9.9 ± 3.8	3.8 ± 1.3
DRB5 vs. *DRB5* (4)	4.3 ± 4.4	10.8 ± 3.7	2.7 ± 1.9	0.8 ± 0.6	2.7 ± 1.6	0.6 ± 0.4
Overall mean (32)	5.6 ± 3.9	26.9 ± 3.3***	9.7 ± 2.2	3.8 ± 0.7*	7.4 ± 1.7	2.7 ± 0.6**

Tests of the hypothesis that $d_S = d_N$: * $p < 0.05$; ** $p < 0.01$; *** $p < 0.001$.
D1 = α_1 or β_1 domain; D2 = α_2 or β_2 domain.

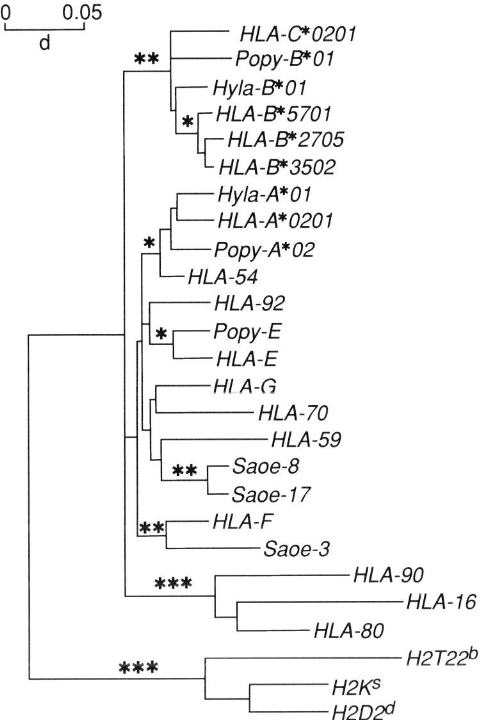

Figure 3.5. Minimum evolution tree (Rzhetsky and Nei, 1992) of class I MHC genes from primate and mouse, based on the numbers of nucleotide substitutions per site (d). Prefixes indicate species as follows: *HLA*, human; *H2*, mouse (*Mus musculus*); *Hyla*, gibbon (*Hylobates lar*); *Popy*, orangutan (*Pongo pygmaeus*); *Saoe*, cottontop tamarin (*Saguinus oedipus*). Tests of the hypothesis that the length of a given internal branch = 0: $^{*}p < 0.05$; $^{**}p < 0.01$; $^{***}p < 0.001$.

class Ib loci (which have a more restricted expression), and non-expressed pseudogenes and gene fragments. The function of the class Ib loci remains mysterious, although it is now known that at least some of them can present antigens (Pamer *et al.*, 1992). When a phylogenetic tree of class I genes is constructed from mammals of different orders, a remarkable pattern is seen: in every case, the class Ib genes of one order cluster with the class Ia genes of that order (Hughes and Nei, 1989a). This pattern of relationship is illustrated by a phylogenetic tree of selected class I genes from primates and the mouse (*Figure 3.5*). All the primate genes cluster together, being separated from the rodent genes by a strongly supported internal branch.

This type of clustering pattern can be explained by either of the following two hypotheses: (i) the class Ib genes have arisen independently by gene duplication in different orders of mammals (Hughes and Nei, 1989a); or (ii) some process of interlocus recombination or 'gene conversion' has acted to homogenize the class I loci within each order of mammals ('concerted evolution'; Rada *et al.* 1990).

Rada *et al.* (1990) compared certain class I sequences from the rat with certain mouse sequences. Because the rat sequences were more similar to each other (as measured by the overall percentage of nucleotide similarity) than to any of the mouse sequences analysed, these authors claimed that a process of concerted evolution has homogenized the class I genes within these two species (which separated 15–25 million years ago). More complete analysis revealed orthologous relationships between certain class I genes of mouse and rat (Hughes, 1991a). Therefore, the hypothesis of concerted evolution in the class I MHC does not seem well supported in this case (Hughes, 1991a). Orthologous relationships are now known among both class Ia and class Ib loci of Old World and New World primates. Interestingly, the class Ia loci of a New World monkey, the cottontop tamarin (*Saguinus oedipus*), appear to be closely related to the human class 1b locus *HLA-G* (Watkins *et al.*, 1990; *Figure 3.5*).

Such findings suggest that the class I MHC evolved by a process of repeated gene duplication, reduction in level of expression of certain duplicated genes, silencing of other duplicated genes, and finally deletion of genes. This would explain why orthologous relationships are found among class I loci of mammals of the same order but not among mammals of different orders, since over a long enough period of time there would be a turnover of class I loci. On this hypothesis, the class Ia and class Ib loci do not necessarily constitute mutually exclusive groups over time (Watkins *et al.*, 1990). Rather, both class Ib loci and class I pseudogenes evolve from duplicated class Ia loci (Hughes, 1995; Hughes and Nei, 1989a).

In some cases the class I MHC may have evolved by duplication of segments containing more than one gene. The truncated pseudogene *HLA-80* is closely linked to *HLA-A* (about 7 kb distant; Geraghty *et al.*, 1992), while the related pseudogene *HLA-16* gene is linked to *HLA-54* (*Figure 3.2*). This pattern may be explained by a tandem duplication involving the ancestor of *HLA-80* and *HLA-16* and the ancestor of *HLA-A* and *HLA-54*. A similar, earlier tandem duplication may have given rise to *HLA-90* and *HLA-75*, which have a linkage pattern similar to that of *HLA-80* and *HLA-A* or *HLA-16* and *HLA-54* (*Figure 3.2*).

Many authors have suggested that after gene duplication MHC genes may interact through interlocus recombination events, which are usually hypothesized to have occurred through a 'gene-conversion' mechanism. To test for patterns of interlocus recombination, Hughes (1995) constructed separate phylogenetic trees for different regions of a set of human class Ia genes, class Ib genes and class I pseudogenes. The regions analysed were: (i) the 5′ flanking region (5′FR), (ii) the 5′ exons and introns (5′EI), consisting of exons 1–3 and introns A–B, (iii) intron C, which is the longest intron in the HLA class I gene, and (iv) the 3′ exons and introns (exons 3–7, introns D–F) (3′EI). These were chosen because preliminary analyses suggested recombination among these regions but did not provide strong evidence of recombination within these regions.

In the 5′FR, the genes analysed formed two major clusters, which were separated by statistically significant internal branches (*Figure 3.6a*): (i) a cluster

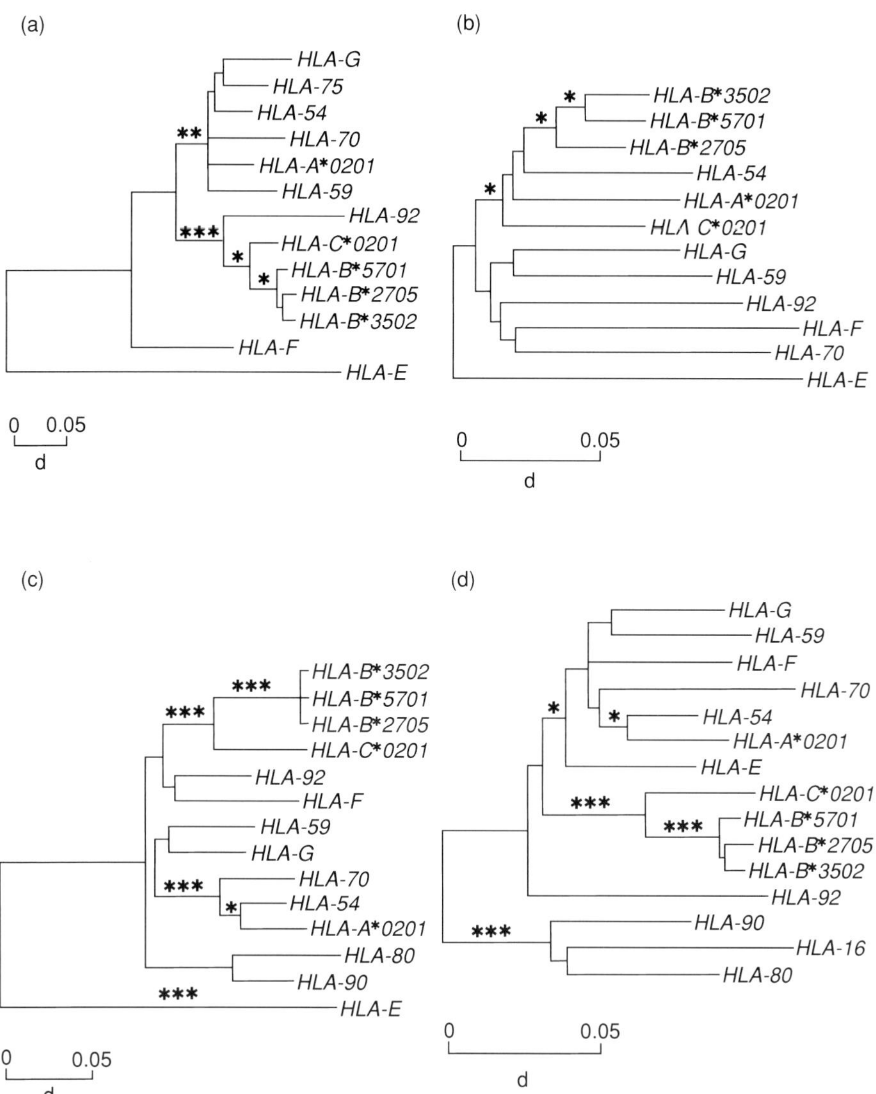

Figure 3.6. Phylogenetic trees of different regions of *HLA* class I genes, based on the numbers of nucleotide substitutions per site (d). Regions from which phylogenies are constructed are: (a) 5′ flanking region (5′FR); (b) exons 1-3 and introns A–B (5′EI); (c) intron C; and (d) exons 4–7 and introns D–F (3′EI). Tests of the hypothesis that the length of an internal branch = 0: $^*p < 0.05$; $^{**}p < 0.01$; $^{***}p < 0.001$.

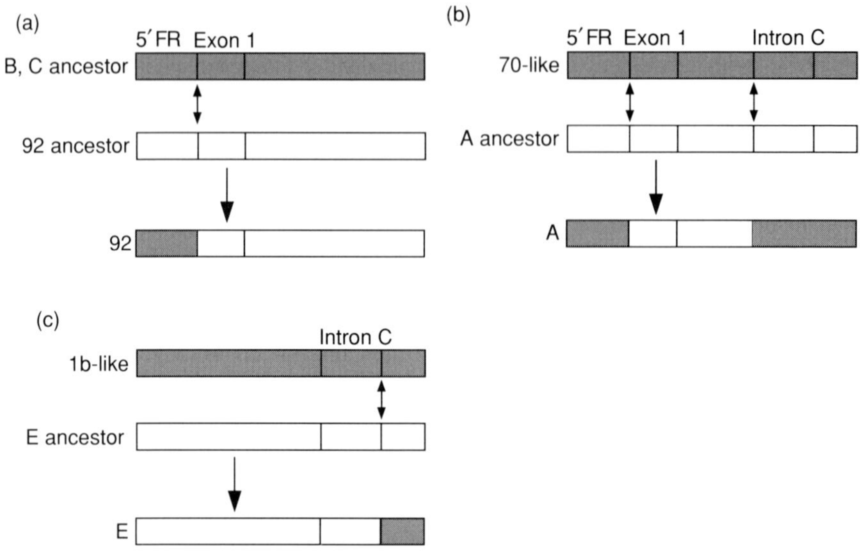

Figure 3.7. Some hypothetical recombination events in the history of the *HLA* class I region (see text for details). Double-headed arrows indicate recombination points.

of *HLA-B* and *HLA-C* alleles along with the class I pseudogene *HLA 92*; and (ii) *HLA-A* and *HLA-G* and all other class I pseudogenes analysed. *HLA-E* and *HLA-F* fell outside both of these clusters. Thus, in the 5′FR, *HLA-92* is closely related to the class Ia *B* and *C* loci. In other regions, however, the pattern of relationship of *HLA-92* is quite different. In the 5′EI, *HLA-A*, *HLA-B* and *HLA-C* cluster together, apart from class Ib genes and pseudogenes, including *HLA-92* (*Figure 3.6b*). In intron C (*Figure 3.6c*) and the 3′EI (*Figure 3.6d*), the relationship of *HLA-92* to other alleles is not clearly resolved, but in neither of these regions does *HLA-92* show the strong pattern of clustering with *HLA-B* and *HLA-C* seen in the 5′FR. Therefore, it seems likely that the 5′FR of *HLA-92* is derived from a gene related to the ancestor of *HLA-B* and *HLA-C*, although the rest of the *HLA-92* gene is not closely related to these loci (*Figure 3.7a*).

Hughes and Nei (1988, 1989b) observed that *HLA-A* alleles are much more divergent from *HLA-B* and *HLA-C* alleles in exons 4–5 than in exons 2-3. They attributed this pattern to a past recombinational event in the history of the *HLA-A* locus; observing similarity between *HLA-A* alleles and the class Ib locus *HLA-E* in exons 4–5, Hughes and Nei (1989b) suggested that the ancestor of *HLA-A* had recombined with an *E*-like locus.

In the present analyses, *HLA-A* clusters with the other class Ia loci in the 5′EI (*Figure 3.6b*). In fact, this same pattern was seen in a tree based on introns A and B only (data not shown), indicating that this clustering pattern is not simply due to convergent evolution among class Ia loci in the PBRs encoded in exons 2–3. By contrast, in the 5′FR, in intron C and in the 3′EI (*Figures 3.6a,c,d*), *HLA-A* clusters with a group of class Ib and pseudogene loci, which includes *HLA-70*, *HLA-G* and *HLA-59*. In intron C, the clustering of *HLA-A* and the closely related *HLA-54* with *HLA-70* is supported by a branch that received strong statistical support (*Figure 3.6c*). Thus, these results suggest that the recombination event in the history of the *HLA-A* locus may have involved the donation of exons 1–3 and introns A–B by a class-Ia-like locus to a locus more closely related to *HLA-70* than to *HLA-E* (*Figure 3.7b*).

HLA-E in the 3′EI clusters with *HLA-A*, *HLA-F*, *HLA-G*, *HLA-59* and *HLA-70* (*Figure 3.6d*). However, in other regions, it is much more divergent, clustering outside all other class I genes. Thus, *HLA-E* itself appears to be a recombinant, with exons 1–3 and introns A–C originating from a highly divergent gene and exons 4–8 and introns D–G originating from a gene similar to most other *HLA* class Ib and pseudogene loci (*Figure 3.7c*).

Such analyses indicate that interlocus recombination has been a recurrent feature in the evolutionary history of the *HLA* class I region. This is true of other multi-gene families (e.g. Hughes, 1991b, 1992), and there is no indication that the rate of such events is particularly high in the case of the *HLA* class I.

3.5 Evolution of class II loci

Whereas class I MHC loci do not show orthologous relationships among different orders of mammals, orthologous relationships are generally found among mammalian class II loci (Hughes and Nei, 1990). This pattern is illustrated by a phylogenetic tree based on the conserved domain 2 amino acid sequences of α and β chain genes (*Figure 3.8*). Since the gene duplication that gave rise to separate α and β chains occurred early in the history of the MHC, the α chain tree serves to root the β chain tree and vice versa. An example of a cluster of orthologous genes is that containing *DR* β chain genes from primates (human and tamarin), Rodentia (the *Eb* gene of mouse), Carnivora (dog) and Artiodactyla (pig and sheep) (*Figure 3.8*). Similarly, orthologues of the *DP* and *DQ* region genes are found in non-human mammals (*Figure 3.8*). Therefore, the class II gene regions seem to have arisen prior to the divergence of the orders of placental mammals.

Like the class I MHC, the class II MHC includes both polymorphic and monomorphic loci as well as a number of pseudogenes. Certain α chain loci are monomorphic (such as human *DRA*) but form heterodimers with highly polymorphic β chain loci. In addition, certain class II loci seem analogous with class I 'non-classical' loci in that they are monomorphic and lack a known function. In the mouse the *Na* and *Ob* genes encode a nearly monomorphic heterodimer (Karlsson and Peterson, 1992).

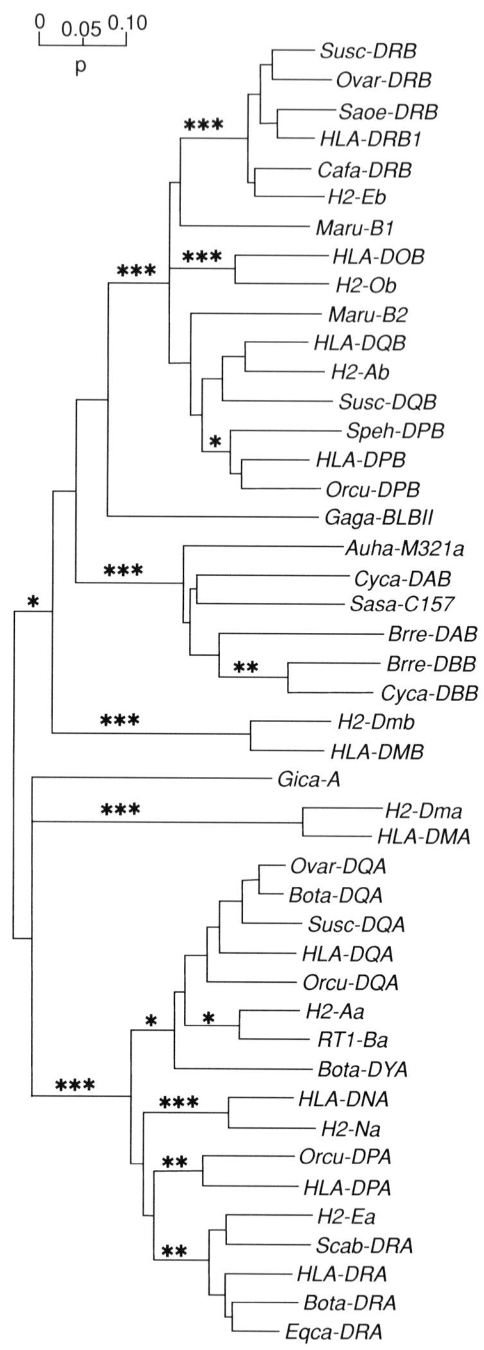
0
0.05
0.10
p
***
***
***
*
***
*
***
**
***
***
*
*
***
***
**
**
Susc-DRB
Ovar-DRB
Saoe-DRB
HLA-DRB1
Cafa-DRB
H2-Eb
Maru-B1
HLA-DOB
H2-Ob
Maru-B2
HLA-DQB
H2-Ab
Susc-DQB
Speh-DPB
HLA-DPB
Orcu-DPB
Gaga-BLBII
Auha-M321a
Cyca-DAB
Sasa-C157
Brre-DAB
Brre-DBB
Cyca-DBB
H2-Dmb
HLA-DMB
Gica-A
H2-Dma
HLA-DMA
Ovar-DQA
Bota-DQA
Susc-DQA
HLA-DQA
Orcu-DQA
H2-Aa
RT1-Ba
Bota-DYA
HLA-DNA
H2-Na
Orcu-DPA
HLA-DPA
H2-Ea
Scab-DRA
HLA-DRA
Bota-DRA
Eqca-DRA

Homologues of these genes (*DNA* and *DOB*) exist in humans but in this case the heterodimer may not be expressed because of defects in the *DNA* mRNA (Trowsdale and Kelly, 1985). The *DMA* and *DMB* genes in humans (*Ma* and *Mb* in the mouse) constitute another pair of genes encoding an almost monomorphic heterodimer (Cho *et al.*, 1991; Kelly *et al.*, 1991).

Figure 3.8 illustrates a contrast between monomorphic class II loci and class Ib loci; unlike the latter, the monomorphic class II loci seem to have diverged early in mammalian evolution. Both human *DOB* and *DNA* cluster with their mouse homologues *H2-Ob* and *H2-Na*, respectively (*Figure 3.8*). More remarkably, *DMA* clusters outside all other mammalian class II β chain genes, and *DMB* clusters outside all class II β chain genes. In each case, this pattern is strongly supported (*Figure 3.8*). In the case of class II β chain genes, chicken and bony fish sequences are available and the *DMB* genes cluster outside both of these, although here this branching pattern is not very strongly supported. The tree thus suggests that the *DMA/DMB* heterodimer may be a very ancient feature of the vertebrate class II MHC, having arisen by a tandem duplication of α and β chain genes possibly before tetrapods diverged from bony fishes. Such an ancient divergence is consistent with evidence that the DM heterodimer plays a role in facilitating antigen binding by class II molecules (Sanderson *et al.*, 1994).

3.6 Conclusions

Only 25 years ago, the function of MHC molecules was unknown. A great deal of progress has been made in our understanding of the MHC since that time, both as a result of immunological experiments that have revealed the function of the MHC and of genetic dissection of the MHC region of selected species. Jan Klein in particular championed the view that the MHC could only be understood in an evolutionary perspective: "The study of MHC evolution will probably be a slow process, which will not be marked by spectacular successes. Yet, if we want to comprehend the MHC, we must conquer the last frontier"

Figure 3.8. Phylogenetic tree of class II MHC α_2 and β_2 domains, based on the proportion of amino acid difference (p). Prefixes indicate species as follows: *HLA*, human; *H2*, mouse (*Mus musculus*); *Auha*, cichlid fish (*Aulonocara hansbaenschi*); *Bota*, bovine (*Bos taurus*); *Brre*, zebrafish (*Brachydanio rerio*); *Cafa*, dog (*Canis familiaris*); *Cyca*, carp (*Cyprinus carpio*); *Eqca*, horse (*Equus caballus*); *Gaga*, chicken (*Gallus gallus*); *Gici*, nurse shark (*Ginglymostoma cirratum*); *Maru*, red-necked wallaby (*Macropus rufogriseus*); *Orcu*, rabbit (*Oryctalagus cuniculus*); *Ovar*, sheep (*Ovis aries*); *Saoe*, cottontop tamarin (*Saguinus oedipus*); *Sasa*, salmon (*Salmo salar*); *Scab*, tassel-eared squirrel (*Sciurus aberti*); *Speh*, mole rat (*Spalax ehrenbergi*); *Susc*, pig (*Sus scrofa*); *RT1*, rat (*Rattus norvegicus*). Tests of the hypothesis that the length of an internal branch = 0: $*p < 0.05$; $**p < 0.01$; $***p < 0.001$.

(Klein, 1986, p. 716). Clearly, in the past decade, the process of conquest has begun. Some of the findings have been summarized here while other exciting results have been passed over due to space limitations. But, as so often in biology, new discoveries have raised new questions and the attempt to answer these in turn will challenge the ingenuity of a new generation of MHC researchers.

Acknowledgement

My research on the evolution of the MHC is supported by grants from the National Institutes of Health.

References

Bjorkman PJ, Parham P. (1990) Structure, function, and diversity of class I major histocompatibility complex molecules. *Annu. Rev. Biochem.* **59**: 253–288.

Bjorkman PJ, Saper MA, Samraoui B, Bennett WS, Strominger JL, Wiley DC. (1987a) Structure of the human class I histocompatibility antigen, HLA-A2. *Nature* **329**: 506–512.

Bjorkman PJ, Saper MA, Samraoui B, Bennett WS, Strominger JL, Wiley DC. (1987b) The foreign antigen binding site and T cell recognition regions of class I histocompatibility antigens. *Nature* **329**: 512–518.

Brown JH, Jardetzky T, Saper MA, Samraoui B, Bjorkman PJ, Wiley DC. (1988) A hypothetical model of the foreign antigen binding site of class II histocompatibility molecules. *Nature* **332**: 845–850.

Brown JH, Jardetzky TS, Gorga JC, Stern LJ, Urban RG, Strominger JL, Wiley DC. (1993) Three dimensional structure of the human class II histocompatibility antigen HLA-DR1. *Nature* **364**: 33–39.

Cho S, Attaya M, Monaco JJ. (1991) New class II-like genes in the mouse MHC. *Nature* **353**: 573–576.

Doherty PC, Zinkernagel R. (1975) Enhanced immunologic surveillance in mice heterozygous at the H-2 gene complex. *Nature* **256**: 50–52.

Garber TL, Hughes AL, Letvin NL, Templeton JW, Watkins DI. (1993) Sequence and evolution of cattle MHC class I cDNAs: concerted evolution has not taken place in cattle. *Immunogenetics* **38**: 11–20.

Geraghty DE, Koller BH, Hansen JA, Orr HT. (1992) The HLA class I gene family includes at least six genes and twelve pseudogenes and gene fragments. *J. Immunol.* **149**: 1934–1936.

Gyllensten UB, Erlich HA. (1989) Ancient roots for polymorphism at the HLA-DQa locus in primates. *Proc. Natl Acad. Sci. USA* **86**: 9986–9990.

Hedrick PW, Thompson G. (1983) Evidence for balancing selection at HLA. *Genetics* **104**: 449–456.

Hill AVS, Allsopp CEM, Kwiatkowski D *et al.* (1991) Common West African HLA antigens are associated with protection from severe malaria. *Nature* **352**: 595–600.

Howard JC. (1987) MHC organization of the rat: evolutionary considerations. In: *Evolution and Vertebrate Immunity* (eds G Kelsoe and DH Schulze). University of Texas Press, Austin, TX, pp. 397–411.

Hughes AL. (1991a) Independent gene duplications, not concerted evolution, explain relationships among class I MHC genes of murine rodents. *Immunogenetics* **33**: 367–373.

Hughes AL. (1991b) Evolutionary origin and diversification of the mammalian CD1 antigen genes. *Mol. Biol. Evol.* **8**: 185–201.

Hughes AL. (1992) Coevolution of the vertebrate integrin α- and β-chain genes. *Mol. Biol. Evol.* **9**: 216–234.

Hughes AL. (1995) Origin and evolution of *HLA* class I pseudogenes. *Mol. Biol. Evol.* **12**: 247–258.

Hughes AL, Nei M. (1988) Pattern of nucleotide substitution at major histocompatibility complex class I loci reveals overdominant selection. *Nature* **335**: 167–170.

Hughes AL, Nei M. (1989a) Nucleotide substitution at major histocompatibility complex class II loci: evidence for overdominant selection. *Proc. Natl Acad. Sci. USA* **86**: 958–962.

Hughes AL, Nei M. (1989b) Evolution of the major histocompatibility complex: independent origin of nonclassical class I genes in different groups of mammals. *Mol. Biol. Evol.* **6**: 559–579.

Hughes AL, Nei M. (1990) Evolutionary relationships of class II major-histocompatibility-complex genes in mammals. *Mol. Biol. Evol.* **7**: 491–514.

Karlsson L, Peterson PA. (1992) The α chain of H-2O has an unexpected location in the major histocompatibility complex. *J. Exp. Med.* **176**: 477–483.

Kelly AP, Monaco JJ, Cho S, Trowsdale J. (1991) A new human HLA class II-related locus DM. *Nature* **353**: 571–576.

Klein J. (1986) *Natural History of the Major Histocompatibility Complex.* John Wiley & Sons, New York.

Klein J, Figueroa F. (1986) Evolution of the major histocompatibility complex. *CRC Crit. Rev. Immunol.* **6**: 295–386.

Lawlor DA, Ward FE, Ennis PD, Jackson AP, Parham P. (1988) HLA-A,B polymorphisms predate the divergence of humans and chimpanzees. *Nature* **335**: 268–271.

Lawlor DA, Zemmour J, Ennis PD, Parham P. (1990) Evolution of class I MHC genes and proteins: from natural selection to thymic selection. *Annu. Rev. Immunol.* **8**: 23–64.

Mayer WE, Jonker D, Klein D, Ivanyi P, van Seventer G, Klein, J. (1988) Nucleotide sequence of chimpanzee MHC class I alleles: evidence for trans-species mode of evolution. *EMBO J.* **7**: 2765–2774.

Monaco JJ. (1992) A molecular model of MHC class-I-restricted antigen processing. *Immunol. Today* **13**: 173–178.

Nei M, Gojobori T. (1986) Simple methods for estimating the numbers of synonymous and nonsynonymous nucleotide substitutions. *Mol. Biol. Evol.* **3**: 418–426.

O'Rourke AM, Mescher MF. (1993) The role of CD8 in cytotoxic T lymphocyte function. *Immunol. Today* **14**: 183–188.

Pamer EG, Wang C-R, Flaherty L, Fischer Lindahl K, Beran MJ. (1992) H-2M3 presents a *Listeria monocytogenes* peptide to cytotoxic T lymphocytes. *Cell* **70**: 215–223.

Rada C, Lorenzi R, Powis SJ, van den Bogaerde J, Parham P, Howard JC. (1990) Concerted evolution of class I genes in the major histocompatibility complex of murine rodents. *Proc. Natl Acad. Sci. USA* **87**: 2167–2171.

Rammensee H-G, Friede T, Stevanovic S. (1995) MHC ligands and peptide motifs: first listing. *Immunogenetics* **41**: 178–228.

Rzhetsky A, Nei M. (1992) A simple method for estimating and testing minimum evolution trees. *Mol. Biol. Evol.* **9**: 945–967.

Sanderson F, Kleijmeer MJ, Kelly A, Verwoerd D, Tulp A, Neefjes JJ, Genze HJ, Trowsdale J.(1994) Accumulation of HLA-DM, a regulator of antigen presentation, in MHC class II compartments. *Nature* **266**: 1566–1569.

Takahata N, Nei M. (1990) Allelic genealogy under overdominant and frequency-dependent selection and polymorphism of major histocompatibility complex loci. *Genetics* **124**: 967–978.

Trowsdale J. (1995) 'Both man and bird and beast': comparative organization of MHC genes. *Immunogenetics* **41**: 1–17.

Trowsdale J, Kelly A. (1985) The human HLA class II α chain gene DZα is distinct from genes in the DP, DQ, and DZ subregions. *EMBO J.* **9**: 2231–2237.

Tulp A, Verwoerd D, Dobberstein B, Ploegh H, Pleters J. (1994) Isolation and characterization of the intracellular MHC class II compartment. *Nature* **369**: 120–126.

Watkins DI, Chen ZW, Hughes AL, Evans MG, Tedder TF, Letvin NL. (1990) Evolution of the MHC class I genes of a New World primate from ancestral homologues of human non-classical genes. *Nature* **346**: 60–63.

Watkins DI, Garber TL, Chen ZW, Toukatly G, Hughes AL, Letvin NL. (1991) Unusually limited nucleotide sequence variation of the expressed major histocompatibility complex class I genes of a New World primate species (*Saguinus oedipus*). *Immunogenetics* **33**: 79–89.

4

Evolution of the G-protein-coupled receptor superfamily

Shozo Yokoyama and William T. Starmer

4.1 Introduction

Transmembrane signalling pathways have generated a great deal of attention because they modulate many physiological and pharmacological events. Despite their functional diversities, the signalling systems are remarkably similar, possessing three common components: a receptor, an effector, and a heterotrimeric guanine-nucleotide-binding protein (or G protein). The G-protein-coupled receptors (GPCRs) are integral membrane proteins involved in the transmission of signals across membranes (*Figure 4.1*). After binding a suitable ligand, GPCRs undergo one or more conformational changes that trigger interactions between the receptor and the G protein. These initiate the intracellular signalling response and modulate the activity of distinct effector systems such as adenylyl cyclase, phospholipases, cGMP phosphodiesterases and ion channels. GPCRs also undergo desensitization, becoming refractory to further stimulation after the initial response despite the continued presence of a stimulus of constant intensity. Phosphorylation is a critical event in regulating this process (*Figure 4.1*; see also Dohlman *et al.*, 1991; Gilman, 1987; Kaziro *et al.*, 1991; Stryer and Bourne, 1986).

One striking common feature of diverse GPCRs is the existence of seven hydrophobic segments with distinctive sequence patterns, implying that they diverged from a common ancestor and share the same basic tertiary structure. Using extensive sequence data from the GPCR superfamily, we can now start answering basic evolutionary questions. For example, how are these diverse GPCRs evolutionarily related? How did paralogous GPCR genes, derived

Human Genome Evolution, edited by M. Jackson, T. Strachan and G. Dover.

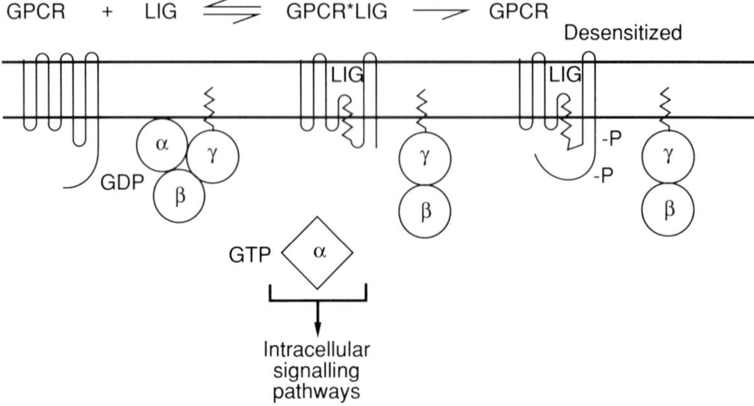

Figure 4.1. Activation and desensitization of a G-protein-coupled receptor. The receptor (GPCR) interacts with a ligand (LIG) to stabilize the activated form of the receptor (GPCR*). This catalyses the exchange of α-subunit-associated GDP for GTP. The G-protein α subunit can then dissociate to initiate intracellular signalling. Desensitization of the receptor involves phosphorylation (-P) and can involve other proteins (see Section 4.3.2). Modified with permission from Simon *et al.* (1991) *Science* **252**: 802–808. ©1991 American Association for the Advancement of Science.

from independent gene duplication events, achieve similar functions? How did orthologous genes derived from speciation events result in similar proteins achieving different functions? Since these questions are closely related to the functions of GPCRs, such knowledge in turn helps us to understand the mechanisms involved in the functional differentiation of different GPCRs. In particular, evolutionary analyses can identify changes that are potentially responsible for the important process of functional differentiation. Effects of such changes on some GPCRs have been evaluated using site-directed mutagenesis (Baldwin 1994; Strader *et al.*, 1994) and the feasibility of such mutagenesis experiments for the GPCR superfamily provides a unique opportunity to combine evolutionary and genetic approaches for the rigorous study of molecular mechanisms involved during functional differentiation.

This chapter reviews the results of molecular evolutionary studies of GPCRs whose ligands include neurotransmitters, neuropeptides and peptide hormones. Before going into detail, however, the general features of GPCRs, G proteins and their interactions will be reviewed.

4.2 G-protein α subunits

G proteins are multi-subunit proteins, consisting of α, β and γ subunits. When guanosine diphosphate (GDP) is bound by a G protein, the Gα subunit associates with the Gβγ subunits to form a membrane-bound Gαβγ complex.

When guanosine triphosphate (GTP) is bound, the Gα subunit (Gα-GTP) dissociates from the Gβγ subunits. The Gα-GTP then alters the activity of the target enzyme or channel. The role of Gβγ is to present Gα-GTP to the activated receptor. The role of GPCRs in this system is to serve as catalysts for the action of G proteins. The Gβγ subunits are structurally similar and functionally interchangeable, but the Gα subunit is unique for each G protein (Gilman, 1987; Kaziro *et al.*, 1991; Simon *et al.*, 1991; Stryer and Bourne, 1986). Thus, specificity in the signalling pathways seems to be achieved by the Gα subunits as a result of interactions with distinct GPCRs and effectors (Conklin and Bourne, 1993).

4.2.1 Evolutionary relationship of Gα subunits

Currently, about 60 different Gα subunit genes have been characterized in eukaryotes. The topology and branch lengths of a phylogenetic tree for the Gα subunit genes have been evaluated by applying the neighbour-joining method (Saitou and Nei, 1987) to the numbers of nucleotide substitutions at the first and second positions within each codon. The levels of bootstrap support were determined by bootstrap analysis with 1000 replications (CLUSTAL V; Higgins *et al.*, 1992). According to the neighbour-joining tree, the Gα subunits may be classified into three groups: the GPA (G-protein α subunits), Gα-I and Gα-II groups (*Figure 4.2*). The GPA group consists of Gα subunits from fungi, plants and slime mould. The Gα-I group consists of inhibitory G_i, transducin G_t and other (G_o and G_x) subunits, while the Gα-II group consists of stimulatory G_s, G_q and G_{12} subunits. This tree topology is quite different from those of Simon *et al.* (1991), Strathman and Simon (1991) and Wilkie *et al.* (1992), where the inhibitory G_i and stimulatory G_q subunits are most closely related, with the common ancestor of these subunits and those of other stimulatory subunits (G_{12} and G_s) diverging about the same time. The difference seems to have occurred because these authors assumed the rate of amino acid replacement was constant, while *Figure 4.2* was obtained without the assumption of rate constancy. As described below, *Figure 4.2* clearly shows that the Gα subunits evolved at different rates and that the rate constancy assumption does not seem to hold for these molecules.

The function of the Gα subunits within a given cluster are often similar (Wilkie and Yokoyama, 1994). For instance, the G_s subunits activate adenylyl cyclase whereas the G_q subunits mediate pertussis-toxin-resistant activation of phospholipase Cβ. G_s and G_q subunits cluster with the G_{12} subunits, whose signalling properties remain to be clarified. G_i subunits regulate a more diverse set of effectors but their functions require fatty acid acylation and are often inhibited by pertussis-toxin-catalysed ADP ribosylation. Interestingly, transducins t1 and t2, which detect visual signals, and gustducin, which mediates taste, belong to the same G_t cluster, suggesting a common evolutionary origin of the two signalling systems.

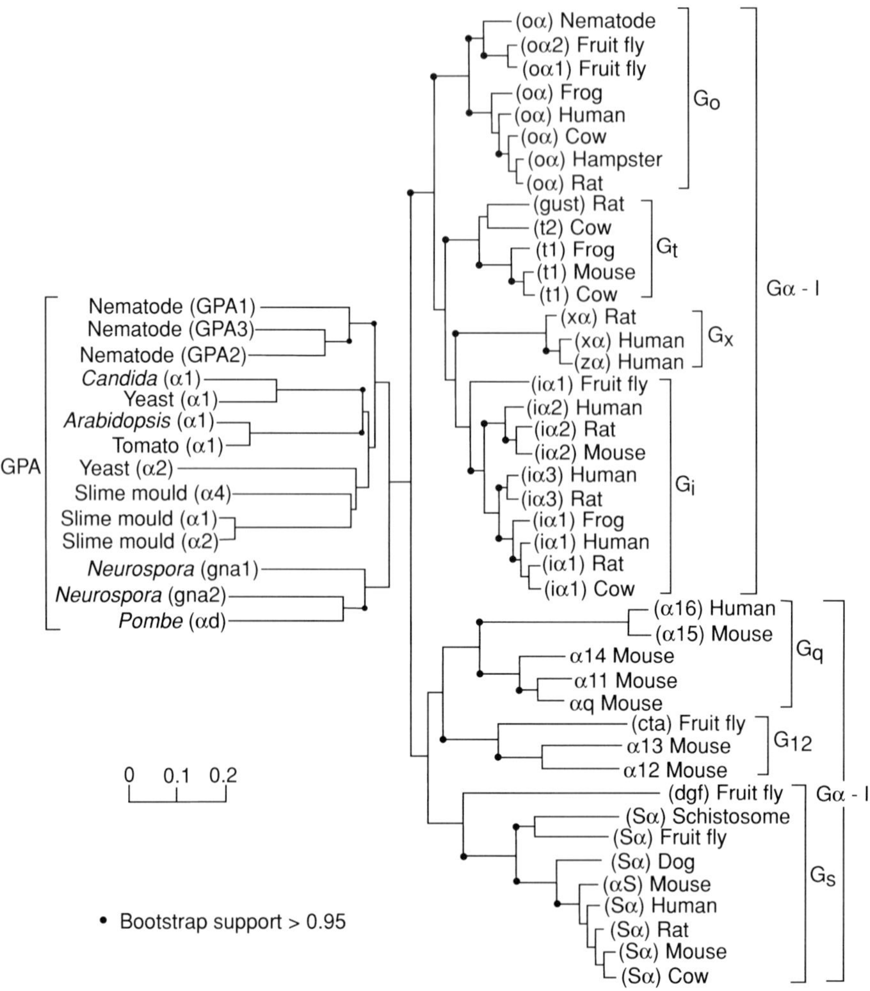

Figure 4.2. Phylogenetic tree of G-protein α subunits. Branch length was measured by the number of nucleotide substitutions per site. Reproduced from *Molecular Biology and Function of Carrier Proteins – 4th Annual Symposium – Society of General Physiologists* (1994), pp. 249–270, by copyright permission of The Rockefeller University Press.

4.2.2 Evolutionary rates

Figure 4.2 shows that significantly fewer nucleotide changes have accumulated in the Gα-I group than in the Gα-II group. Furthermore, within the G_q cluster, the α11, α14 and αq subunits are evolving at a much slower rate than the α15 and α16 subunits. Interestingly, α15 and α16 are expressed specifically in haematopoietic cells whereas α11, α14 and αq are widely expressed. The faster evolutionary rate of α15 and α16 may be explained either by a relaxation of selective constraints or by functional adaptation to the haematopoietic cell environment (Wilkie and Yokoyama, 1994).

It is also of interest to point out that during mammalian evolution the rates of synonymous and non-synonymous substitutions for the Gα subunit genes have been estimated to be $1.77–5.63 \times 10^{-9}$ per site per year and $0.008–0.067 \times 10^{-9}$ per site per year, respectively (Yokoyama and Starmer, 1992). The latter rates are similar to those for histone genes, suggesting that the Gα subunits have equally important biological functions (Yokoyama and Starmer, 1992).

4.3 Some features of GPCRs

In *Table 4.1*, GPCRs are separated into five different groups according to their endogenous ligands: (i) peptides and peptide hormones; (ii) neurotransmitters; (iii) other regulatory factors; (iv) sensory stimuli; and (v) others with unknown ligands (orphans). It is clear that the lengths of GPCRs vary considerably, ranging from the 297 amino acid residue human adrenocorticotrophin (Acth-Hs) to the 764 residue human thyrotrophin (Tsh-Hs) (*Table 4.1*).

4.3.1 Protein structure

As already noted, a striking feature of the amino acid sequences of GPCRs is the presence of seven stretches of relatively hydrophobic amino acids representing possible membrane-spanning domains. This profile is remarkably similar to that of bacteriorhodopsin, a non-G-protein-coupled light-driven proton pump from *Halobacterium halobium*. When analysed by electron microscopy and high-resolution electron diffraction, bacteriorhodopsin is seen as seven α-helical transmembrane segments arranged in a bundle perpendicular to the plane of the lipid bilayer (Harrison and Unwin, 1975). Since the seven hydrophobic domains in each GPCR are similar in size and of sufficient length to span the lipid bilayer, bacteriorhodopsin-like topographies have been proposed for the bovine rhodopsin (Hargrave, 1982; Ovchinnikov *et al.*, 1982) and for the hamster β2-adrenergic receptor (Dixon *et al.*, 1986). Recently, the map of bovine rhodopsin has been determined by two-dimensional electron crystallography to a resolution of 9 Å, which confirmed the existence of seven transmembrane helices (Schertler *et al.*, 1993). Similarly, a model for the hamster β2-adrenergic receptor having seven transmembrane helices has also been supported by proteolytic mapping and immunolocalization of antipeptide antibody-binding sites (Dohlman *et al.*, 1991).

All of these analyses strongly suggest that most GPCRs can be considered to have seven transmembrane domains (Baldwin, 1993; Dohlman *et al.*, 1991). *Figure 4.3* shows a schematic representation of bovine rhodopsin in the membrane with seven transmembrane segments, TM I–TM VII (Hargrave *et al.*, 1983; Nathans, 1992). According to this model, loops I–II, III–IV and V–VI and the carboxyl terminus are located on the cytoplasmic surface of the

Table 4.1. G-protein-coupled receptors categorized according to endogenous ligands

Receptor	Amino acid number	Organism	Notation	Accession number
Peptide and peptide hormones				
Adrenocorticotrophin	297	Human	Acth-Hs	X65633
Angiotensin 1	359	Human	Ang1-Hs	M93394
Angiotensin 2	363	Mouse	Ang2-Mm	S67465
Angiotensin 2	359	Human	Ang2-Hs	Z11162
Arginine vasopressin	394	Rat	Avr-Rn	Z11690
Antidiuretic hormone	371	Human	Adhr-Hs	Z11687
Bradykinin	391	Human	Bk2A-Hs	M88714
C5a anaphylatoxin	350	Human	C5ana-Hs	X58674
Cholecystokinin	444	Rat	Cck-Rn	M88096
Endothelin	426	Rat	Endo-Rn	M60786
Endothelin	442	Human	Endo-Hs	M74921
Follicle-stimulating hormone	695	Human	Fsh-Hs	M65085
N-formylpeptide R26	350	Human	Fmlp26-Hs	M60627
N-formylpeptide I	353	Human	FmlpI-Hs	M76673
N-formylpeptide II	351	Human	FmlpII-Hs	M76672
Gastrin	453	Dog	Gstr-Cf	M87834
Gastrin-releasing peptide	384	Human	Grp-Hs	M73481
Gonadotrophin-releasing hormone	327	Mouse	Gorh-Mm	M93108
Interleukin 8-1	350	Human	IL8-1-Hs	M68932
Interleukin 8-2	355	Human	IL8-2-Hs	M73969
Luteinizing hormone/human chorionic gonadotrophin	699	Human	Lh/Hcg-Hs	M63108
Melanocortin	360	Human	Mc-Hs	L06155
Melanocyte-stimulating hormone	317	Human	Msh-Hs	X65634
Neuromedin B	390	Human	Nmedin B-Hs	M73482
Neuromedin K	465	Human	NmedinK-Hs	X65176
Neuropeptide Y	384	Human	NpepY-Hs	M84755
Neuropeptide Y	353	Bovine	NpepY-Bt	M86739
Neurotensin	418	Human	Ntensin-Hs	X70070
Opioid	440	Human	Opioid-Hs	M84605
Oxytocin	389	Human	Oxy-Hs	X64878
Somatostatin 1	391	Human	Sstat1-Hs	M81829
Somatostatin 2	369	Human	Sstat2-Hs	M81830
Substance P	407	Human	SubsP-Hs	M74290
Substance K	398	Human	SubsK-Hs	M60284
Thyrotrophin	764	Human	Tsh-Hs	M31774
Thyrotrophin-releasing hormone	393	Mouse	Trf-Mm	M59811
Neurotransmitters				
Adenosine A1	326	Rat	AdnA1-Rn	M64299
Adenosine A2A	409	Human	AdnA2A-Hs	M97370
Adenosine A2B	332	Human	AdnA2B-Hs	M97759
Adenosine A3	317	Sheep	AdnA3-Oa	S65334
Adrenaline α1A	501	Human	Adrα1A-Hs	M76446
Adrenaline α1B	515	Rat	Adrα1B-Rn	X51585
Adrenaline α1C	466	Bovine	Adrα1C-Bt	J05426
Adrenaline α2A	450	Human	Adrα2A-Hs	M18415
Adrenaline α2B	450	Human	Adrα2B-Hs	M34041
Adrenaline α2C	461	Human	Adrα2C-Hs	J03853
Adrenaline β1	477	Human	Adrβ1-Hs	J03019
Adrenaline β2	413	Human	Adrβ2-Hs	J02960
Adrenaline β3	402	Human	Adrβ3-Hs	M29932
Dopamine D1	446	Human	DopaD1-Hs	X58987
Dopamine D2	443	Human	DopaD2-Hs	M29066
Dopamine D2B	444	Rat	DopaD2B-Rn	X14028
Dopamine D3	454	Rat	DopaD3-Rn	M69194

Table 4.1. *(continued)*

Receptor	Amino acid number	Organism	Notation	Accession number
Dopamine D4	467	Human	DopaD4-Hs	X58497
Dopamine D5	477	Human	DopaD5-Hs	X58454
Histamine H1	491	Bovine	HisH1-Bt	D10197
Histamine H2	359	Human	HisH2-Hs	M64799
5-Hydroxytryptamine 1a	421	Human	5Ht1a-Hs	X13556
5-Hydroxytryptamine 1b	390	Human	5Ht1b-Hs	M89478
5-Hydroxytryptamine 1b	386	Rat	5Ht1b-Rn	X62944
5-Hydroxytryptamine 1c	458	Human	5Ht1c-Hs	M81778
5-Hydroxytryptamine 1d	377	Human	5Ht1d-Hs	M89955
5-Hydroxytryptamine 1e	365	Human	5Ht1e-Hs	Z11166
5-Hydroxytryptamine 2a	456	Human	5Ht2a-Hs	M86841
5-Hydroxytryptamine 2b	479	Rat	5Ht2b-Rn	X66842
Muscarinic receptor 1	460	Human	Musc1-Hs	M35128
Muscarinic receptor 2	466	Human	Musc2-Hs	X15264
Muscarinic receptor 3	479	Human	Musc3-Hs	X15265
Muscarinic receptor 4	590	Human	Musc4-Hs	X15266
Muscarinic receptor 5	532	Human	Musc5-Hs	M80333
Other regulatory factors				
Cannabinoid	472	Human	Cann-Hs	X54937
Prostaglandin E	365	Mouse	PgE-Mm	D10204
Platelet-activating factor	342	Human	Paf-Hs	M76674
Thrombin	425	Human	Thr-Hs	M62424
Thromboxane A2	341	Rat	TxA2-Rn	D21158
Sensory stimuli				
Olfactory stimulus	313	Dog	Olf-Cf	X64996
Olfactory stimulus 071	314	Human	Olf071-Hs	X64994
Olfactory stimulus 07J	320	Human	Olf07J-Hs	X64995
Olfactory stimulus F3	333	Rat	OlfF3-Rn	M64376
Olfactory stimulus F5	313	Rat	OlfF5-Rn	M64377
Olfactory stimulus F6	311	Rat	OlfF6-Rn	M64378
Olfactory stimulus F8	312	Rat	OlfF8-Rn	M64387
Olfactory stimulus F12	317	Rat	OlfF12-Rn	M64381
Olfactory stimulus I3	310	Rat	OlfI3-Rn	M64385
Olfactory stimulus I7	327	Rat	OlfI7-Rn	M64386
Olfactory stimulus I9	314	Rat	OlfI9-Rn	M64388
Olfactory stimulus I14	312	Rat	OlfI14-Rn	M64391
Olfactory stimulus I15	314	Rat	OlfI15-Rn	M64392
Rhodopsin	348	Human	Rhod-Hs	K02281
Green opsin	351	Chicken	Gops-Gg	M88178
Blue opsin	361	Chicken	Bops-Gg	M92037
Violet opsin	347	Chicken	Vops-Gg	M92039
Green opsin	355	Fish	G103-Af	M60938
Red opsin	357	Fish	R007-Af	M90075
Red opsin	364	Human	Rops-Hs	M13300
Orphan				
Aorta cells (RTA)	343	Rat	Rta-Rn	M35297
Endothelial cells	381	Human	Edg-Hs	M31210
Liver cells	395	Rat	G10D-Rn	L09249
Mas oncogene	325	Human	Mas-Hs	M13150
Mas-related	378	Human	MasX-Hs	S78653
Pituitary cells	320	Rat	Cgpc-Rn	X61496
T cells	423	Mouse	Gpc-Mm	M80610
Thyroid cells	362	Dog	Rdc1-Cf	X14048

Af, *Astyanax fasciatus* (Mexican cave fish); Bt, *Bos taurus* (cow); Cf, *Canis familiaris* (dog); Hs, *Homo sapiens* (human); Mm, *Mus musculus* (mouse); Oa, *Ovis aries* (sheep); Rn, *Rattus norvegicus* (rat);

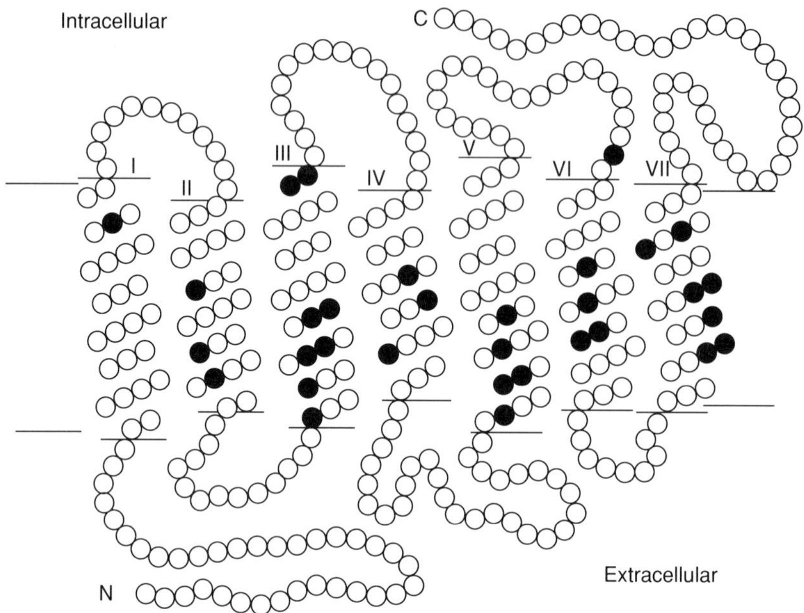

Figure 4.3. Model for the bovine rhodopsin structure (Hargrave and McDowell, 1992). The seven hydrophobic transmembrane segments are denoted I–VII. Filled circles show sites of mutations that affect the function of opsins or the binding of agonists or antagonists. For specific amino acid changes, see Table 1 in Baldwin (1994) and references therein. C, carboxyl terminal; N, amino terminal.

intact disk membrane, while the amino-terminal sequence and loops II–III, IV–V, and VI–VII bind to the outside surface of the plasma membrane. Compared with rhodopsin, the hamster β2-adrenergic receptor has a much longer carboxyl terminus and longer loop V–VI (Strader *et al.*, 1994).

4.3.2 Structure–function relationship

GPCRs, their stimuli, interacting Gα subunits and effectors are shown in *Table 4.2*, together with an amazingly wide range of physiological responses that can be incurred.

Ligand binding. A wide range of extracellular signalling molecules binds to specific GPCRs. The binding of the ligand promotes an interaction between the receptor and a G protein on the intracellular side of the membrane. Thus, one of the major functions of most GPCRs is to recognize specific hormones and regulatory molecules present in the extracellular environment. In this sense, phototransduction is unique since the ligand, the retinal chromophore, which is either 11-*cis* retinal (vitamin A1 aldehyde) or 3-dehydroretinal (vitamin A2 aldehyde), is covalently bound to opsin. The combined protein, known as a visual pigment, is responsible for the capture of photons and the initiation of visual excitation.

Many experiments have been designed to demonstrate which sites in GPCR molecules contribute to ligand-binding. Recently, Baldwin (1994) has reviewed sites of mutations that affect the function of opsins or the binding of agonists or antagonists. The sites of such mutations suggest that the transmembrane region may be distinguished into two parts: the intracellular third of the transmembrane region and the central and extracellular third of this region (see filled circles in *Figure 4.3*). The sites in the upper third of *Figure 4.3*, towards the intracellular side of the transmembrane region, are probably too far from the extracellular side of the membrane to be part of a ligand-binding pocket. The mutations in this region, however, may affect agonist binding by altering the ease with which the conformational transition between high- and low-affinity forms takes place, or they may affect coupling by altering the conformation of the G-protein binding site (Kjelsberg *et al.*, 1992; Kosugi *et al.*, 1992; Parma *et al.*, 1993; Ren *et al.*, 1993; Savarese *et al.*, 1992). The sites in the middle and extracellular thirds seem to affect antagonist binding. They are unlikely to alter G-protein binding directly because the sites are too far from the intracellular surface (Baldwin, 1994).

The mutation analyses also show that when the endogenous ligand is a small molecule, the GPCR lacks a significant amino-terminal extracellular domain and the ligand binds in a pocket that is located in the extracellular half of the transmembrane domain. Typical GPCRs of this type are adrenergic and muscarinic acetylcholine receptors. On the other hand, human follistatin, (Fsh-Hs), luteinizing hormone/human chorionic gonadotrophin (Lh/Hcg-Hs) and Tsh-Hs (*Table 4.2*) bind to the relatively large glycoprotein hormones and contain a large amino-terminal extracellular domain. This domain has been established as the ligand-binding site using chimeric receptors (Braun *et al.*, 1991; Nagayama *et al.*, 1991). Some larger neuropeptides are also expected to bind the extracellular loops and/or the amino terminus because the ligands are too large to be accommodated entirely in a pocket between the seven helices. For example, the amino-terminal sequence and loop II–III of human substance SubsP-Hs have been shown to be important for ligand binding (Fong *et al.*, 1992a,b).

Coupling to G proteins. The interactions between GPCRs and G proteins in the intracellular domain have been examined by constructing chimeric receptors and site-directed mutants. The important regions in binding and interacting with G proteins include loop III–IV, loop V–VI and the carboxyl-terminal region near transmembrane segment VII. For example, loop V–VI and the carboxyl-terminal region of the adrenalin receptor Adrα2-Hs confer the specificity for coupling to G_i and G_o proteins, as demonstrated by replacing different segments of Adrα2-Hs with the equivalent segments of Adrβ2-Hs (which couples to G_s) (Cheung *et al.*, 1991; Liggett *et al.*, 1991).

Table 4.2. GPCRs, their G proteins, effectors, and some physiological responses[a]

Receptor	Stimulus	G protein	Effector[b]	Physiological responses
Peptides/peptide hormones				
Acth,Mc,Msh	Melanocortins	G_s	Activation of AC	Regulation of the synthesis and release of glucocorticoids
Ang 1,2	Angiotensin	G_q	Activation of PP	Stimulation of vasoconstriction, retention of salt and water
Avr	Vasopressin	G_q	Activation of PP	Regulation of the water content in the body
Adhr	Vasopressin	G_s	Formation of cAMP	Stimulation of water reabsorption in the urinary tract
Bk2A	Bradykinin	G_q	Activation of PP	Contraction of smooth muscles
C5ana	C5a anaphylatoxin	G_q	Activation of PP	Accumulation of phagocytic cells at the site of injury or infection
Cck	Cholecystokinin	G_q	Activation of PP	Gall bladder contraction, pancreatic enzyme secretion, secretion/absorption in the gastrointestinal tract
Endo	Endothelin	G_q	Activation of PP	Regulation of cardiovascular system
Fsh	Follicle-stimulating hormone	G_s	Activation of AC	Regulation of oestrogen secretion in female, gametogenesis in male
Gstr	Gastrin	G_q	Activation of PP	Stimulation of acid secretion in stomach
Grp	Bombesins	G_q	Activation of PP	Contraction of smooth muscles
Gorh	Gonadotrophin-releasing hormone	G_q	Activation of PP	Regulation of reproduction
IL8-1,-2	Interleukin 8		Stimulation of PP	Inflammatory response
Lh/Hcg	Luteinizing hormone/human chorionic gonadotrophin	G_s	Activation of AC	Stimulation of ovulation, stimulation of testosterone
Fmlp26,I,II	Formyl-methyonil Peptides	G_i/G_o	Activation of PP, inhibition of AC	Accumulation of phagocytic cells at the site of injury or infection
NmedinB	Neuromedin B	G_q	Activation of PP	Contraction of smooth muscles
Nmedin K	Neuromedin K	G_q	Activation of PP	Regulation of dopaminergic neurones
Opioid	Opioid peptide	G_i/G_o	Activation of K^+, inhibition of AC	Regulation of sensory function, neuroendocrine activity, respiration, mood and gut motility
NpepY	Neuropeptide Y	G_i/G_o	Inhibition of AC	Stimulation of food intake, anxiety, facilitation of learning and memory, regulation of the cardiovascular and neuroendocrine systems
Ntensin	Neurotensin	G_q	Activation of PP	Contraction of smooth muscles
Oxy	Oxytocin	G_q	Activation of PP	Contraction of uterine smooth muscle
Sstat1,2	Somatostatin	G_i/G_o	Inhibition of AC	Inhibition of growth hormone secretion, regulation of smooth muscle contraction, locomotion, and cognitive function
SubsP,K	Substance P and K	G_q	Activation of PP	Contraction of smooth muscles
Tsh	Thyrotrophin	G_s	Activation of AC	Stimulation of iodine uptake, synthesis of thyroid hormones
Trf	Thyrotrophin-releasing hormone	G_q	Activation of PP	Stimulate synthesis and release of thyroid-stimulating hormone

Neurotransmitters				
AdnA1	Adenosine	G_i/G_o	Inhibition of AC and Ca^{2+}, activation of K^+	Vasodilation, bronchoconstriction, immunosuppression, reduction of motor activity, depression of respiration
AdnA2A	Adenosine	G_s	Activation of AC	
AdnA2B	Adenosine	G_s	Stimulation of cAMP	
AdnA3	Adenosine	G_i/G_o	Inhibition of AC	
Adrα1A,B,C	Adrenaline	G_q	Activation of PP	Regulation of cardiovascular system
Adrα2A	Adrenaline	G_i/G_o	Inhibition of AC and Ca^{2+}, activation of K^+	
Adrα2B	Adrenaline	G_i/G_o	Inhibition of AC and Ca^{2+}	
Adrα2C	Adrenaline	G_i/G_o	Inhibition of AC	
Adrβ1,2,3	Adrenaline	G_s	Activation of AC	
DopaD1,D5	Dopamine	G_s	Stimulation of AC	Regulation of movement
DopaD2,D3,D4	Dopamine	G_i/G_o	Inhibition of AC and Ca^{2+}, activation of K^+	Cognitive and emotional responses, vasodilation
HistH1	Histamine	G_q	Activation of PP	Mediator of inflammation and allergy; regulation of gastric acid from parietal cells
HistH2	Histamine	G_s	Formation of cAMP	
5Ht1a,b,d,e	Serotonin	G_i/G_o	Inhibition of AC, activation of K^+	Contraction of smooth muscles, induction of endothelin-dependent vasodilation
5Ht1c,2a,2b	Serotonin	G_q	Activation of PP	
Musc1,3,5	Acetylcholine	G_q	Activation of PP	Contraction of skeletal and smooth muscles
Musc2,4	Acetylcholine	G_i/G_o	Inhibition of AC and Ca^{2+}	Inhibition of cardiac muscles
Other regulatory factors:				
Cann	Cannabinoid	G_i/G_o	Inhibition of AC	Hallucinations, memory deficit
PgE	Prostaglandins	G_i/G_o	Inhibition of AC	Mediator in the cardiovascular and immune systems, and in pain
Paf	Platelet-activating factor		Activation of PP	Mediator in allergic and inflammatory conditions
Thr	Thrombin		Activation of PP, inhibition of AC	Stimulation of aggregation and secretion in blood platelets
TxA2	Thromboxanes	G_q	Activation of PP	Regulation of the interaction of platelets with the vascular endothelium
Sensory stimuli				
Olf	Odour	G_s/G_{olf}	Formation of cAMP, activation of PP	Odour detection
Opsin	Light	G_t	Stimulation of cGMP	Visual excitation

[a]For more details, see Watson and Arkinstall (1994).
[b]AC, adenylyl cyclase; Ca^{2+}, Ca^{2+} channel; cGMP, cGMP phosphodiesterase; K^+, K^+ channel; PP, phosphoinositide pathway.

Mutations at the first and second residues in loop III–IV of rhodopsin have been shown to affect activation of transducin, G_t (Cohen *et al.*, 1993; Franke *et al.*, 1992; Min *et al.*, 1993) and the inositol triphosphate (IP3) response (Ohyama *et al.*, 1992). Similarly, charged residues at either end of loop V–VI in the muscarinic receptor Musc3-Hs (Kunkel and Peralta, 1993) and the thyrotrophin receptor Tsh-Hs (Parma *et al.*, 1993) are important in its interaction with G proteins.

Receptor desensitization. A general property of the signal–response system is that prolonged stimulation of a cell often results in reduced responsiveness to subsequent challenges by the same stimulus (*Figure 4.1*). Dohlman *et al.* (1991) distinguished the diminished role of receptors to agonist stimulation, termed receptor desensitization, into two types: homologous desensitization; and heterologous desensitization. The former affects only the agonist-activated receptor, where prolonged agonist exposure results in an uncoupling of the agonist-occupied receptor from its G protein and effector system. In the latter, prolonged agonist exposure results not only in a reduced response from the occupied receptor but also in a reduced response from other receptors which are coupled to the same G protein and effector system. For example, heterologous desensitization of the adrenalin receptor Adrβ2 involves phosphorylation of the receptor by the cAMP-dependent protein kinase and follows any stimulus that raises cAMP levels in the cell, while homologous desensitization of the Adrβ2 receptor occurs only at high levels of agonist (Dohlman *et al.*, 1991).

Rhodopsin is one of the most extensively studied GPCRs with respect to desensitization. It undergoes a light-dependent and ATP-dependent phosphorylation. The enzyme responsible for this activity, rhodopsin kinase, uses only the photoactivated form of rhodopsin as a substrate (Kuhn, 1978). Rhodopsin is phosphorylated at multiple serine and threonine sites in the carboxyl terminus (Wilden and Kuhn, 1982). Although phosphorylation of photoexcited rhodopsin does not completely prevent its ability to activate transducin, it allows the binding of arrestin to photoexcited rhodopsin and formation of this complex prevents the further activation of transducin (Wilden *et al.*, 1986). After the photoactivated conformation of phosphorylated rhodopsin relaxes, a phosphatase is able to dephosphorylate the receptor. After the rebinding of chromophore 11-*cis* retinal, a rhodopsin molecule is again capable of participating in the phototransduction process (Hargrave and McDowell, 1992).

The β-adrenergic receptors are phosphorylated by a related but different enzyme: β-adrenergic kinase (Benovic *et al.*, 1987b). Similarly, a protein homologous to arrestin, β-arrestin, has been shown to terminate receptor signalling by the β-adrenergic receptor (Benovic *et al.*, 1987a). Although our knowledge of receptor desensitization is mostly limited to rhodopsin and β-adrenergic receptors, this mechanism may be common to other G-protein receptors (Savarese and Fraser, 1992).

4.4 Phylogenetic and functional relationships of GPCRs

Despite their variable lengths, GPCR amino acid sequences can be aligned reasonably well around transmembrane segments I–VII. The alignment shows that length variation occurs in three major regions: the carboxyl terminus; the amino terminus; and loop V–VI. Compared with rhodopsin (*Figure 4.3*), the three longest peptide receptors (those for follistatin, luteinizing hormone/human chorionic gonadotrophin and thyrotrophin) all have much longer amino-terminal sequences, while several neurotransmitter receptors (AdnA2-Hs, Dopa D1-Hs, Dopa D5-Hs, Adrα1A-Hs, and Adrα1C-Bt) have much longer carboxyl-terminal sequences. Similarly, the loop V–VI region of adrenergic, dopamine, histamine and muscarinic acetylcholine receptors are much longer than that of rhodopsin.

In constructing the phylogenetic tree for the GPCRs listed in *Table 4.1*, only the region between transmembrane segment I and VII has been considered, which corresponds to residues 37–311 of rhodopsin. The topology and branch lengths of the phylogenetic tree were evaluated using the neighbour-joining method based on the number of Poisson-corrected amino acid replacements per site for the pairwise comparisons (*Figure 4.4*).

Figure 4.4 shows that GPCRs may be distinguished into two major groups: the peptide receptors (P1–7 clusters); and neurotransmitter receptors (N1–4 clusters), with a rather poor bootstrap support of 0.58. Two types of sensory receptors, opsins (S1 cluster) and olfactory receptors (S2 cluster) are distantly related to each other and are more closely related to the peptide receptors. Different clusters in *Figure 4.4* are reasonably consistent with those of Mountjoy *et al.* (1992), where a total of 39 GPCRs were considered. As we will see, the evolutionary relationships of the GPCRs often reflect those of the Gα subunits discussed in Section 4.2.

4.4.1 Neurotransmitter receptors (N1 and N2 clusters)

Adenosines are ubiquitous modulators of numerous physiological activities, particularly within the cardiovascular and nervous systems. AdnA2 receptors stimulate adenylyl cyclase (AC) through G_s while AdnA3 receptors inhibit adenylyl cyclase through G_i and G_o (Williams, 1987). Interestingly, the two types of receptors are evolutionarily distinguished in the N1 cluster (*Figure 4.4*).

In the N2 cluster, three types of adrenergic receptors (Adrα1, Adrα2 and Adrβ3), serotonin receptors (5Ht1a, b, d and e), dopamine receptors (DopaD1 and DopaD5), and the histamine receptor (HisH2) are almost equally distantly related (*Figure 4.4*). The catacholamines adrenaline (epinephrine) and noradrenaline (norepinephrine) mediate various physiological effects in different tissues by binding to different subtypes of adrenergic receptors (Dohlman *et al.*, 1991). The serotonin receptors mediate a number of peripheral effects of serotonin, including the contraction of a number of smooth muscles, endothelium-dependent coronary artery relaxation, and

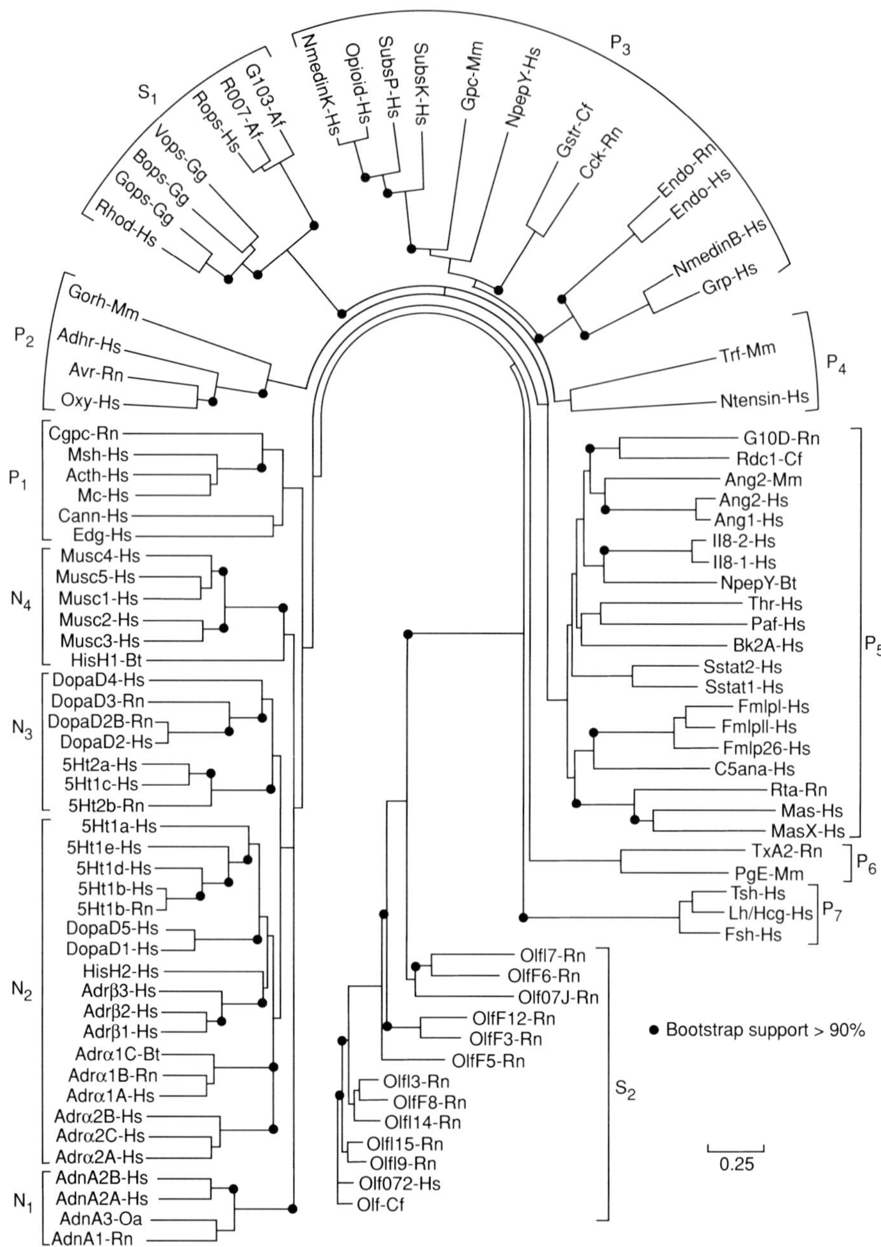

Figure 4.4. Phylogenetic tree of GPCRs. Branch length was measured by the number of amino acid replacements per site. Two-letter prefixes indicate the species of origin as follows: Af, *Astynaxas fasciatus* (Mexican cave fish); Bt, *Bos taurus* (Cow); Cf, *Canis familiaris* (dog); Gg, *Gekko gekko* (gecko); Hs, *Homo sapiens* (human); Mm, *Mus musculus* (mouse); Oa, *Ovis aries* (sheep); Rn, *Rattus norvegicus* (rat); N, neurotransmitter receptors; P, peptide receptors; S, sensory receptors.

prejunctional inhibition of sympathetic norepinephrine release (Hamblin and Metcalf, 1991). The dopaminergic system in brain function is important because of its association with neurological and psychiatric disorders such as Parkinson's disease and schizophrenia (Sunahara *et al.*, 1991). Similarly, histamine receptors regulate gastric acid secretion and have an important effect on the treatment of peptic ulcer disease (Ganz *et al.*, 1991). The Adrβ1–3 and dopamine D1 and D5 receptors stimulate adenylyl cyclase through G_s and the Adrα1 receptors activate the phosphoinositide pathway through G_q, while the 5Ht1 and Adrα2 receptors inhibit adenylyl cyclase through G_i and G_o (*Table 4.2*).

4.4.2 Neurotransmitter receptors (N3 and N4 clusters)

The N3 cluster consists of serotonin receptors 5Ht1c, 5Ht2 and dopamine receptors DopaD2, D3 and D4. The serotonin receptors stimulate phosphoinositide hydrolysis through G_q (Foguet *et al.*, 1992), while the dopamine receptors inhibit adenylyl cyclase through G_i and G_o (*Table 4.2*). In *Figure 4.4*, the two groups of serotonin and dopamine receptors belong to the N2 and N3 clusters. However, the bootstrap supports for the N2 and N3 clusters are only 0.26 and 0.29, respectively. Thus, the closer relationships implied in *Figure 4.4* between the 5Ht1a, b, d and e receptors and the DopaD1 and D5 receptors (in the N2 cluster) and between the 5Ht1c, 2a and 2b receptors and the DopaD2, D3 and D4 receptors (in the N3 cluster) are not as firm as they may seem.

The N4 cluster contains various muscarinic acetylcholine receptors and the histamine receptor HisH1. Among the muscarinic acetylcholine receptors, two groups (Musc1, 4 and 5, and Musc2 and 3) are clearly distinct (*Figure 4.4*). Functionally, however, Musc1, 3 and 5 activate the phosphoinositide pathway through G_q, while Musc2 and 4 inhibit adenylyl cyclase (*Table 4.2*). Thus, it is likely that the activation of the phosphoinositide pathway or inhibition of adenylyl cyclase by the muscarinic receptors must have been achieved independently after the divergence of the two evolutionarily distinct muscarinic acetylcholine subgroups. Histamine receptors H1 and H2 are present in clusters N4 and N2, respectively (*Figure 4.4*). Since the bootstrap support for the N4 cluster is only 0.23, the classification of H1 and H2 into the two separate clusters is questionable.

4.4.3 Peptide receptors (P1 and P2 clusters)

The P1 cluster contains receptors such as those for melanocortin peptides (Mc), adrenocorticotrophic hormone (Acth), melanocyte-stimulating hormone (Msh) and cannabinoid peptide (Cann). Both Acth and Msh activate adenylyl cyclase (*Table 4.2*), while Cann inhibits it. The two groups can also be distinguished evolutionarily. The P2 cluster contains oxytocin (Oxy), arginine vasopressin (Avr), antidiuretic hormone (Adhr), and gonadotrophin-releasing hormone (Gorh) receptors. Oxy stimulates contraction of uterine smooth muscle and milk secretion. Avr has diverse functions including the inhibition of diuresis, counteraction of smooth muscle, stimulation of liver glycolysis and modulation of adrenocorticotropic hormone release from the pituitary (Morel *et al.*, 1992). Antidiuretic hormone binds to Adhr or renal-collecting tubule cells, stimulates adenylyl cyclase and promotes the cAMP-mediated incorporation of water pores into the luminal surface of these cells (Birnbaumer *et al.*, 1992), where the close pharmacological and structural relationships among the neurohypophyseal hormone receptors Adhr, Avr, and Oxy have been noted (see *Table 4.2*). *Figure 4.4* shows that they are also evolutionarily closely related.

4.4.4 Peptide receptors (P3 and P4 clusters)

The P3 cluster includes a number of neuropeptide receptors that control various physiological responses in neuroendocrine systems such as HCl release, gastrin release, vasoconstriction, satiety and mitogenic responses. The family of closely related tachykinin peptides consists of three distinct peptides: substance P; substance K (known also as neurokinin A); and neuromedin K (or neurokinin B). These possess receptors denoted Subs P, Subs K and Nmedin K, respectively (*Figure 4.4*; see also Takahashi *et al.*, 1992). Neuropeptide Y is one of the most abundant neuropeptides in the mammalian nervous system and exhibits a diverse range of important physiological activities, including effects on psychomotor activity, food intake, regulation of central endocrine secretion, and potent vasoactive effects on the cardiovascular system (Herzog *et al.*, 1992). *Figure 4.4* shows that the human neuropeptide Y (NpepY-Hs) is evolutionarily much more closely related to the tachykinin receptors Subs P, Subs K and NmedinK than to the bovine neuropeptide Y (NpepY-Bt)(see the P5 cluster).

Gastrin is an important stimulant of acid secretion by gastric parietal cells and is structurally similar to the peptide hormone cholecystokinin (CCK) (Kopin *et al.*, 1992). The two receptors Gstr and Cck activate the phosphoinositide pathway through G_q (*Table 4.2*) and are evolutionarily closely related (*Figure 4.4*). Endothelin binding to the GPCR (Endo receptor) is followed by an increase in intracellular Ca^{2+}, which is preceded by an increase in inositol phosphates (Lin *et al.*, 1991). Bombesin is a tetradecapeptide originally purified from the skin of the European frog *Bombina bombina*. Two known mammalian counterparts to the amphibian bombesin peptides, gastrin-releasing peptide and neuromedin B, seem to be important in the pathogenesis and maintenance of some human, small-cell, lung carcinoma tumours (Corjay *et al.*, 1991). The receptors Endo, Grp and NmedinB are evolutionarily and functionally closely related (*Figure 4.4*).

The P4 cluster includes thyrotrophin-releasing factor (Trf) and neurotensin (Ntensin) receptors. Thyrotrophin-releasing hormone is a regulatory molecule that functions as a releasing factor in the anterior pituitary gland and as a neurotransmitter or neuromodulator in the central and peripheral nervous system (Straub *et al.*, 1990), while neurotensin is found in the central nervous system and gastrointestinal tissues, where it modulates intracellular levels of cGMP, cAMP and inositol phosphate (Vita *et al.*, 1993).

All receptors in the P3 and P4 clusters, with the exceptions of Opioid, NpepY and Gpc receptors, have the similar function of activating the phosphoinositide pathway, while Opioid and NpepY are known to inhibit adenylyl cyclase (*Table 4.2*).

4.4.5 Peptide receptors (P5, P6 and P7 clusters)

An orphan receptor, Rdc1, is expressed in the heart, kidney and thyroid (Libert *et al.*, 1989) while another orphan receptor, G10D, is most prominently expressed in the lung, liver and adrenal gland (Harrison *et al.*, 1993). The two receptors Rdc1 and G10D are evolutionarily closely related. Angiotensin II receptors consist of type 1 (Ang1) and type 2 (Ang2) receptors. These receptors activate phosphodiesterase C (*Table 4.2*), which results in the mobilization of intracellular Ca^{2+} and the activation of protein kinase C or they inhibit adenylyl cyclase activity (Nakajima *et al.*, 1993).

Interleukin-8 receptors IL8-1 and IL8-2 are the major mediators of the inflammatory response. They activate inositide-specific phospholipase C and interact with endogenous pertussis-toxin-sensitive G proteins to release free $G\beta\gamma$ subunits that then specifically activate β2 isoforms of phospholipase C (Wu *et al.*, 1993). Somatostatin is a tetradecapeptide that is widely distributed in the body and acts in brain, pituitary gland, gut, exocrine and endocrine pancreas, adrenals, thyroid and kidney to inhibit the release of many hormones and other secretory proteins. Furthermore, it functions as a neuropeptide affecting the electrical activity of neurons (Yamada *et al.*, 1992). Somatostatin receptor isoforms 1 (Sstat 1) and 2 (Sstat 2) are evolutionarily closely related to each other (*Figure 4.4*).

The P5 cluster also contains phospholipase-C-coupled paracrine or autocrine receptors, including receptors for formylated peptides (Fmlp) and for C5a anaphylatoxin (C5ana), which mediate the immune response. For example, C5ana seems to specifically interact with $G_{\alpha 16}$ (α16 in *Figure 4.2*), a haematopoeitic-specific G protein (Amatruda *et al.*, 1993). P5 also includes a group of orphan receptors, such as that for mas oncogene protein (Mas) and another orphan receptor, Rta, which may be activated by proteases.

Prostanoids, prostaglandins and thromboxanes have important physiological roles in the cardiovascular and immune systems and in pain sensation of peripheral tissues (Watson and Arkinstall, 1994). The thrombone receptor TxA2 and the prostaglandin receptor PgE in the P6 cluster are evolutionarily closely related. The P7 cluster consists of three closely related glycoprotein hormone (gonadotrophin) receptors: Lh/Hcg; Tsh; and Fsh. Luteinizing hormone (LH) and human chorionic gonadotrophin (hCG) elicit their biological actions through the same receptor, Lh/Hcg. LH is released from the anterior pituitary whereas hCG is released by the placenta in pregnancy. LH not only stimulates ovulation in females but also stimulates Leydig cells to secrete testosterone. Similarly, follicle-stimulating hormone (FSH) is released from the anterior pituitary. It acts on the ovaries to promote the development of follicles and is the major hormone regulating secretion of oestrogens in females (in males, the equivalent hormone supports gametogenesis). Thyroid-stimulating hormone (TSH) is also synthesised in the anterior pituitary. It acts on the thyroid gland, stimulating iodine uptake, as well as the synthesis and release of thyroid hormones. The P6 and P7 clusters seem to be most closely related to the S2 cluster (*Figure 4.4*).

4.4.6 Sensory receptors (S1 and S2 clusters)

In phototransduction, photoexcited opsins in the S1 cluster open G_t binding sites and an active subunit of transducin activates the effector cGMP phosphodiesterase (*Table 4.2*). Ligands and signalling characteristics of olfactory receptors in the S2 cluster have not been established. However, olfactory signalling may be modulated by adenylyl cyclase via G_s or G_{olf} (Reed, 1992) or by the stimulation of IP_3 (inositol triphosphate) by phospholipase C (Raming *et al.*, 1993) (see also *Table 4.2*). The S1 cluster seems to be related to the P3 and P4 clusters, while the S2 cluster is more closely related to the P6 and P7 clusters. Although the evolutionary analysis of Gα subunits suggested that vision and taste may have had a common origin, the olfactory system might have developed independently from these two sensory systems.

4.5 Adaptive evolution of the S1 (opsin) cluster

A major function of visual pigments is measured by the wavelength of maximal absorption (λ_{max}). The λ_{max} has been determined for a large number of vertebrate visual pigments (Lythgoe, 1972). In humans, the visual pigments in rods (rhodopsins) mediate vision in dim light and have a λ_{max} of 496 nm. Three types of visual pigments present in cone cells have λ_{max} values of 420, 530 and 558 nm (Dartnall *et al.*, 1983), which are sufficient to mediate vision across the entire spectrum of visible colour.

A central unanswered question in phototransduction is how visual pigments achieve a wide range of λ_{max}, from approximately 350 to 560 nm, using the same chromophore. Several authors approached this problem using site-directed mutagenesis of the gene encoding bovine rhodopsin. The majority of mutants have been generated to test the role of electrostatic interactions between charged amino acids in spectral tuning (Nathans, 1990a, b; Sakmar *et al.*, 1989; Weitz and Nathans, 1993; Zhukovsky and Oprian, 1989). In these analyses, it has been found that one amino acid change from glutamic acid (E) to glutamine (Q) at the 113th residue (E113Q) shifts λ_{max} from 500 nm to 380 nm (Nathans, 1990b; Sakmar *et al.*, 1989; Zhukovsky and Oprian, 1989). Several other mutants also shift λ_{max} of the visual pigment by 7–33 nm

(Weitz and Nathans, 1993) but most of these mutants are not found in nature and the significance of such amino acid changes in the regulation of absorption spectra is not clear.

In 1990, using an alternative approach, we proposed that the three amino acid changes, alanine to serine (A180S), phenylalanine to tyrosine (F277Y) and alanine to threonine (A285T), are important in modifying the green-sensitive opsin into red-sensitive opsin (Yokoyama and Yokoyama, 1990; see also Neitz *et al.*, 1991). This proposal, based on an evolutionary argument expounded below, was tested by Chan *et al.* (1992) who showed that F277Y and A285T do indeed have major effects, resulting in a shift of λ_{max} by a total of 24 nm towards red (see also Asenjo *et al.*, 1994). By analysing opsins in different species and comparing their evolutionary relationships and the direction of their λ_{max} shifts, potentially important amino acid changes can be identified which may modify the wavelength of absorption. Identification of functionally important amino acid changes is of great interest because they have actually occurred and become fixed in different species. Such evolutionary analyses throw new light on our understanding of mechanisms for the regulation of wavelength absorption by a visual pigment (Yokoyama, 1995).

The rooted phylogenetic tree for some representative vertebrate opsins is shown in *Figure 4.5*, where branch lengths are measured by the number of Poisson-corrected amino acid replacements per site using the neighbour-joining method (Yokoyama, 1995). Opsins are evolutionarily distinguished into rhodopsin (RH1), rhodopsin-like (RH2), short-wavelength-sensitive (SWS), and long-wavelength-sensitive or middle-wavelength-sensitive (LWS/MWS) clusters (Yokoyama, 1994). The bootstrap values support the formation of the RH1 and LWS/MWS clusters: both are 1.0 and highly reliable values. However, those for the RH2 and SWS cluster are 0.97 and 0.53, respectively. When the LWS/MWS cluster is excluded from the analysis, the support for the SWS cluster becomes much higher (Yokoyama, 1994).

Knowing the λ_{max} values associated with the opsins in *Figure 4.5*, the time and direction of the functional changes of different opsins can be identified. In the rhodopsin (RH1) cluster, for example, the UV-sensitive protein Zebrafish 360 (an opsin with a λ_{max} value of 360nm) is most closely related to the rhodopsins in fish, implying that the ancestor of zebrafish achieved its UV vision using one of the duplicate rhodopsin genes. It should be noted that many other vertebrates, including other species of fishes, birds, amphibians, reptiles and mammals, possess UV vision (Jacobs, 1992). The duplication of rhodopsin genes in the ancestor of zebrafish seems to have occurred long after the divergence between fish and the other vertebrates (*Figure 4.5*). This strongly suggests that UV vision in zebrafish and in many other vertebrates has evolved independently (Yokoyama, 1995).

In the rhodopsin-like (RH2) cluster, the blue-shift of the opsin Gecko 467 of *Gekko gekko* must have occurred after its divergence from the other members of this cluster. Four opsins, Goldfish 441, Chicken 455, Human 420, and Chicken 415, seem to have achieved their SWS evolutionary status after their

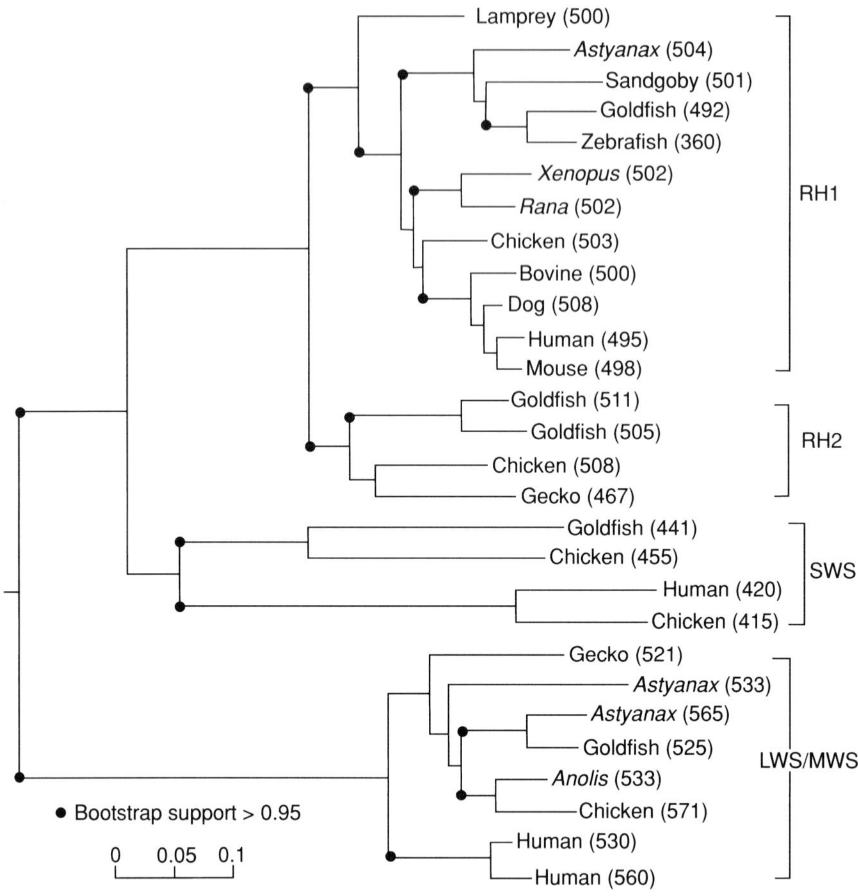

Figure 4.5. Phylogenetic tree of vertebrate opsins based on the number of amino acid replacements. LWS/MWS, long-wavelength-sensitive or medium-wavelength-sensitive clusters; RHI, rhodospin clusters; RH2, rhodopsin-like clusters; SWS, short-wavelength-sensitive clusters of opsins. Values in parentheses denote λ_{max} values. Modified from Yokoyama (1995) in *Molecular Biology and Evolution*, with permission from The University of Chicago Press.

common ancestor diverged from that of the RH1 and RH2 members. The λ_{max} values of Human 420 and Chicken 415 are much lower than those of Goldfish 441 and Chicken 455, indicating a further functional differentiation between the two sets of SWS opsins.

In the LWS/MWS cluster, Goldfish 525, Astyanax 563 of *Astyanax fasciatus* (Mexican blind cave fish), Anolis 561 of *Anolis carolinensis* (American chameleon), Chicken 571 and Human 560 have the red opsin-specific amino acids serine-180, tyrosine-277 and threonine-285, while Gecko 521, Astyanax 533 and Human 530 have the green-specific amino acids alanine-180, phenyl-

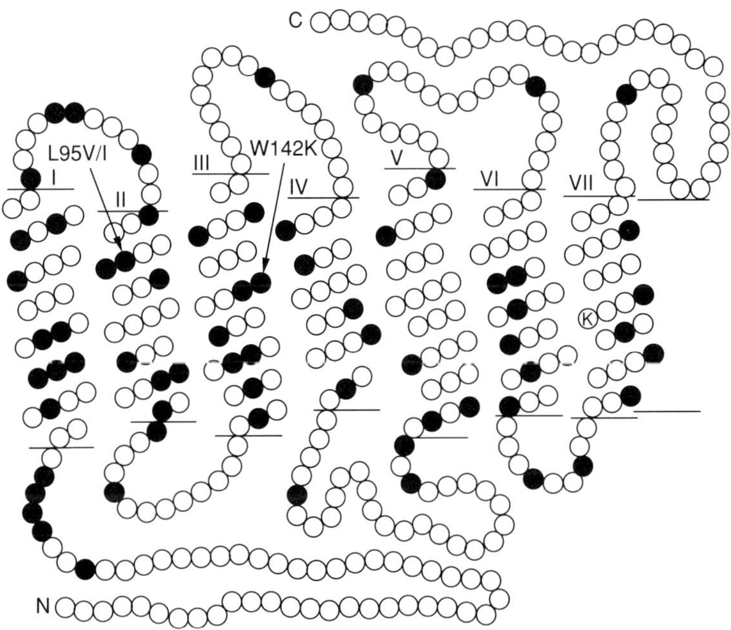

Figure 4.6. Amino acid changes which may be important in the blue shift in λ_{max}. The model is of long-wavelength-sensitive or middle-wavelength-sensitive (LWS/MWS) opsins (Nathans, 1992). I–VII denote seven transmembranes.

alanine-277 and alanine-285 (Yokoyama and Yokoyama, 1990). Goldfish 525 is supposed to encode the red opsin in goldfish (Johnson *et al.*, 1993). However, the red visual pigment in goldfish detected by microspectrophotometry absorbs light maximally at 579 nm when 11-*cis* retinal predominates and at 625 nm when 3-dehydroretinal predominates (Tsin *et al.*, 1981). Johnson *et al.* (1993) suggest that Goldfish 525 may be encoded by a variant of the gene for red opsin. This discrepancy remains to be resolved.

Amino acid changes associated with a potential blue-shift are shown as filled circles in *Figure 4.6* (for more details, see Yokoyama, 1995). Out of 66 amino acid changes suggested, 49 changes reside inside or very close to the transmembrane regions. Since the opsin and chromophore is important in regulating the wavelength of absorption, it is reasonable to assume that these 49 amino acid changes might have been important in regulating λ_{max}.

Two out of the 49 amino acid changes deserve special comments. When the sequence encoding Zebrafish 360 is compared with the most closely related Goldfish 492 sequence, 27 amino acids differ. Only one of these amino acid differences, trytophan to lysine (W142K), is unique to the zebrafish; thus, the major step in the development of Zebrafish 360 seems to be this particular amino acid change, as pointed out by Robinson *et al.* (1993). This hypothesis, however, remains to be tested empirically. Another amino acid change, leucine to valine or isoleucine (L95V/I), is unique to the SWS opsin, where L95 has been conserved not only among all vertebrate opsins but also among other GPCRs, including muscarinic, adrenergic, olfactory and dopamine receptors. Amino acid changes at this highly conserved residue suggest that they may be important in the blue-shift of visual pigments.

The amino acid changes suggested from the 'electrostatic interaction' and the 'evolutionary' models discussed above do not overlap. The amino acid changes suggested in the evolutionary model do not involve charge differences and provide a new set of mutants to consider. Importantly, the polymorphic amino acids have been evolutionarily accepted by organisms, actually exist in nature, and may be responsible for shifts in λ_{max}. These wild-type opsins and their mutant forms (derived by site-directed mutagenesis) can be expressed in cultured cells, reconstituted with 11-*cis* retinal and measured for the absorption spectra of the visual pigments (Chan *et al.*, 1992; Nathans, 1990a, b; Oprian *et al.*, 1991; Sakmar *et al.*, 1989; Weitz and Nathans, 1993; Yokoyama *et al.*, 1995; Zhukovsky and Oprian, 1989). Thus, the effects of these mutations can be experimentally tested rigorously and provide an excellent opportunity to investigate the molecular mechanism underlying the adaptive evolution of opsins.

4.6 Summary

G-protein-mediated signal transduction systems have been found in a diverse array of eukaryotes, generating an amazing range of physiological responses. The broad distribution and sequence conservation of G proteins and GPCRs imply that genes encoding the components of G-protein signalling evolved at an early stage of eukaryotic evolution. Here, we have considered the evolution of two major superfamilies, involved in these signalling systems, Gα subunits and GPCRs.

Among heterodimeric G proteins, G$\beta\gamma$ subunits are structurally similar and functionally interchangeable but Gα subunits in vertebrates seem to be grouped into two clusters, one cluster consisting of G_i, G_t, G_o and G_x subunits and another cluster consisting of G_s, G_q and G_{12} subunits. Most members in the former cluster have evolved at much slower rates than those in the latter cluster.

GPCRs may be distinguished into peptide receptors and neurotransmitter receptors. Opsins and olfactory receptors seem to be related more to the peptide receptors than to the neurotransmitter receptors. Although details of the tree topology obtained are not completely clear, many branches reflect similar characteristics in physiological responses and signalling pathways. Interactions between GPCRs, Gα subunits, agonists and antagonists have been studied using mutants generated by site-directed mutagenesis. Such analyses suggest that the sites towards the intracellular side of the transmembrane region of receptors are responsible for agonist binding and that those in the central and extracellular regions of the protein affect antagonist binding. By constructing the evolutionary tree for 28 opsins with known function (λ_{max}), a total of 66 amino acid changes have been shown to correlate with the directions of λ_{max}-shift and might have been important in determining the λ_{max} of a visual pigment. Among these amino acid changes, three changes are already proven to be responsible for the differentiation of the green-sensitive visual pigment from the red-sensitive visual pigment.

In conclusion, evolutionary analyses of the GPCRs and Gα subunits open a new direction in which to study the molecular mechanisms involved in acquiring new functions and adaptive evolution.

Acknowledgements

Comments by Ruth Yokoyama were greatly appreciated. This work was supported by NIH grant GM42379.

References

Amatruda TT, Gerard NP, Gerard C, Simon MI. (1993) Specific interactions of chemoattractant factor receptors with G-proteins. *J. Biol. Chem.* **268**: 10139–10144.

Asenjo A, Rim J, Oprian DD. (1994) Molecular determinants of human red/green color discrimination. *Neuron* **12**: 1131–1138.

Baldwin JM. (1993) The probable arrangement of the helices in G protein-coupled receptors. *EMBO J.* **12**: 1693–1703.

Baldwin JM. (1994) Structure and function of receptors coupled to G proteins. *Curr. Opin. Cell Biol.* **6**: 180–190.

Benovic JL, Kuhn H, Weyand I, Codina J, Caron MG, Lefkowitz RJ. (1987a) Functional desensitization of the isolated β-adrenergic receptor by the β-adrenergic receptor kinase: potential role of an analog of the retinal protein arrestin (48 kDa protein). *Proc. Natl Acad. Sci. USA* **84**: 8879–8882.

Benovic JL, Mayer JF, Staniszewski C, Lefkowitz RJ, Caron MG. (1987b) Purification and characterization of the β-adrenergic receptor kinase. *J. Biol. Chem.* **262**: 9026–9032.

Birnbaumer M, Seibold A, Gilbert S, Ishido M, Barberis C, Antaramian A, Brabet P, Rosenthal W. (1992) Molecular cloning of the receptor for human antidiuretic hormone. *Nature* **357**: 333–335.

Braun T, Schofield PR, Sprengel R. (1991) Amino-terminal leucine-rich repeats in gonadotropin receptors determine hormone selectivity. *EMBO J.* **10**: 1885–1890.

Chan T, Lee M, Sakmar TP. (1992) Introduction of hydroxyl-bearing amino acids causes bathochromic spectral shifts in rhodopsin. Amino acid substitutions responsible for red–green color pigment spectral tuning. *J. Biol. Chem.* **267**: 9478–9480.

Cheung AH, Huang R-RC, Graziano MP, Strader CD. (1991) Specific activation of Gs by synthetic peptide corresponding to an intracellular loop of the β-adrenergic receptor. *FEBS Lett.* **279**: 277–280.

Cohen GB, Yang T, Robinson PR, Oprian DD. (1993) Constitutive activation opsin: Influence of change at position 134 and size at position 296. *Biochemistry* **32**: 6111–6115.

Conklin BR, Bourne HR. (1993) Structural elements of Gα subunits that interact with Gβγ, receptors, and effectors. *Cell* **73**: 631–641.

Corjay MH, Dobrzanski DJ, Way JM, Viallet J, Shapira H, Worland P, Sausville EA, Battey JF. (1991) Two distinct bombesin receptor subtypes are expressed and functional in human carcinoma cells. *J. Biol. Chem.* **266**: 18771–18779.

Dartnall HJA, Bowmaker JK, Mollon JD. (1983) Microspectrophotometry of human photoreceptors. In: *Colour Vision* (ed. JD Mollon and LT Sharpe). Academic Press, New York, pp. 69–80.

Dixon RAF, Kobilka BK, Strader DJ *et al.* (1986) Cloning of the gene and cDNA for mammalian α-adrenergic receptor and homology with rhodopsin. *Nature* **321**: 75–79.

Dohlman HG, Thorner J, Caron MG, Lefkowitz RJ. (1991) Model systems for the study of seven-transmembrane-segment receptors. *Annu. Rev. Biochem.* **60**: 653–688.

Foguet M, Hoyer D, Pardo LA, Parekh A, Kluxen FW, Kalkman HO, Stuhmer W, Lubbert H. (1992) Cloning and functional characterization of the rat stomach undus serotonin receptor. *EMBO J.* **11**: 3481–3487.

Fong TM, Huang R-RC, Strader CD. (1992a) Localization of agonist and antagonist binding domains of the human neurokinin-1 receptor. *J. Biol. Chem.* **267**: 25664–25667.

Fong TM, Yu H, Huang R-RC, Strader CD. (1992b) The extracellular domain of the neurokinin-1 receptor is required for high-affinity binding of peptides. *Biochemistry* **31**: 11806–11811.

Franke RR, Sakmar TP, Graham RM, Khorana HG. (1992) Structure and function of rhodopsin: studies of the interaction between the rhodopsin cytoplasmic domain and transducin. *J. Biol. Chem.* **267**: 14767–14774.

Ganz I, Munzert G, Tashiro T, Schaffer M, Wang L, Del Valle J, Yamada T. (1991) Molecular cloning of the human histamine H2 receptor. *Biochem. Biophys. Res. Commun.* **178**: 1386–1392.

Gilman, AG. (1987) G proteins: transducers of receptor-generated signals. *Annu. Rev. Biochem.* **56**: 615–649.

Hamblin MW, Metcalf MA. (1991) Primary structure and functional characterization of a human 5–HT_{1D}-type serotonin receptor. *Mol. Pharmacol.* **40**: 143–148.

Hargrave PA. (1982) Rhodopsin chemistry, structure, and topology. *Prog. Retinal Eye Res.* **1**: 1–51.

Hargrave PA, McDowell JH. (1992) Rhodopsin and phototransduction: a model system for G protein-linked receptors. *FASEB J.* **6**: 2323–2331.

Hargrave PA, McDowell JH, Curtis DR, Wang JK, Juszczak E, Fing S-L, Mohana Rao JK, Argos P. (1983) The structure of bovine rhodopsin. *Biophys. Struct. Mech.* **9**: 235–244.

Harrison R, Unwin PNT. (1975) Three dimensional model of purple membrane obtained by electron microscopy. *Nature* **257**: 28–32.

Harrison JK, Barker CM, Lynch KR. (1993) Molecular cloning of a novel rat G protein-coupled receptor gene expressed prominently in lung, adrenal, and liver. *FEBS Lett.* **318**: 17–22.

Herzog H, Hort YJ, Ball HJ, Hayes G, Shine J, Sekbie LA. (1992) Cloned human neuropeptide Y receptor couples to two different second messenger systems. *Proc. Natl Acad. Sci. USA* **89**: 5794–5798.

Higgins DS, Bleasby AJ, Fuchs R. (1992) CLUSTAL V: improved software for multiple sequence alignment. *Comput. Appl. Biosci.* **8**: 189–191.

Jacobs GH. (1992) Ultraviolet vision in vertebrates. *Am. Zool.* **32**: 544–554.

Johnson RL, Grant KBG, Zankel TC, Boehm MF, Merbs SL, Nathans J, Nakanishi K. (1993) Cloning and expression of goldfish opsin sequences. *Biochemistry* **32**: 208–214.

Kaziro Y, Itoh H, Kozasa T, Nakafuku M, Satoh T. (1991) Structure and function of signal-transducing GTP-binding proteins. *Annu. Rev. Biochem.* **60**: 349–400.

Kopin AS, Lee Y-M, McBride EW, Miller LJ, Lu M, Lin HY, Kolakowski LF, Beinborn M. (1992) Expression, cloning and characterization of the canine parietal cell gastrin receptor. *Proc. Natl Acad. Sci. USA* **89**: 3605–3609.

Kjelsberg MA, Cotecchia S, Ostrowski J, Caron MG, Lefkowitz RJ. (1992) Constitutive activation of the α1b-adrenergic receptor by all amino acid substitutions at a single site. Evidence for a region which constraints receptor activation. *J. Biol. Chem.* **267**: 1430–1433.

Kosugi S, Okajima F, Ban T, Hidaka A, Shenker A, Kohn LD. (1992) Mutation of alanine 623 in the third cytoplasmic loop of the rat thyrotropin (TSH) receptor results in a loss in the phosphoinositide but not cAMP signal induced by TSH and receptor autoantibodies. *J. Biol. Chem.* **267**: 24153–24156.

Kuhn H. (1978) Light-regulated binding of rhodopsin kinase and other proteins to cattle photoreceptor membranes. *Biochemistry* **17**: 4389–4395.

Kunkel MT, Peralta EG. (1993) Amino acids required for signal transduction by the M3 muscarinic acetylcholine receptor. *EMBO J.* **12**: 3809–3815.

Libert F, Parmentier M, Lefort A, Dinsart C, van Sande J, Maenhaut C, Simons MJ, Dumont JE, Vassart G. (1989) Selective amplification and cloning of four new members of the G protein-coupled receptor family. *Science* **244**: 569–572.

Liggett SB, Caron MG, Lefkowitz RJ, Hnatowich M. (1991) Coupling of a mutated form of the human β2–adrenergic receptor to Gi and Gs: requirement for multiple cytoplasmic domains in the coupling process. *J. Biol. Chem.* **266**: 4816–4821.

Lin HY, Kaji EH, Winkel GK, Ives HE, Lodish HF. (1991) Cloning and functional expression of a vascular smooth muscle endothelin 1 receptor. *Proc. Natl Acad. Sci. USA* **88**: 3185–3189.

Lythgoe JN. (1972) List of vertebrate visual pigments. In: *Photochemistry of Vision, Handbook of Sensory Physiology* VII/1 (ed. HJA Dartnall). Springer-Verlag, Berlin, pp. 604–624 .

Min KC, Zvyaga TA, Cypress AM, Sakmar TP. (1993) Characterization of mutant rhodopsins responsible for autosomal dominant retinitis pigmentosa. Mutations on the cytoplasmic surface affect transducin activation. *J. Biol. Chem.* **268**: 9400–9404.

Morel A, O'Carroll A-M, Brownstein MJ, Lolait SJ. (1992) Molecular cloning and expression of a rat Vla arginine vasopressin receptor. *Nature* **356**: 523–526.

Mountjoy KG, Robbins LS, Mortrud MT, Cone RD. (1992) The cloning of a family of genes that encode the melanocortin receptors. *Science* **257**: 1248–1251.

Nagayama Y, Wadsworth HL, Chazenbalk GD, Russo D, Seto P, Rapoport B. (1991) Thyrotropin-luteinizing hormone/chorionic gonadotropin receptor extracellular domain chimeras as probes for thyrotropin receptor function. *Proc. Natl Acad. Sci. USA* **88**: 902–905.

Nakajima M, Mukoyama M, Pratt RE, Horiuchi M, Dzau VJ. (1993) Cloning of cDNA and analysis of the gene for mouse angiotensin II type 2 receptor. *Biochem. Biophys. Res. Commun.* **197**: 393–399.

Nathans J. (1990a) Determinants of visual pigment absorbance: role of charged amino acids in the putative transmembrane segments. *Biochemistry* **29**: 937–942.

Nathans J. (1990b) Determinants of visual pigment absorbance: identification of the retinylidene Schiff's base counterion in bovine rhodopsin. *Biochemistry* **29**: 9746–9752.

Nathans J. (1992) Rhodopsin: structure, function, and genetics. *Biochemistry* **31**: 4923–4931.

Neitz M, Neitz J, Jacobs GH. (1991) Spectral tuning of pigments underlying red-green color vision. *Science* **252**: 971–974.

Ohyama K, Yamano Y, Chaki S, Kondo T, Inagami T. (1992) Domains for G-protein coupling in angiotensin II receptor type I: studies by site-directed mutagenesis. *Biochem. Biophys. Res. Commun.* **189**: 677–683.

Oprian DD, Asenjo AB, Lee N, Pelletier SL. (1991) Design, chemical synthesis, and expression of genes for the three human color vision pigments. *Biochemistry* **30**: 11367–11372.

Ovchinnikov YA, Abdulaev NG, Feigina MY *et al.* (1982) The complete amino acid sequence of visual rhodopsin. *Bioorg. Khim.* **8**: 1011–1014.

Parma J, Duprez L, van Sande J, Cochaux P, Gervy C, Mockel J, Dumont J, Vassart G. (1993) Somatic mutations in the thyrotropin receptor gene cause hyperfunctioning thyroid adenomas. *Nature* **365**: 649–651.

Raming K, Kriegr J, Strotmann J, Boekhoff I, Kubick S, Baumstark C, Breer H. (1993) Cloning and expression of odorant receptors. *Nature* **361**: 353–356.

Reed RR. (1992) Signaling pathways in odorant detection. *Neuron* **8**: 205–209.
Ren Q, Kurose H, Lefkowitz RJ, Cotecchia S. (1993) Constitutively active mutants of the α2-adrenergic receptor. *J. Biol. Chem.* **268**: 16483–16487.
Robinson J, Schmitt EA, Harosi FI, Reece RJ, Dowling JE. (1993) Zebrafish ultraviolet visual pigment: Absorption spectrum, sequence, and localization. *Proc. Natl Acad. Sci. USA* **90**: 6009–6012.
Saitou N, Nei M. (1987) The neighbor-joining method: a new method for reconstructing phylogenetic trees. *Mol. Biol. Evol.* **4**: 406–425.
Sakmar TP, Franke RR, Khorana HG. (1989) Glutamic acid-113 serves as the retinylidene Schiff base counterion in bovine rhodopsin. *Proc. Natl Acad. Sci. USA* **86**: 8309–8313.
Saverese TM, Fraser CM. (1992) *In vitro* mutagenesis and the search for structure-function relationships among G protein-coupled receptors. *Biochem. J.* **283**: 1–19.
Saverese TM, Wang C-D, Fraser CM. (1992) Mutagenesis of the rat m1 muscarinic acetylcholine receptor: role of conserved cysteins in receptor function. *J. Biol. Chem.* **267**: 11439–11448.
Schertler GFX, Villa C, Henderson R. (1993) Projection structure of rhodopsin. *Nature* **362**: 770–772.
Simon MI, Strathman MP, Gautum N. (1991) Diversity of G proteins in signal transduction. *Science* **252**: 802–808.
Strader CD, Fong TM, Tota MR, Underwood D, Dixon RAF. (1994) Structure and function of G protein-coupled receptors. *Annu. Rev. Biochem.* **63**: 101–132.
Strathman M, Simon MI. (1991) Gα12 and Gα13 subunits define a fourth class of G protein α subunits. *Proc. Natl Acad. Sci. USA* **88**: 5582–5586.
Straub RE, Frech GC, Joho RH, Gershengorn MC. (1990) Expression cloning of a cDNA encoding the mouse pituitary thyrotropin-releasing hormone receptor. *Proc. Natl Acad. Sci. USA* **87**: 9514–9518.
Stryer L, Bourne HR. (1986) G proteins: a family of signal transducers. *Annu. Rev. Cell Biol.* **2**: 391–419.
Sunahara RK, Guan H-C, O'Dowd BF *et al.* (1991) Cloning of the gene for a human dopamine D5 receptor with higher affinity for dopamine D1. *Nature* **350**: 614–619.
Takahashi K, Tanaka A, Hara M, Nakanishi S. (1992) The primary structure and gene organization of human substance P and neuromedin K receptors. *Eur. J. Biochem.* **204**: 1025–1033.
Tsin ATC, Liebman PA, Beatty DD, Drazymala R. (1981) Rod and cone visual pigments in the goldfish. *Vision Res.* **21**: 943–946.
Vita N, Laurent P, Lefort S, Chalon P, Dumont X, Kaghad M, Gully D, Le Fur G, Ferrara P, Caput D. (1993) Cloning and expression of a complementary DNA encoding a high affinity human neurotensin receptor. *FEBS Lett.* **317**: 139–142.
Watson S, Arkinstall S. (1994) *The G-protein Linked Receptor Facts Book.* Academic Press, New York.
Weitz CJ, Nathans J. (1993) Rhodopsin activation: effects of the metarhodopsin I - metarhodopsin II equilibrium of neutralization or introduction of charged amino acids within putative transmembrane segments. *Biochemistry* **32**: 14176–14182.
Wilden U, Kuhn H. (1982) Light-dependent phosphorylation of rhodopsin: the number of phosphorylation sites. *Biochemsitry* **21**: 3014–3022.
Wilden U, Hall SW, Kuhn H. (1986) Phosphodiesterase activation by photoexcited rhodopsin is quenched when rhodopsin is phosphorylated and binds the intrinsic 48-kDa protein of rod outer segments. *Proc. Natl Acad. Sci. USA* **83**: 1174–1178.
Wilkie TM, Yokoyama S. (1994) Evolution of the G protein alpha subunit multigene family. In: *Molecular Evolution of Physiological Processes* (ed. DM Fambrough). Rockefeller University Press, New York, pp. 249–270.
Wilkie TM, Gilbert DJ, Olsen AS *et al.* (1992) Evolution of the mammalian G protein α subunit multigene family. *Nature Genetics* **1**: 85–91.
Williams M. (1987) Purine receptors in mammalian tissues: pharmacology and functional significance. *Annu. Rev. Pharmacol. Toxicol.* **27**: 315–345.

Wu D, LaRosa GJ, Simon MI. (1993) G protein-coupled signal transduction pathways for interleukin-8. *Science* **261**: 101–103.

Yamada Y, Post SR, Wang K, Tager HS, Bell GI, Seino SZ. (1992) Cloning and functional characterization of a family of human and mouse somatostatin receptors expressed in brain, gastrointestinal tract, and kidney. *Proc. Natl Acad. Sci. USA* **89**: 251–255.

Yokoyama S. (1994) Gene duplication and evolution of the short wavelength-sensitive visual pigments in vertebrates. *Mol. Biol. Evol.* **11**: 32–39.

Yokoyama S. (1995) Amino acid replacements and wavelength absorption of visual pigments in vertebrates. *Mol. Biol. Evol.* **12**: 53–61.

Yokoyama S, Starmer WT. (1992) Phylogeny and evolutionary rates of G protein α subunit genes. *J. Mol. Evol.* **35**: 230–238.

Yokoyama R, Yokoyama S. (1990) Convergent evolution of the red- and green-like visual pigment genes in fish, *Astyanax fasciatus*, and human. *Proc. Natl Acad. Sci. USA* **87**: 9315–9318.

Yokoyama R, Knox BE, Yokoyama S. (1995) Rhodopsin from the fish, *Astyanax*: role of tyrosine 261 in the red shift. *Invest. Ophthalmol. Visual Sci.* **36**: 939–945.

Zhukovsky EA, Oprian DD. (1989) Effect of carboxylic acid side chains on the absorption maximum of visual pigments. *Science* **246**: 928–930.

5

Evolution of centromeric alpha satellite DNA: molecular organization within and between human and primate chromosomes

Peter E. Warburton and Huntington F. Willard

5.1 Introduction

The concerted evolution of tandemly repeated DNA families is a complex process that involves the interaction of many genetic turnover mechanisms at both the chromosomal and population level (Dover, 1982, 1986). Insight into these processes can be gained by consideration of repeat unit structure and organization, and how these vary both within populations and between related species (Brown *et al.*, 1971; Charlesworth *et al.*, 1994; Dover, 1993; Southern, 1975; Willard, 1991; Willard and Waye, 1987a). The alpha satellite DNA family has been undergoing extensive characterization at the molecular level, having been described at the centromere of every human chromosome and at the chromosome centromeres of many other primate species. Thus, it can be considered as a paradigm for addressing specific questions about the processes of concerted evolution in tandemly repeated DNA families (Willard, 1991; Willard and Waye, 1987a).

5.2 Repeat unit organization of alpha satellite DNA

Alpha satellite DNA is characterized by many levels of hierarchical organization in genomes, from suprachromosomal subfamilies to chromosome-specific

Human Genome Evolution, edited by M. Jackson, T. Strachan and G. Dover.

subsets, to polymorphic variation within these subsets. Thus, experiments that examine these relationships between repeat units at different levels of organization give insight into genetic processes such as mutation, unequal crossing-over and intra- and interchromosomal exchange.

The fundamental repeat unit of all alpha satellite DNA is based on diverged monomers of approximately 171 bp in length. These monomers are organized in distinct linear arrangements that form higher-order repeat units, which can be revealed by restriction enzyme site periodicity and/or sequence analysis (*Figure 5.1*; Willard and Waye, 1987a). Individual monomers can diverge in sequence from each other by up to 35%, both within a higher-order

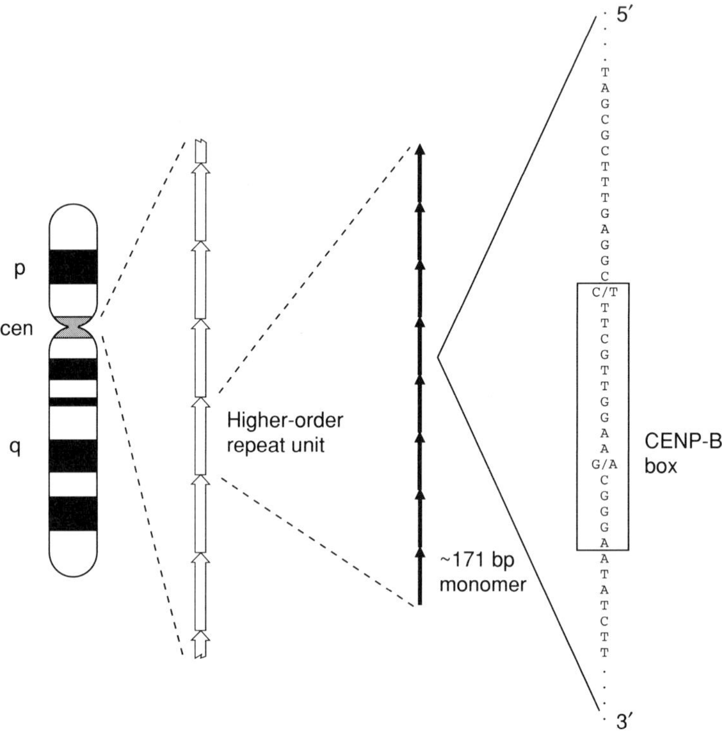

Figure 5.1. Monomeric and higher-order repeat unit organization of alpha satellite DNA. Tandemly repeated monomers of ~171 bp (small black arrows) are the fundamental repeat unit. Within and between chromosomes, monomers can be significantly diverged (~35%) from each other. Some, but not all monomers, contain the 17 bp binding site for the centromere-specific, DNA-binding protein CENP-B, called the CENP-B box. Distinct linear arrangements of diverged monomers form higher-order repeat units (larger open arrows), which are in turn tandemly repeated to form arrays up to several million bp long. Alphoid arrays have been identified at all human and primate centromeres examined so far. Although monomers can be significantly diverged, higher-order repeat units from a particular chromosomal locus are generally highly similar (>95%).

repeat structure and between repeat structures on different chromosomes. However, higher-order repeat units from the same locus are more than 95% similar to each other, and this high sequence identity defines different subsets of alpha satellite repeats (Willard, 1991). Over 33 distinct alpha satellite subsets have been described, with higher-order repeat units ranging in size from as few as two to as many as 35 monomers (*Table 5.1*; Choo *et al.*, 1991). This higher-order repeat organization is consistent with linear sets of diverged monomers becoming the unit of crossing-over during the process of sequence homogenization (Southern, 1975; Willard and Waye, 1987a).

The higher-order repeat units of alpha satellite DNA are organized in a largely chromosome-specific manner (Choo *et al.*, 1991; Willard, 1985). This has been revealed using cloned higher-order repeat units as hybridization probes at high stringency, both to Southern blotted DNA and to metaphase chromosome spreads in *in situ* hybridization experiments (Jorgensen *et al.*, 1986; Waye and Willard, 1985; Willard, 1985; Wolfe *et al.*, 1985). *Table 5.1*

Table 5.1. Higher-order repeat unit structure of alpha satellite subsets

Chromosome	Higher-order repeat unit	Enzyme	Reference
1	1.9 kb (11mer)	*Hin*dIII, *Xba*I, *Sph*I	Choo *et al.* (1991)
2	0.68 kb (4mer)	*Hin*fI, *Xba*I, *Hin*dIII	"
3	2.9 kb (17mer)	*Acc*I, *Hin*dIII, *Pvu*II	"
4	3.2 kb (19mer)	*Msp*I	Maskova *et al.* (1994)
4, 9	1.2 kb (7mer)	*Pst*I, *Rsa*I, *Taq*I	Choo *et al.* (1991)
5, 19	2.25 kb (13mer)	*Eco*RI, *Pst*I, *Rsa*I	"
6	2.9 kb (17mer)	*Bam*HI, *Taq*I	"
7	2.7kb (16mer)	*Hin*dIII, *Taq*I	"
	1.0 kb (6mer)	*Eco*RI	"
8	2.5 kb (15mer)	*Sca*I, *Hin*dIII, *Pst*I	Ge *et al.* (1992)
9	2.7 kb (16mer)	*Pst*I, *Rsa*I, *Msp*I	Choo *et al.* (1991)
10	1.35 kb (8mer)	*Pst*I, *Rsa*I, *Acc*I	"
11	0.85 kb (5mer)	*Xba*I	"
12	1.4 kb (8mer)	*Hin*dIII, *Taq*I, *Pvu*II	"
13, 21	1.9 kb (11mer)	*Hin*dIII, *Taq*I, *Sau*3AI	Greig *et al.* (1993)
	0.68 kb (4mer)	*Eco*RI, *Xba*I	Choo *et al.* (1991)
13, 14, 21	3.95 kb (23mer)	*Rsa*I, *Msp*I	"
14, 22	1.4 kb (8mer)	*Xba*I, *Taq*I	"
15	2.5kb (14mer)	*Rsa*I, *Dra*I	"
	4.5 kb (26mer)	*Acc*I, *Bgl*II, *Bst*NI	"
16	1.7 kb (10mer)	*Sau*3A1, *Eco*RV	"
17	2.7kb (16mer)	*Eco*RI, *Pst*I, *Pvu*II	"
18	1.4 kb (8mer)	EcoRI	Alexandrov *et al.* (1991)
	1.7 kb (10mer)	*Hin*dIII	"
20	1.0 kb (6mer)	*Hin*fI	Baldini *et al.* (1992)
22	2.7 kb (16mer)	*Eco*RI	Choo *et al.* (1991)
X	2.0 kb (12mer)	*Bam*HI, *Pst*I, *Sst*I	"
Y	5.7 kb (33mer)	*Eco*RI, *Pst*I, *Hin*dIII	"

The length of the repeat unit in kilobases (kb), the number of monomers (mer) present in each of the alphoid subsets described, and the restriction enzymes that cut once per higher-order repeat unit are shown. Only subsets for which the higher-order structure has been determined are included; other cloned subsets not fully described are not included here. The work by Choo *et al.* (1991) is the most recent comprehensive survey published, and references to specific subsets are included therein. References to subsets described subsequent to 1991 are also included in the table.

makes it clear that chromosome-specific subsets have been identified for most human chromosomes (Choo *et al.*, 1991; Waye and Willard, 1987; Willard and Waye, 1987a). The high degree of repeat unit similarity within homologous chromosomes compared with the relatively modest repeat unit similarity between non-homologous chromosomes can be used to infer the relative rates of concerted evolution within and between chromosomes. Specifically, it suggests that the rate of processes homogenizing repeat units within homologous chromosomes (such as interhomologue exchange) are more efficient, or occur more frequently, than processes which spread repeats between non-homologous chromosomes (*Figure 5.2*; Charlesworth *et al.*, 1994; Dover, 1986; Ohta and Dover, 1983).

As discussed above, the centromere of each human chromosome is characterized by one or more subsets of distinct higher-order repeat units. The size of these regions can be assessed by pulsed-field gel electrophoresis (PFGE), using restriction enzymes that cut the higher-order repeat units rarely, or not at all, but which cut frequently in surrounding non-satellite DNA (*Figure 5.3*; Tyler-Smith and Brown, 1987; Warburton *et al.*, 1992; Wevrick and Willard, 1989). Such analyses have revealed the presence of up to several thousand repeat units arranged in an apparently uninterrupted fashion and forming arrays of several million base pairs (Jabs *et al.*, 1989; Jackson *et al.*, 1993; Mahtani and

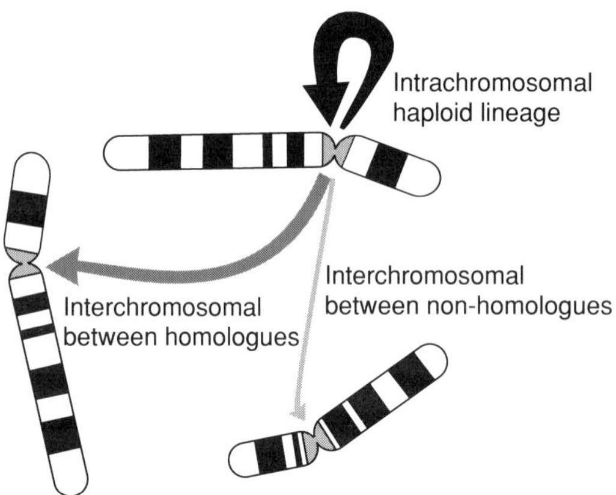

Figure 5.2. Relative rates of chromosomal exchange involved in the evolution of alpha satellite DNA. Larger and darker arrows represent higher rates of sequence exchange. The present-day chromosomal organization of alpha satellite DNA has probably resulted from the relative (qualitative) rates depicted here. Intrachromosomal exchange would result in haplotype-specific variation. Interchromosomal exchange between homologues would result in chromosome specificity. Interchromosomal exchange between non-homologues would result in the emergence of suprachromosomal subfamilies within a species and non-orthologous evolution between species. Adapted from Warburton and Willard (1995) Inter-homologue sequence variation of chromosome 17 alpha satellite DNA: evidence for concerted evolution along haplotypic lineages. *J. Mol. Evol.*, **41**: 1006–1015. © Springer-Verlag.

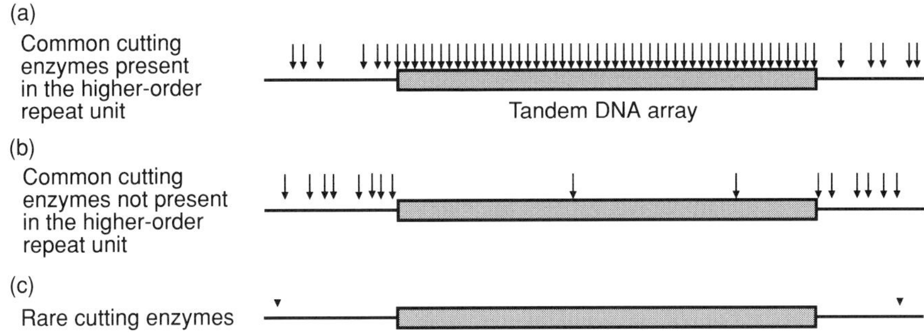

Figure 5.3. Restriction enzyme analysis of alpha satellite DNA. Restriction enzymes can be used both to characterize repeat units and to estimate repeat array length. (a) Restriction enzymes which cut once within the higher-order repeat unit release most or all of the array into fragments characteristic of the array (see *Table 5.1*). (b) Restriction enzymes not present in the higher-order repeat unit will cut in none, or very few, of the tandemly arranged repeats and will cut the array into one or a small number of large fragments containing little or no flanking DNA. (c) Rare cutting enzymes also release the array into large fragments but these can include large amounts of flanking genomic DNA. In general, enzymes which give the patterns of digestion shown in (a) and (b) have a 4–6 bp recognition site with no CpG dinucleotide, and are predicted to cut genomic DNA approximately every 15–50 kb. Rare cutting enzymes have recognition sites 6–8 bp long and contain a CpG dinucleotide. The behaviour of a particular enzyme will depend specifically on the DNA sequence of the particular alpha satellite subset being examined. Adapted from Warburton *et al.* (1992) with permission from Humana Press.

Willard, 1990; Tyler-Smith and Brown, 1987; Warburton and Willard, 1990; Waye and Willard, 1989a; Wevrick and Willard, 1989). Array sizes are highly variable between individuals, with arrays from homologous chromosomes varying in size by a factor of up to 4- to 6-fold (Mahtani and Willard, 1990; Oakey and Tyler-Smith, 1990; Warburton *et al.*, 1992; Wevrick and Willard, 1989). The wide variation in array size between individuals, which are nonetheless stably transmitted through at least three generation pedigrees, provides a valuable source of variation for the genetic mapping of these regions (Jabs *et al.*, 1989; Mahtani and Willard, 1990; Wevrick and Willard, 1989; Willard *et al.*, 1986). Such variation between array sizes is consistent with unequal crossing-over within and between homologous chromosomes.

5.3 Variation between homologous higher-order repeat units

The chromosome-specific subsets of alpha satellite DNA described in *Table 5.1* are not the only alpha satellite repeats present on many chromosomes. In many cases, a variety of higher-order repeat units can be identified in addition to the abundant chromosome-specific repeats. These variant repeats are presumed to have arisen from the chromosome-specific repeats by one or more of several mutational mechanisms. They are also often polymorphic, which means they are useful as genetic markers.

One type of variant repeat can arise by point mutations within the repeat units, and these can be easily detected by digesting genomic DNA with restriction enzymes. For example, the chromosome-X-specific subset of human alpha satellite DNA is predominantly composed of higher-order repeat units of 2.0 kb that are 12 monomers in length. Most or all of these repeat units contain a recognition site for the restriction enzyme *Bam*HI; when these sites are digested most of the array is released into multiple 2.0 kb fragments (Waye and Willard, 1985; Willard *et al.*, 1983; Yang *et al.*, 1982). Other restriction enzymes such as *Xba*I or *Hin*dIII, however, cleave only a subset of the 2.0 kb repeat units, and release a ladder of multiples of 2.0 kb (Durfy and Willard, 1987; Willard *et al.*, 1986). These complex restriction patterns show a high degree of polymorphism between individuals, and are useful for the construction of centromere-based genetic linkage maps (Mahtani and Willard, 1988; Willard *et al.*, 1986). Such sequence polymorphisms have been described for other chromosome-specific alpha satellite DNA subsets (Greig *et al.*, 1989, 1993; Jabs *et al.*, 1984; Maskova *et al.*, 1994; Waye *et al.*, 1987a, b). Individual homologous chromosomes can reveal a remarkable variability in the number of repeats that contain a particular sequence polymorphism. Restriction enzyme digestion of these polymorphic chromosomes can release fragments from a very few repeats, giving faint or no bands on a Southern blot, or from a large proportion of the array, giving very dark bands (Durfy and Willard, 1987; Waye and Willard, 1986a; Waye *et al.*, 1987b).

Another type of polymorphism found in alpha satellite DNA is higher-order repeat units from a particular chromosomal subset that differ by an integral number of monomers but are nonetheless closely related in sequence. Such length variants are presumably generated by unequal crossing-over between existing higher-order repeat units that are misaligned in the register of their constituent monomers, resulting in a precise monomeric deletion or duplication (Southern, 1975; Waye and Willard, 1986a, b). For example, both restriction site periodicity and sequencing has shown that the chromosome-17-specific alpha satellite subset is characterized by multiple repeat units of different lengths, ranging in size from 11 to 16 monomers. Individual chromosomes 17 are characterized by the presence or absence of sets of these repeat-unit length variants, which can be used to define haplotypes (Waye and Willard, 1986 a, b; Willard *et al.*, 1986, 1987).

5.3.1 Variant repeats are clustered within arrays

The long-range organization of variant repeat units within alpha satellite DNA arrays has been analysed using a novel two-dimensional gel electrophoretic technique. This combines the long-range resolution of PFGE (in the first dimension) with the relatively short-range resolution of conventional electrophoresis (in the second dimension), enabling the analysis of the repeat unit variants contained in separate regions of the arrays (the individual PFGE fragments resolved in the first dimension). This technique has shown that variant repeat units (identified using enzymes that define the repeat unit and resolved in the second dimension) are often localized in specific PFGE fragments as opposed to occurring in all PFGE fragments. This indicates the presence of distinct domains of variant repeats within arrays of alpha satellite DNA (Ge *et al.*, 1992; Warburton and Willard, 1990; Warburton *et al.*, 1992). Some of these variant domains contain up to several hundred similar chromosome-specific variant repeat units and extends previous estimates of the size of such domains, previously observed in lambda and cosmid clones, by at least an order of magnitude (Durfy and Willard, 1989; Tyler-Smith, 1987).

The degree of misalignment between individual arrays has been suggested to be an important factor in the rate and extent of homogenization of tandemly repeated DNA (Durfy and Willard, 1989; Ohta, 1980; Smith, 1976). The existence of localized homogeneous domains of variant repeats within alpha satellite arrays suggests that crossing-over occurs between sister chromatids that are misaligned primarily by one or a few repeat units, spreading repeat unit variants locally and not throughout the entire array. Further support for this conclusion comes from the discovery that haplotypic variation in the copy number of variant repeats between homologous chromosomes can range from complete absence to almost complete fixation (Section 5.3.2; Dover, 1986; Warburton and Willard, 1995).

5.3.2 Sequence analysis of variant repeats

Cloning particular variant repeat units from library screens is difficult and labour-intensive, especially for the less abundant variant repeat units, due to the background of highly similar repeat units in the genome. Therefore, to facilitate the isolation and characterization of repeat unit variants, a system based on the polymerase chain reaction (PCR) that can amplify alpha satellite DNA in a chromosome-specific manner was developed (*Figure 5.4*; Warburton *et al.*, 1991). This assay allowed the haplotype-specific amplification of all the major chromosome 17 variants, and thus could be used for rapid PCR-based genotyping of this subset. PCR-amplified repeat units of different lengths are easily isolated and characterized to determine the extent of sequence variation and the molecular basis of putative unequal crossing-over events that gave rise to them. An example of this is shown in *Figure 5.5*, where sequence comparison of a variant 11mer from chromosome 17 with the sequence of the constitutive 16mer indicates that the deletion giving rise to the 11mer is localized within the constitutive repeat (Warburton and Willard, 1992).

The examination of multiple unequal crossing-over events of this kind could, in theory, identify underlying biological principles or constraints involved in synapsis or location of crossing-over within the higher-order repeat

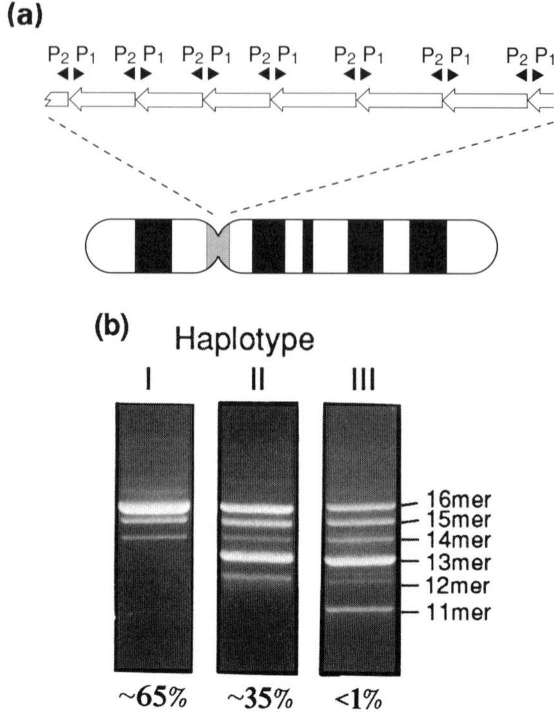

Figure 5.4. RepPCR amplification of alpha satellite DNA. (a) PCR primers that are specific for a particular chromosome-specific, higher-order repeat unit (P1 and P2) can amplify multiple repeat units (large open arrows of varying length) that are representative of the alpha satellite subset – hence the name repPCR. (b) The haplotypic nature of amplified products obtained using repPCR. This diagram shows fractionated PCR products obtained when primers designed for the subset specific for chromosome 17 are used to analyse individual chromosomes 17 in mouse–human somatic cell hybrids. Haplotype I, which is the most common human haplotye (~65%) is characterized by 14mers, 15mers and 16mers. Haplotype II (~35%) contains 12mers and 13mers not seen in haplotype I, while haplotype III (<1%) contains an abundant 11mer not present in either haplotypes I or II. Adapted from Warburton and Willard (1995) Inter-homologue sequence variation of chromosome 17 alpha satellite DNA: evidence for concerted evolution along haplotypic lineages. *J. Mol. Evol.* **41**: 1006–1015. © Springer-Verlag.

unit. Therefore, a total of 13 variant repeat units that belong to the pentamer subfamily were collected and characterized. These include the PCR-amplified, variant chromosome 17 repeat units plus other repeats cloned from cosmids and lambda libraries from human chromosomes 17 and X (Warburton and Willard, 1992; Warburton *et al.*, 1993; Waye and Willard, 1986a, b). The putative unequal crossing-over events that led to these variant repeat units were found to cluster within a distinct region of the 171 bp monomeric repeat unit, as opposed to occurring randomly (*Figure 5.6*). This was proposed to be due to secondary characteristics such as chromatin structure, where the crossing-over was occurring between phased nucleosomes on the 171 bp monomers (Warburton *et al.*, 1993). Furthermore, recognition sites for CENP-B, a protein that is centromere-specific and binds to alpha satellite DNA (Masumoto *et al.*, 1989) were found to be juxtaposed in each misalignment of repeat units (*Figure 5.7*), which suggests a possible role for this protein in the synapsis of these recombination events (Warburton *et al.*, 1993).

The spread of these repeat units through chromosomes and populations will depend on the relative rates of several recombinational processes including intra- and interchromosomal exchange (Charlesworth *et al.*, 1994; Dover, 1986, 1993; Willard, 1991). To examine these processes, the extent of sequence variation between repeat units from within and between different chromosomes 17 in the population was determined. PCR amplification of alpha satellite DNA results in the simultaneous amplification of multiple copies of each repeat unit. Direct sequencing of this PCR-amplified DNA reveals positions of sequence heterogeneity between repeat units appearing as two bands at the same position on a sequencing ladder (Warburton and Willard, 1992). Direct sequence analysis of PCR-amplified alpha satellite DNA from several chromosomes 17 detected no variation between chromosomes from the same haplotype, but distinct variation between chromosomes from different haplotypes (Warburton and Willard, 1992, 1995). Individual repeat units can also be conveniently cloned from the PCR reaction for further analysis. Analysis of multiple individual clones detected extensive haplotype-specific sequence variation, not simply a few rare mutations. Phylogenetic analysis of multiple individual clones indicates that these repeat units on chromosome 17 evolve along haplotypic lineages (Warburton and Willard, 1995). These studies provide further evidence for a high rate of intrachromosomal (sister chromatid) exchange relative to exchange between homologues (*Figure 5.2*).

5.4 Suprachromosomal organization of alpha satellite DNA

Despite the predominantly chromosome-specific organization of alpha satellite DNA subsets, sequence and hybridization studies have revealed the existence of major related subfamilies among different chromosomes (Willard, 1991). Southern blot and *in situ* hybridizations, using repeat units at stringencies lower than that required for chromosome specificity, reveal the relationships between subsets from different chromosomes (Alexandrov *et al.*, 1988). Extensive sequence analysis has allowed the determination of characteristic sequences (and deletions) at particular locations that define a common origin of monomeric units, despite a significant overall divergence between some individual members (Alexandrov *et al.*, 1991; Willard and Waye, 1987b).

To date, four suprachromasomal familes have been identified in the human genome. The repeat units from suprachromosomal subfamily 1, found on human chromosomes 1, 3, 5, 6, 7, 10, 16 and 19, are all based on characteristic *Eco*RI sites containing dimeric or tetrameric repeat units related to the first

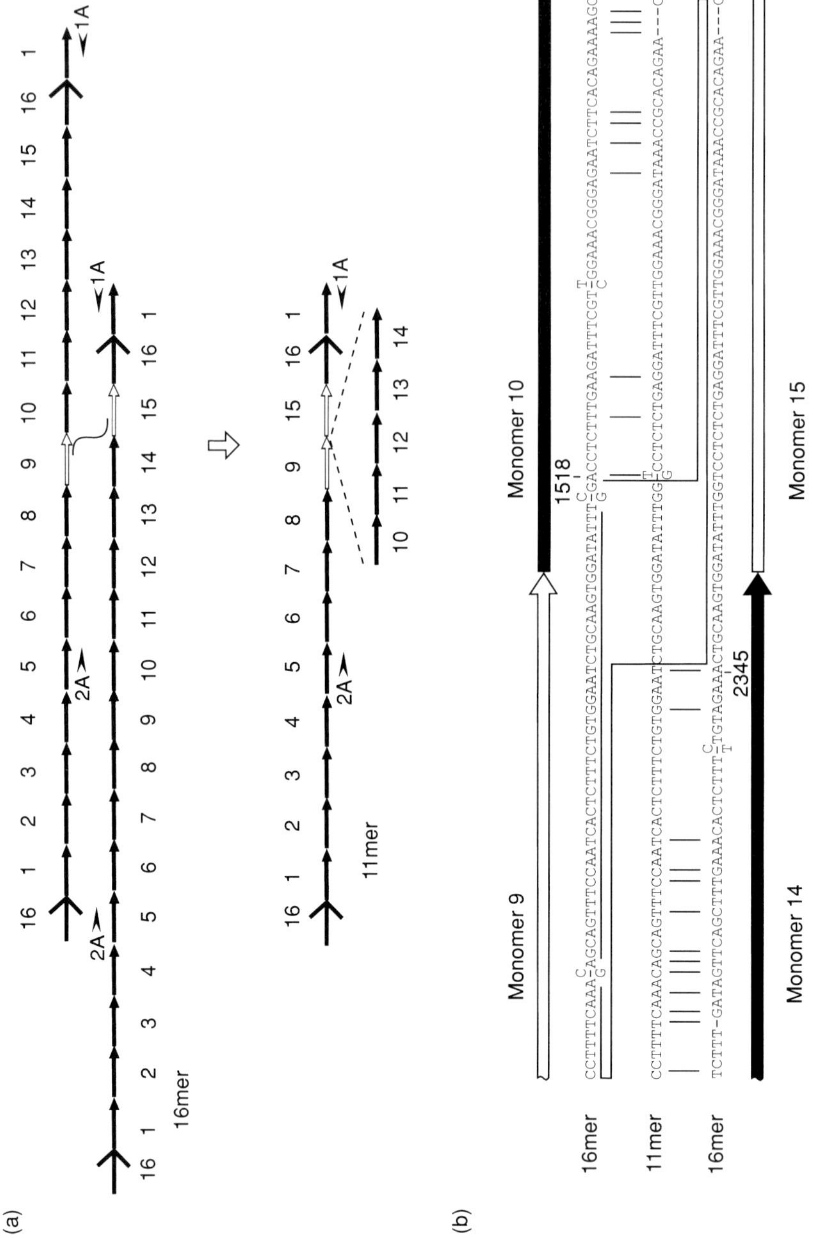
(a)
16mer
11mer
1A
2A
(b)
Monomer 9
Monomer 10
Monomer 14
Monomer 15
1518
2345
16mer
11mer
16mer

Figure 5.5. Determination of unequal crossing-over in a chromosome 17 alpha satellite repeat unit variant. (a) The proposed formation of an 11mer repeat unit variant by unequal crossing-over (or unequal sister chromatid exchange) occurring between two constituent 16mers. The exchange shown occurs at the junctions of monomers 9 and 10 on one 16mer and of monomers 14 and 15 in the other 16mer, resulting in the loss of five monomers (numbers 10–14). The location of the PCR primers 1A and 2A used in this analysis are shown. (b) Sequence analysis which identifies the site of unequal exchange during the formation of the variant 11mer shown in (a). When the sequence of the 11mer (centre) is compared with the sequence of the constitutive 16mer (top and bottom), an abrupt and symmetric breakpoint in similarity is seen between nucleotide positions 1500–1517 inclusive (junction of monomers 9 and 10, top strand) and positions 2346–2363 inclusive (junction of monomers 14 and 15, bottom strand) of the 16mer. Vertical lines indicate the position of mismatches between the 11mer and 16mer sequences. This allows the putative unequal crossing-over event to be localized to a window of 18 bp. Nucleotide positions are as in Waye and Willard (1986b). Adapted from Warburton *et al.* (1993) with permission from the American Society for Microbiology.

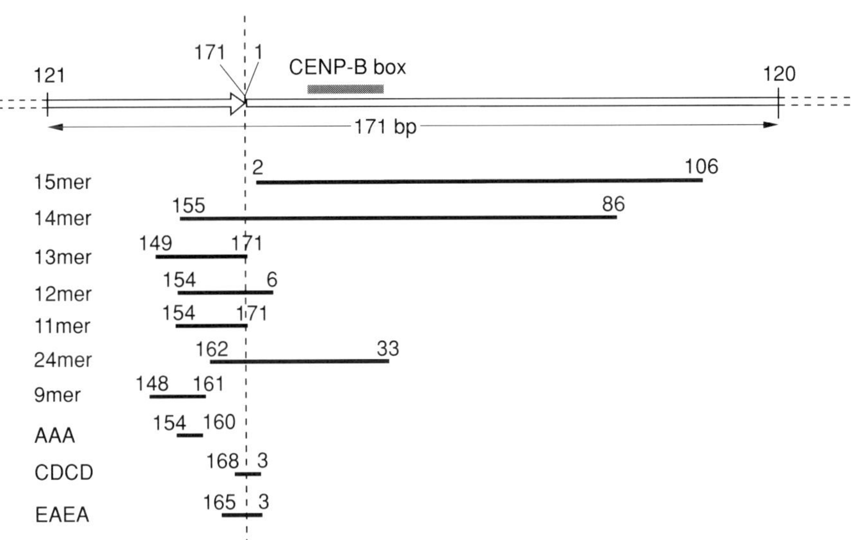

Figure 5.6. Location of windows of recombination in variant repeat units. The putative windows of recombination were determined as shown in *Figure 5.5* for a series of variant repeat units from chromosome 17 and other subsets from the pentameric subfamily. The 9mer and 24mer are low-copy variants isolated from a cosmid library. AAA, CDCD and EAEA refer to monomer homology groups in the pentamer subfamily (see *Figure 5.8*). The location of these recombination windows are shown relative to a 171 bp monomer. Instead of being randomly distributed throughout the 171 bp, the windows appear to cluster in a 20–25 bp region. The location of the CENP-B box is indicated. The large windows seen for the 15mer and 14mer are due to a recombination between highly homologous monomers. Adapted from Warburton *et al.* (1993) with permission from the American Society for Microbiology.

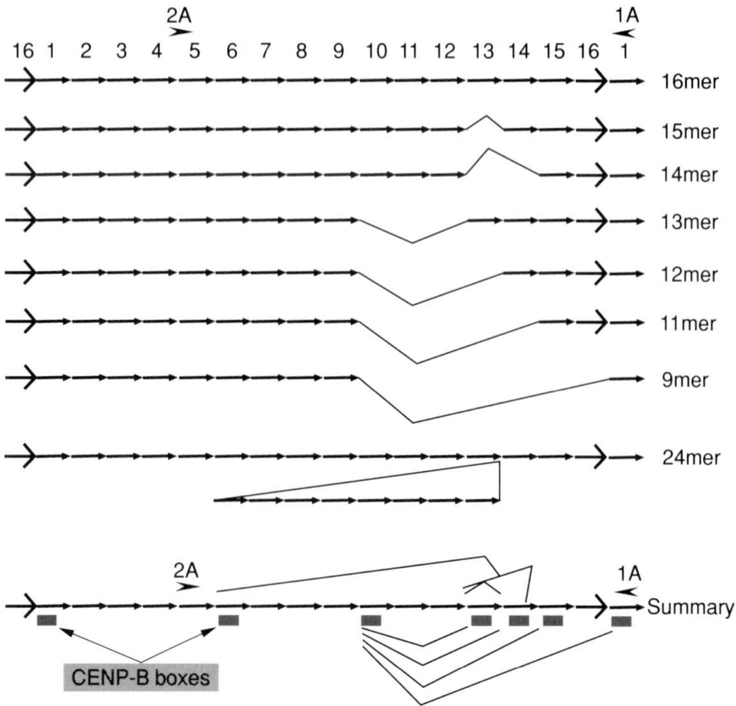

Figure 5.7. Structure of variant repeats for chromosome 17 alpha satellite DNA. The variant repeats are shown as deletions relative to the constitutive 16mer, or as an insertion in the case of the 24mer. At the bottom of the figure is a summary of the deletions, with the location of the CENP-B box shown. In each case, the misalignment giving rise to the recombination event results in the juxtaposition of two CENP-B boxes in close proximity to the cross-over. This suggests that CENP-B may play a role in the determination of recombination events. Adapted from Warburton *et al.* (1993) with permission from the American Society for Microbiology.

described human alpha satellite DNA clone (Alexandrov *et al.*, 1988; Wu and Manuelidis, 1980). Suprachromosomal subfamily 2, which is found on human chromosomes 2, 4, 8, 9, 13, 14, 15, 18, 20, 21 and 22 (Alexandrov *et al.*, 1988), is also characterized by a dimeric organization. These dimers, however, characteristically contain *Xba*I sites and are distinct in origin from the *Eco*RI dimers of subfamily 1 (Gray *et al.*, 1985; Jorgensen *et al.*, 1987a; Thompson *et al.*, 1989). The third suprachromosomal subfamily, which is found on human chromosomes 1, 11, 17 and X, is based on a five monomer repeating pattern and is often referred to as the pentamer subfamily (*Figure 5.8*; Alexandrov *et al.*, 1988; Waye and Willard, 1985; Willard and Waye, 1987b). A fourth suprachromosomal subfamily, with a monomeric organization in which the amplification events are based on individual monomers, as opposed to higher-order repeat units, has also been described (Alexandrov *et al.*, 1993), although the organization of this subfamily within the genome is unclear.

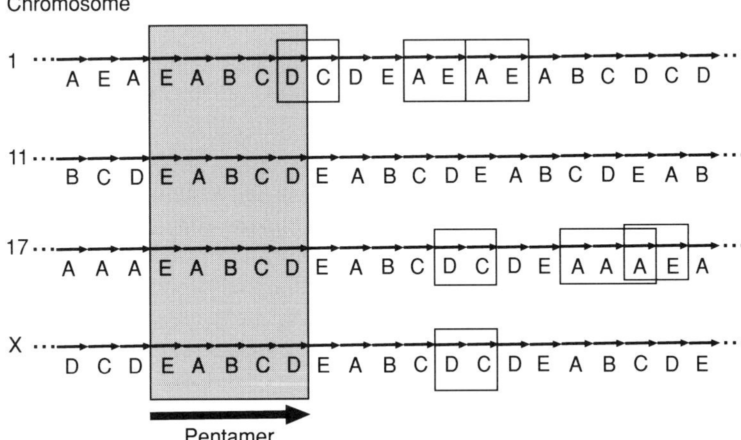

Figure 5.8. Monomeric organization of alpha satellite suprachromosomal subfamily 3 (pentamer subfamily). This subfamily consists of five distinct monomeric homology groups (A–E). Each of the chromosome-specific, higher-order repeat units contains a different organization of these monomeric homology groups. These have presumably arisen by unequal crossing-over of ancestral pentamers which are still found in the repeats (shown in the shaded region). The boxed regions indicate regions that deviate from this ancestral pentamer and represent the location of presumed unequal crossing-over events. Adapted from Willard and Waye (1987b) Chromosome-specific subsets of human alpha satellite DNA: analysis of sequence divergence within and between chromosomal subsets and evidence for an ancestral pentameric repeat. *J. Mol. Evol.* **25**: 207–214. © Springer-Verlag.

Subsets of alpha satellite DNA have been isolated that cannot be clearly classified into any of the major subfamilies, including a subset from chromosome 7 (Waye *et al.*, 1987b) and from the Y chromosome (Tyler-Smith and Brown, 1987; Wolfe *et al.*, 1985). The subset from the Y chromosome is particularly noteworthy, as the centromeric region of the Y chromosome does not have an opportunity to undergo meiotic or interhomologue exchange; thus, this alpha satellite may be evolving essentially in isolation.

5.4.1 Structural relationship between members of a suprachromosomal family

The underlying pentameric organization of suprachromosomal subfamily 3 has been analysed in detail and has provided insight into the structural relationships between the four chromosome-specific arrays which comprise this group (present on chromosomes 1, 11, 17 and X). Sequence analysis has confirmed that the monomeric units which make up the different higher-order repeat units specific for these four chromosomes (11mer, 5mer, 16mer and 12mer, respectively) can be classified into one of five distinct homology groups called A, B, C, D and E (Waye and Willard, 1985, 1986b). Monomers from within a homology group, regardless of chromosomal origin, are more similar to each other than they are to monomers from any of the other homology groups, suggesting a common origin for the members of each homology group (Willard and Waye, 1987b). Despite these relationships, each chromosome-specific subset has higher-order repeat units that are distinguishable by mutations in the organization and primary sequence of monomers. These mutations can be viewed as having accumulated during the generation and fixation of each higher-order repeat unit on a specific chromosome (Willard and Waye, 1987b). These higher-order repeat units have apparently formed by rearrangements of canonical pentamers (ABCDE; *Figure 5.8*), presumably by unequal crossing-over, to give distinct chromosome-specific patterns in which intact pentamers can still be seen (Alexandrov *et al.*, 1988; Waye and Willard, 1985, 1986b; Waye *et al.*, 1987a, b; Willard and Waye, 1987b).

The five monomeric homology groups characteristic of this subfamily allow more detailed analysis of patterns of variation and crossing-over than the dimeric subfamilies. The relationships within the pentameric subfamilies suggest that interchromosomal exchanges (between non-homologues) have occurred (at least ancestrally) such that repeat units have spread to non-homologous chromosomes (*Figure 5.2*). However, the distinct chromosome-specificity and considerable divergence of members of subfamilies from different chromosomes provides further evidence that this process appears to be very infrequent compared with exchanges between homologous chromosomes (Charlesworth *et al.*, 1994; Dover, 1986; Ohta and Dover, 1983; Willard, 1991). In the case of the pentamer subfamily, these interchromosomal changes are likely to have pre-dated the separation of great apes and humans, a theory based on the similarity of higher-order repeat units from homologous chromosomes in these species (Baldini *et al.*, 1991a; Durfy and Willard, 1990; Laursen *et al.*, 1992).

5.5 Subsets present on more than one chromosome and chromosomes with more than one subset

Not all alpha satellite DNA subsets show a strict specificity for a single chromosome but instead appear to be found on more than one chromosome. One such example is a dimeric subset from suprachromosomal subfamily 1 that hybridizes to the centromeres of chromosomes 1, 5 and 19 (Baldini *et al.*, 1989; Hulsebos *et al.*, 1988). No hybridization conditions could be found that eliminated hybridization to any one of these chromosomes (Baldini *et al.*, 1989). These data suggest that relatively recent exchanges have occurred between the centromeres of these chromosomes, leading to a higher degree of homogenization and a lack of distinct chromosome-specific alpha satellite subsets. It would be interesting to examine the sequence of repeat units of known origin from each of these chromosomes, to assess the level (if any) of chromosome-specific sequence changes that may be below the level detectable by *in situ* hybridization.

Given the identification of alpha satellite subsets which map to more than one chromosome, it is not surprising that some human chromosomes are characterized by more than one type of alpha satellite DNA. For example, the centromeric region of chromosome 7 contains two distinct types of alpha satellite: a 6 monomer higher-order repeat belonging to subfamily 1 and a 16 monomer repeat not classified into a subfamily (*Table 5.1*; Waye *et al.*, 1987b). The origin of two unrelated alpha satellite arrays at the same centromere is unclear, but could be ancestral or have arisen by interchromosomal transfer. The organization of these two arrays has been further examined by PFGE (Wevrick and Willard, 1991) and by extended chromatin fibre *in situ* hybridization (Haaf and Ward, 1994), which showed that they are physically distinct and separated by approximately 1000 kb of DNA. This physical separation would help to explain the apparent lack of homogenization between the two subsets despite being on the same chromosome (Waye *et al.*, 1987b). Chromosome 1 also contains two types of alpha satellite DNA from subfamilies 1 and 3 (Alexandrov *et al.*, 1988; Waye *et al.*, 1987a;), while chromosome 18 has two distinct subsets from subfamily 2 that are as different from each other as they are from other members of subfamily 2 on other chromosomes (Alexandrov *et al.*, 1991). Chromosome 15 also contains two distinct subsets of alpha satellite DNA, both belonging to subfamily 2 (*Table 5.1*; Choo *et al.*, 1990).

5.5.1 The acrocentric chromosomes: a complex pattern of subset organization

The human acrocentric chromosomes, numbers 13, 14, 15, 21 and 22, contain the tandemly arranged ribosomal DNA repeat units on their short arms. These chromosomes frequently form Robertsonian translocations such that the entire long arm, and sometimes portions of the short arm, are transferred. This is thought to occur by exchanges between the large amount of satellite DNA in the short arm of these chromosomes when they are organized in the nucleolus (Gravholt *et al.*, 1992; Sullivan *et al.*, 1996; Therman *et al.*, 1989). The alpha satellite DNA from these chromosomes, which all belong to subfamily 2, reveals a complex organization with some subsets being chromosome-specific and others being dispersed on one or more chromosomes (Choo, 1990; Choo *et al.*, 1991). Specific subsets have been found on chromosome 22 (McDermid *et al.*, 1986; Rocchi *et al.*, 1994), chromosome 14 (Waye *et al.*, 1988) and, as mentioned above, two distinct subsets have been identified which are specific for chromosome 15 (Choo *et al.*, 1990).

Many other subsets are found on more than one acrocentric chromosome, with different subsets being identified on 13 and 21 (*Table 5.1*; Devilee *et al.*, 1986; Ikeno *et al.*, 1994; Jorgensen *et al.*, 1987a) and 14 and 22 (*Table 5.1*; Jorgensen *et al.*, 1988). At least four distinct subsets have been reported to be shared by chromosomes 13, 14 and 21 (*Table 5.1*; Choo *et al.*, 1988, 1989, 1991; Trowell *et al.*, 1993).

The lack of chromosome specificity displayed by the acrocentric alpha satellite subsets suggests that relatively frequent interchromosomal exchanges have taken place. This has been attributed to the close association of these non-homologous chromosomes in the nucleolus (Choo, 1990). Such exchanges are presumed to be tolerated because the short arms of these chromosomes contain functionally similar ribosomal DNA genes. It is interesting to note that a higher frequency of translocations occur between chromosomes 13 and 14 and between 14 and 21 (Therman *et al.*, 1989; Warburton, 1991), which has been postulated to be due to an inversion of one or more of the shared subsets of homologous alpha satellite DNA on chromosome 14 relative to chromosomes 13 and 21 (Choo, 1990; Trowell *et al.*, 1993). Sequence analysis of 11mer subset shared between chromosomes 13 and 21 shows some rare polymorphisms that make it possible to distinguish these two chromosomes (Greig *et al.*, 1993; Warren *et al.*, 1990), suggesting that intrachromosomal homogenization is also taking place.

5.6 Alpha satellite DNA in non-human primates

Since its original identification in the African green monkey (Maio, 1971), alpha satellite DNA with its characteristic restriction pattern of 170 bp ladders has been described in many primate species ranging from Old and New World monkeys to higher primates and humans (Alves *et al.*, 1994; Donehower and Gillespie, 1979; Fanning *et al.*, 1993; Maio *et al.*, 1981; Musich *et al.*, 1980; Prassolov *et al.*, 1986). The wide range of alpha satellite DNA throughout primate species provides an opportunity to examine aspects of concerted evolution over time by defining relationships between the repeat units in different species. It may also provide data relevant to primate phylogeny. Initial investigations have involved using available clones in hybridization experiments and the sequencing of randomly cloned monomers and dimers from non-human primates.

5.6.1 Hybridization data

There is sufficient sequence similarity between the alpha satellite DNA from higher primates (orangutan, gorilla, chimpanzee and human) to allow cross-hybridization on Southern blots or in *in situ* hybridization experiments using conditions of reduced stringency. A human clone from suprachromosomal subfamily 2 cross-hybridizes to all four of these species at low stringency but does not cross-hybridize to DNA from Old or New World monkeys or to the gibbon (Gray *et al.*, 1985). An alpha satellite DNA repeat cloned from the gorilla was reported to hybridize *in situ* under high stringency to a subset of chromosomes from gorilla, human, orangutan and, less abundantly, to chromosomes from chimpanzee (Baldini *et al.*, 1991b). This clone hybridized to at least 10 gorilla chromosomes and to the human centromeres in subfamily 2. It is possible that the stringency conditions used allowed significant cross-hybridization which may account for the lack of species-specific patterns of hybridization. In a related study, a large collection of human-derived alpha satellite DNA clones, as well as a gorilla-derived probe, showed species specificity at high stringency conditions, and cross-reaction at reduced stringency – a result more consistent with the expectations of a system undergoing concerted evolution (Waye and Willard, 1989b). These somewhat contradictory studies illustrate that subtle differences in hybridization conditions may give misleading results. However, the results may also be due to differences in the evolutionary history of the particular repeats used. More informative results require the nucleotide sequence analysis of related repeats from different species.

5.6.2 Sequence data

The first clones of alpha satellite DNA obtained from the genomes of non-human primates using restriction site periodicities were cloned randomly and were unable to take into account the chromosomal origin of the clones. None of the sequences obtained in this way bear an obvious relationship to any of the three major human suprachromosomal subfamilies (subfamilies 1–3; Alexandrov *et al.*, 1991), although a sequence from the African green monkey is known to belong to the monomeric subfamily 4 (Alexandrov *et al.*, 1993). The interpretation of sequence comparisons between species is also unclear. For example, an alpha satellite DNA dimer from the rhesus monkey was reported to be 98% similar to a baboon alpha satellite DNA sequence (Singer and Donehower, 1979), 80% similar to an alpha satellite DNA sequence from the African green monkey, and 71% similar to alpha satellite DNA from human subfamilies 1 and 2 (Pike *et al.*, 1986). The significance of these results, while consistent with the accepted relationships of these species as deduced from palaentological studies (Donehower and Gillespie, 1979), is debatable without a knowledge of the chromosomal origin of these sequences. It is well-known that alpha satellite DNA repeats from different chromosomes in the human genome can be diverged from one another by 35% or more (Alexandrov *et al.*, 1991; Looijenga *et al.*, 1992). It has also been established that repeats from different primate species, such as human, gorilla and chimpanzee, can be diverged by only 10% or less (Baldini *et al.*, 1991a; Durfy and Willard, 1990). Thus, interpretation of sequence similarities between repeat units of unknown origin could lead to false conclusions about relationships between species, since variation between monomeric units can be greater within a species than between species.

5.7 Primate alpha satellite DNA: orthologous and non-orthologous patterns of evolution

Given the chromosome-specific nature of alpha satellite DNA, it would be most useful to understand the relationships between repeat units on known homologous chromosomes from related species: these relationships have been investigated using the genomes of human, chimpanzee and gorilla. The alpha satellite subsets specific for the X chromosomes of the gorilla, chimpanzee and human have repeat units that show identical 2.0 kb *Bam*HI, 12 monomer periodicity, although in the gorilla the more prominent repeat units are 2.7 kb and 3.2 kb long (Durfy and Willard, 1990; Laursen *et al.*, 1992). Each repeat unit belongs to suprachromosomal subfamily 3 (pentamer subfamily), and the arrangement of monomers belonging to the five known homology groups (*Figure 5.8*), is co-linear for each 2.0 kb repeat (Durfy and Willard, 1990; Laursen *et al.*, 1992). In a related study, a chimpanzee alpha satellite DNA clone belonging to suprachromosomal subfamily 3 was shown by *in situ* hybridization to be on chimpanzee (*Pan paniscus*) chromosome 19, which is homologous to human chromosome 17 (Baldini *et al.*, 1991a). Limited sequence analysis was suggestive of a conservation of the monomeric pattern found on human chromosome 17 (Baldini *et al.*, 1991a). These patterns of conservation of higher-order repeat units in chimpanzee, gorilla and human suggests that at least the pentamer family pre-dates the split of these species, estimated to be about five million years ago, and that the alpha satellite DNA has been evolving along these chromosomal lineages since this split occurred (Jorgensen *et al.*, 1992; Laursen *et al.*, 1992).

Similar to the situation on human acrocentric chromosomes, the subsets from chimpanzee acrocentric chromosomes 15 and 23 (homologous to human chromosomes 14 and 22, respectively) are indistinguishable by hybridization (Jorgensen *et al.*, 1992). The subsets from chimpanzee acrocentric chromosomes 14 and 22 (homologous to human chromosomes 13 and 21, respectively) are also closely related (Jorgensen *et al.*, 1987b, 1992). However, unlike the situation described above for the X chromosomes, the subsets from the homologous chromosomes between the species are not related, with the chimpanzee sequences belonging to suprachromosomal subfamily 1 and the human sequences to subfamily 2 (Jorgensen *et al.*, 1992). Thus, these alpha satellite DNA subsets do not appear to be evolving in an orthologous manner, which may be related to the lability of the alpha satellite DNA found on acrocentric chromosomes (Section 5.5.1).

Comparative *in situ* hybridization has revealed several other human chromosome-specific subsets that do not hybridize specifically to their corresponding homologous chromosomes in chimpanzees and gorillas (Baldini *et al.*, 1992; D'Aiuto *et al.*, 1993). Recently, a large study using most of the available human chromosome-specific subsets as *in situ* probes against chimpanzee and gorilla chromosomes support the conclusion that, in general, subsets are not found on their corresponding homologous chromosomes (Archidiacono *et al.*, 1995). Another study made use of PCR primers specific for human chromosome 17 (Warburton *et al*, 1991) to amplify by PCR a chromosome-specific subset from chimpanzee (*Pan troglodytes*) that corresponds most closely to that on human chromosome 17 in sequence and repeat unit structure (Warburton *et al.*, 1996). Comparative *in situ* hybridization reveals, however, that this subset is not found on the corresponding homologous chimpanzee chromosome (19) but instead hybridizes specifically to chimpanzee chromosome 12, which corresponds to human chromosome 2p. This provides an example of a chimpanzee subset that has not undergone orthologous evolution (Warburton *et al.*, 1996). These studies revealed complex patterns of hybridization between the homologous chromosomes from the different species, with even an apparent lack of conservation of chromosomes belonging to the different subfamilies. The basis for which chromosome-specific subsets do, or do not, undergo orthologous evolution is poorly understood but may reflect differential rates of stochastic interchromosomal exchanges for certain non-homologous chromosomes. However, given these orthologous and non-orthologous patterns of homogenization, the analysis of primate alpha satellite DNA evolution requires even more care and detailed interpretation, and a comparative description of even more primate subsets will be required before a good understanding of these processes can be realized.

5.8 Primate alpha satellite DNA: sequence variation within and between species

Despite the similarities in organization between repeat units from the X chromosomes of higher primates, they nevertheless show patterns of sequence divergence within and between species consistent with predictions of concerted evolution (*Table 5.2*). A large study of 34 independent human 2.0 kb repeats showed an average within-species sequence divergence of 1.35% (Durfy and Willard, 1989). Additional, more limited, data showed less than 0.5% divergence between 2.0 kb repeat units from the chimpanzee X chromosome (Laursen *et al.*, 1992). Sequence analysis of several repeat units from the subset isolated from chimpanzee chromosome 12 showed a mean sequence variation of 0.34% (Warburton *et al.*, 1996), while the mean variation between repeat units specific for human chromosome 17 was 1.78% (*Table 5.2*; Warburton and Willard, 1995). Thus, on both chimpanzee chromosomes 12 and X there appears to be a significantly lower level of intrachromosomal variation than on the corresponding subsets in humans, which may suggest that applying a constant molecular clock to such DNA species may not be valid (Laursen *et al.*, 1992; Warburton *et al.*, 1996).

The sequence divergence between species is, as expected, significantly higher. The divergence between gorilla and human 2.0 kb repeats from the X chromosome is 9–10%, between chimpanzees and humans it is approximately 7% (*Table 5.2*), and between chimpanzees and gorillas approximately 8% (Durfy and Willard, 1990; Laursen *et al.*, 1992). This is consistent with the sequences undergoing concerted evolution on their respective X chromosomes since the divergence of these species. However, other aspects may be less consistent with some predictions of concerted evolution. For example, the divergence between the alpha satellite DNA from chimpanzee chromosome 12 and

Table 5.2. Alpha satellite DNA sequence comparisons within and between chromosomes of human and chimpanzee

	HSA 17	HSA X	PTR X	PTR 12
HSA 17	**1.78%**	–	–	–
HSA X	15.2%	**1.35%**	–	–
PTR X	14.5%	7.0%	**0.43%**	–
PTR 12	8.4%	13.6%	18.5%	**0.34%**

The mean sequence divergence between alpha satellite repeat units from the indicated chromosomal subsets of chimpanzees (*Pan troglodytes*, PTR) and humans (*Homo sapiens*, HSA) are shown. Numbers along the main diagonal (in bold type) are the mean sequence divergences between repeat units from within each of the subsets. Sequence comparisons are taken from Warburton and Willard (1995), and Warburton *et al.* (1996), and references therein.

human chromosome 17 is 8.4% (Warburton *et al.*, 1996) while the divergence between this chimpanzee repeat and the chimpanzee chromosome X repeat is 18.5% (*Table 5.2*). Thus, the sequence is in fact more closely related to sequences from a different species than it is to sequences from the same species. The analysis of the evolution of alpha satellite DNA in both higher and lower primates, although clearly complex and requiring extensive further research, should prove to be an extremely interesting area of research and might eventually provide insight into mechanisms and patterns of chromosomal evolution.

5.9 Conclusions

By detailed characterization of the human and primate alpha satellite DNA family, many aspects of the evolutionary processes that led to the organization of these DNA families can be further understood (Willard, 1991). The ongoing study of sequence variation and long-range organization of alpha satellite repeat units from different species, chromosomes and individual homologues is providing the basis for relative assessment of various fundamental genetic parameters such as mutation and intra- and interchromosomal exchange (Charlesworth *et al.*, 1994; Dover, 1993). Perhaps large tandemly repeated DNA families no longer have to be considered as arcane intellectual and genetic obstacles, often dismissed as 'junk' DNA, but instead may be realized as a highly informative system that will add greatly to the understanding of the organization, structure and evolution of the human and other complex genomes.

References

Alexandrov IA, Mitkevich SP, Yurov YB. (1988) The phylogeny of human chromosome-specific alpha satellites. *Chromosoma (Berlin)* **96**: 443–453.

Alexandrov IA, Mashkova TD, Akopian TA, Medvedev LI, Kisselev LL, Mitkevich SP, Yurov YB. (1991) Chromosome-specific alpha satellites: two distinct families on human chromosome 18. *Genomics* **11**: 15–23.

Alexandrov IA, Medvedev LI, MashkovaTD, Kisselev LL, Romanova LY, Yurov YB. (1993) Definition of a new alpha satellite suprachromosomal family characterized by monomeric organization. *Nucleic Acids Res.* **21**: 2209–2215.

Alves G, Seuanez HN, Fanning T. (1994) Alpha satellite DNA in neotropical primates (Platyrrhini). *Chromosome* **103**: 262–267.

Archidiacono N, Annotate R, Marzella R, Finelli P, Lonoce A, Rocchi M. (1995) Comparative mapping of human alphoid sequences in great apes using fluorescence *in situ* hybridization. *Genomics* **25**: 477–484.

Baldini A, Smith DI, Rocchi M, Miller OJ, Miller DA. (1989) A human alphoid DNA clone from the *Eco*RI dimeric family: genomic and internal organization and chromosomal assignment. *Genomics* **5**: 822–828.

Baldini A, Miller DA, Miller OJ, Ryder OA, Mitchell AR. (1991a) A chimpanzee-derived chromosome-specific alpha satellite DNA sequence conserved between chimpanzee and human. *Chromosome* **100**: 156–161.

Baldini A, Miller DA, Shridhar V, Rocchi M, Miller OJ, Ward, DC. (1991b) Comparative mapping of a gorilla-derived satellite DNA clone on great ape and human chromosomes. *Chromosome* **101**: 109–114.

Baldini A, Archidiacono N, Carbone R, Bolino A, Shridhar V, Miller OJ, Miller DA, Ward DC, Rocchi M. (1992) Isolation and comparative mapping of a human chromosome 20-specific alpha satellite DNA clone. *Cytogenet. Cell Genet.* **59**: 12–16.

Brown DD, Wensink PC, Jordan E. (1971) A comparison of the ribosomal DNAs of *Xenopus laevis* and *Xenopus mulleri*: the evolution of tandem genes. *J Mol. Biol.* **63**: 57–73.

Charlesworth B, Sniegowski P, Stephan W. (1994) The evolutionary dynamics of repetitive DNA in eukaryotes. *Nature* **371**: 215–220.

Choo KH. (1990) Role of acrocentric cen-pter satellite DNA in Robertsonian translocations and chromosomal non-disjunction. *Mol. Biol. Med.* **7**: 437–449.

Choo KH, Vissel B, Brown R, Filby RG, Earle E. (1988) Homologous alpha satellite DNA sequences on human acrocentric chromosomes with selectivity for chromosomes 13, 14, and 21: implications for recombination between nonhomologues and Robertsonian translocations. *Nucleic Acids Res.* **16**: 1273–1284.

Choo KH, Vissel B, Earle E. (1989) Evolution of α-satellite DNA on human acrocentric chromosomes. *Genomics* **5**: 332–344.

Choo KH, Earle E, Vissel B, Filby RG. (1990) Identification of two distinct subfamilies of alpha satellite DNA that are highly specific for human chromosome 15. *Genomics* **7**: 143–151

Choo KH, Vissel B, Nagy A, Earle E, Kalitsis P. (1991) A survey of the genomic distribution of alpha satellite DNA on all the human chromosomes, and derivation of a new consensus sequence. *Nucleic Acids Res.* **19**: 1179–1182.

D'Aiuto L, Annotate R, Marzella R, Archidiacono N, Rocchi M. (1993) Cloning and comparative mapping of a human chromosome 4-specific alpha satellite DNA sequence. *Genomics* **18**: 230–235.

Devilee P, Cremer T, Slagboom P, Bakker E, Scholl HP, Hager HD, Stevenson AFG, Cornelisse CJ, Pearson PL. (1986) Two subsets of human alphoid repetitive DNA show distinct preferential localisation in the pericentromeric regions of chromosomes 13, 18, and 21. *Cytogenet. Cell Genet.* **41**: 193–201.

Donehower L, Gillespie D. (1979) Restriction site periodicities in highly repetitive DNA of primates. *J. Mol. Biol.* **134**: 805–834.

Dover GA. (1982) Molecular drive: a cohesive mode of species evolution. *Nature* **299**: 111–117.

Dover GA. (1986) Molecular drive in multigene families: how biological novelties arise, spread and are assimilated. *Trends Genet.* **2**: 159–165

Dover GA. (1993) Evolution of genetic redundancy for advanced players. *Curr. Opin. Genet. Dev.* **3**: 902–910.

Durfy SJ, Willard HF. (1987) Molecular analysis of a polymorphic domain of alpha satellite from the human X chromosome. *Am. J. Hum. Genet.* **41**: 391–401.

Durfy SJ, Willard HF. (1989) Patterns of intra- and interarray sequence variation in alpha satellite DNAfrom the human X chromosome: evidence for short-range homogenization of tandemly repeated DNA sequences. *Genomics* **5**: 810–821.

Durfy SJ, Willard HF. (1990) Concerted evolution of primate alpha satellite DNA: evidence for an ancestral sequence shared by gorilla and human X chromosome alpha satellite. *J. Mol. Biol.* **216**: 555–566.

Fanning TG, Seuanez HN, Forman L. (1993) Satellite DNA sequences in the New World primate *Cebus apella* (Platyrrhini, Primates). *Chromosome* **102**: 306–311.

Ge Y, Wagner MJ, Siciliano M, Wells DE. (1992) Sequence, higher order repeat structure, and long-range organization of alpha satellite DNA specific to human chromosome 8. *Genomics* **13**: 585–593.

Gravholt CH, Friedrich U, Caprani M, Jorgensen AL. (1992) Breakpoints in Robertsonian translocations are localized to satellite III DNA by fluorescence *in situ* hybridization. *Genomics* **14**: 924–930.

Gray KM, White JW, Costanzi C, Gillespie D, Schroeder WT, Calabretta B, Saunders GF. (1985) Recent amplification of an alpha satellite DNA in humans. *Nucleic Acids Res.* **13**: 521–535.

Greig GM, England SB, Bedford HM, Willard HF. (1989) Chromosome-specific alpha satellite DNA from the centromere of chromosome 16. *Am. J. Hum. Genet.* **45**: 862–872.

Greig GM, Warburton PE, Willard HF. (1993) The organization and evolution of an alpha satellite subset shared by chromosomes 13 and 21. *J. Mol. Evol.* **37**: 464–475.

Haaf T, Ward DC. (1994) Structural analysis of alpha satellite DNA and centromere proteins using extended chromatin and chromosomes. *Hum. Mol. Genet.* **3**: 697–709.

Hulsebos T, Schonk D, vanDalen I, Coerwinkel-Driessen M, Schepens J, Ropers HH, Wieringa B. (1988) Isolation and characterisation of alphoid DNA sequences specific for the pericentric regions of chromosomes 4, 5, 9, and 19. *Cytogenet. Cell Genet.* **47**: 144–148.

Ikeno M, Masumoto H, Okazaki T. (1994) Distribution of CENP-B boxes reflected in CREST centromere antigenic sites on long-range alpha satellite DNA of human chromosome 21. *Hum. Mol. Genet.* **3**:1245–1257.

Jabs EW, Wolf SF, Migeon BR. (1984) Characterisation of a cloned DNA sequence that is present at centromeres of all human autosomes at the X chromosome and shows polymorphic variation. *Proc. Natl Acad. Sci. USA* **81**: 4884–4888.

Jabs EW, Goble CA, Cutting GR. (1989) Macromolecular organization of human centromeric regions reveals high-frequency, polymorphic macro DNA repeats. *Proc. Natl Acad. Sci. USA* **86**: 202–206.

Jackson MS, Slijepcevic P, Ponder BAJ. (1993) The organization of repetitive sequences in the pericentromeric region of human chromosome 10. *Nucleic Acids Res.* **21**: 5865–5874.

Jorgensen AL, Bostock CJ, Bak AL. (1986) Chromosome-specific subfamilies within human alphoid repetitive DNA. *J. Mol. Biol.* **187**: 185–196.

Jorgensen AL, Bostock CJ, Bak AL. (1987a) Homologous subfamilies of human alphoid repetitive DNA on different nucleolus organizing chromosomes. *Proc. Natl Acad. Sci. USA* **84**: 1075–1079.

Jorgensen AL, Jones C, Bostock CJ, Bak AL. (1987b) Different subfamilies of alphoid repetitive DNA are present on the human and chimpanzee homologous chromosomes 21 and 22. *EMBO J.* **6**: 1691–1696.

Jorgensen AL, Kolvraa S, Jones C, Bak AL. (1988) A subfamily of alphoid repetitive DNA shared by the NOR-bearing human chromosomes 14 and 22. *Genomics* **3**: 100–109.

Jorgensen AL, Laursen HB, Jones C, Bak AL. (1992) Evolutionarily different alphoid repeat DNA on homologous chromosomes in human and chimpanzee. *Proc. Natl Acad. Sci. USA* **89**: 3310–3314.

Laursen HB, Jorgensen AL, Jones C, Bak AL. (1992) Higher rate of evolution of X chromosome alpha repeat DNA in human than in the great apes. *EMBO J.* **11**: 2367–2372.

Looijenga LHJ, Oosterhuis JW, Smit VTHBM, Wessels JW, Mollevanger P, Devilee P. (1992) Alpha satellite DNAs on chromosome 10 and 12 are both members of the dimeric suprachromosomal subfamily, but display little identity at the nucleotide sequence level. *Genomics* **13**: 1125–1132.

Mahtani MM, Willard HF. (1988) A primary genetic map of the pericentromeric region of the human X chromosome. *Genomics* **2**: 294–301.

Mahtani MM, Willard HF. (1990) Pulsed-field gel analysis of alpha satellite DNA at the human X chromosome centromere: high frequency polymorphisms and array size estimate. *Genomics* **7**: 607–613.

Maio JJ. (1971) DNA strand reassociation and polyribonucleotide binding in the African green monkey, *Cercopithecus aethiops*. *J. Mol. Biol.* **56**: 579–595.

Maio JJ, Brown FL, Musich PR. (1981) Toward a molecular palaeontology of primate genomes I: the Hind III and Eco RI dimer families of alphoid DNAs. *Chromosome* **83**: 103–125.

Maskova TD, Akopian TA, Romanova LY, Mitkevich SP, Yurov YB, Kisselev LL, Alexandrov IA. (1994) Genomic organization, sequence and polymorphism of the human chromosome 4-specific alpha satellite DNA. *Gene* **140**: 211–217.

Masumoto H, Masukata H, Muro Y, Nozaki N, Okazaki T. (1989) A human centromere antigen (CENP-B) interacts with a short specific sequence in alphoid DNA, a human centromeric satellite. *J. Cell. Biol.* 1963–1973.

McDermid HE, Duncan AMV, Higgins MJ, Hamerton JL, Rector E, Brasch KR, White BN. (1986) Isolation and characterisation of an α-satellite repeated sequence from human chromosome 22. *Chromosome* **94**: 228–234.

Musich PR, Brown FL, Maio JJ. (1980) Highly repetitive component alpha and related alphoid DNAs in man and monkeys. *Chromosoma (Berlin)* **80**: 331–348.

Oakey R, Tyler-Smith C. (1990) Y chromosome DNA haplotyping suggests that most European and Asian men are descended from one of two males. *Genomics* **7**: 325–330.

Ohta T. (1980) Evolution and variation of multigene families. In: *Lectures in Biomathematics*, Vol. 37. Springer-Verlag, New York.

Ohta T, Dover G. (1983) Population genetics of multigene families that are dispersed into two or more chromosomes. *Proc. Natl Acad. Sci. USA* **80**: 4079–4083.

Pike LM, Carlisle A, Newell C, Hong SB, Musich PR. (1986) Sequence and evolution of Rhesus monkey alphoid DNA. *J. Mol. Evol.* **23**: 127–137.

Prassolov VS, Kuchino Y, Neomoto K, Nishimara S. (1986) Nucleotide sequence of the Bam HI repetitive sequence, including the Hind III fundamental unit, as a possible mobile element from the Japanese monkey *Macaca fuscata*. *J. Mol. Evol.* **23**: 200–204.

Rocchi M, Archidiacono N, Annotate R, Finelli P, D'Aiuto L, Carbone R, Lindsay E, Baldini A. (1994) Cloning and comparative mapping of a recently evolved human chromosome 22-specific alpha satellite DNA. *Somatic Cell Mol. Genet.* **20**: 443–448.

Singer D, Donehower L. (1979) Highly repeated DNA of the baboon: organization and sequence homologies to highly repeated DNA of the African green monkey. *J. Mol. Biol.* **134**: 835–842.

Smith GP. (1976) Evolution of repeated DNA sequences by unequal crossover. *Science* **191**: 528–535.

Southern EM. (1975) Long range periodicities in mouse satellite DNA. *J. Mol. Biol.* **94**: 51–69.

Sullivan BA, Jenkins LS, Karson EM, Lana-Cox J, Schwartz S. (1996) Evidence for structural heterogeneity from molecular cytogenetic analysis of dicentric Robertsonian translocations. *Am J. Hum. Genet.*, in press.

Therman E, Susman B, Denniston C. (1989) The nonrandom participation of human acrocentric chromosomes in Robertsonian translocations. *Ann. Hum. Genet.* **53**: 49–65.

Thompson JD, Sylvester JE, Gonzalez IL, Costanzi CC, Gillespie D. (1989) Definition of a second dimeric subfamily of human alpha satellite DNA. *Nucleic Acids Res.* **17**: 2769–2782.

Trowell HE, Nagy A, Vissel B, Choo KH. (1993) Long-range analyses of the centromeric regions of human chromosomes 13, 14 and 21: identification of a narrow domain containing two key centromeric DNA elements. *Hum. Mol. Genet.* **2**: 1639–1649.

Tyler-Smith C. (1987) Structure of the repeated sequences in the centromeric region of the human Y chromosome. *Development* **101**: 93–100.

Tyler-Smith C, Brown WRA. (1987) Structure of the major block of alphoid satellite DNA on the human Y chromosome. *J. Mol. Biol.* **195**: 457–470.

Warburton D. (1991) *De novo* balanced chromosome rearrangements and extra marker chromosomes identified at prenatal diagnosis: clinical significance and distribution of breakpoints. *Am. J. Hum. Genet.* **49**: 995–1013.

Warburton PE, Willard HF. (1990) Genomic analysis of sequence variation in tandemly repeated DNA: evidence for localized sequence domains within arrays of alpha satellite DNA. *J. Mol. Biol.* **216**: 3–16.

Warburton PE, Willard HF. (1992) PCR amplification of tandemly repeated DNA: analysis of intra- and interchromosomal sequence variation and homologous crossing-over in human alpha satellite DNA. *Nucleic Acids Res.* **20**: 6033–6042.

Warburton PE, Willard HF. (1995) Inter-homologue sequence variation of chromosome 17 alpha satellite DNA: evidence for concerted evolution along haplotypic lineages. *J. Mol. Evol.* **41**: 1006–1015.

Warburton PE, Greig GM, Haaf T, Willard HF. (1991) PCR amplification of chromosome-specific alpha satellite DNA: definition of centromeric STS markers and polymorphic analysis. *Genomics* **11**: 324–333.

Warburton PE, Wevrick R, Mahtani MM, Willard HF. (1992) Pulsed field and two-dimensional gel electrophoresis of long arrays of tandemly repeated DNA: analysis of human centromeric alpha satellite. In: *Pulsed-field Gel Electrophoresis* (eds M Burmeister and L Ulanovsky). Humana Press, Totowa, NJ, pp. 299–317.

Warburton PE, Waye JS, Willard HF. (1993) Non-random localisation of unequal crossing-over in tandemly repeated DNA: implications for higher-order structural characteristics within centromeric heterochromatin. *Mol. Cell. Biol.* **13**: 6520–6529.

Warburton PE, Haaf T, Gosden J, Lawson D, Willard HF. (1996) Characterisation of a chromosome-specific chimpanzee alpha satellite subset: evolutionary relationship to subsets on human chromosomes. *Genomics* **33**: 220–228.

Warren AC, Bowcock AM, Farrier LA, Antonarakis SE. (1990) An alpha satellite DNA polymorphism specific for the centromeric region of chromosome 13. *Genomics* **7**: 110–114.

Waye JS, Willard HF. (1985) Chromosome-specific alpha satellite DNA: nucleotide sequence analysis of the 20 kilobasepair repeat from the human X chromosome. *Nucleic Acids Res.* **13**: 2731–2743.

Waye JS, Willard HF. (1986a) Molecular analysis of a deletion polymorphism in alpha satellite of human chromosome 17: evidence for homologous unequal crossing-over and subsequent fixation. *Nucleic Acids Res.* **14**: 6915–6927.

Waye JS, Willard HF. (1986b) Structure, organization, and sequence of alpha satellite DNA from human chromosome 17: evidence for evolution by unequal crossing-over and an ancestral pentamer repeat shared with the human X chromosome. *Mol. Cell. Biol.* **6**: 3156–3165.

Waye JS, Willard HF. (1987) Nucleotide sequence heterogeneity of alpha satellite repetitive DNA: a survey of alphoid sequences from different human chromosomes. *Nucleic Acids Res.* **15**: 7549–7569.

Waye JS, Willard HF. (1989a) Chromosome specificity of satellite DNAs: short- and long-range organization of a diverged dimeric subset of human alpha satellite from chromosome 3. *Chromosome* **97**: 475–480.

Waye JS, Willard HF. (1989b) Concerted evolution of alpha satellite DNA: evidence for species specificity and a general lack of sequence conservation among alphoid sequences of higher primates. *Chromosome* **98**: 273–279.

Waye JS, Durfy SJ, Pinkel D, Kenwrick S, Patterson M, Davies KE, Willard HF. (1987a) Chromosome-specific alpha satellite DNA from human chromosome 1: hierarchical structure and genomic organization of a polymorphic domain spanning several hundred kilobase pairs of centromeric DNA. *Genomics* **1**: 43–51.

Waye JS, Greig GM, Willard HF. (1987b) Detection of novel centromeric polymorphisms associated with alpha satellite DNA from human chromosome 11. *Hum. Genet.* **77**: 151–155.

Waye JS, England SB, Willard HF. (1987c) Genomic organization of alpha satellite DNA on human chromosome 7: evidence for two distinct alphoid domains on a single chromosome. *Mol. Cell. Biol.* **7**: 349–356.

Waye JS, Mitchell AR, Willard HF. (1988) Organization and genomic distribution of '82H' alpha satellite DNA. *Hum. Genet.* **78**: 27–32.

Wevrick R, Willard HF. (1989) Long-range organization of tandem arrays of alpha satellite DNA at the centromeres of human chromosomes: high frequency array-length polymorphism and meiotic stability. *Proc. Natl Acad. Sci. USA* **86**: 9394–9398.

Wevrick R, Willard HF. (1991) Physical map of the centromeric region of human chromosome 7: relationship between two distinct alpha satellite arrays. *Nucleic. Acids Res.* **19**: 2295–2301.

Willard HF. (1985) Chromosome-specific organization of human alpha satellite DNA. *Am. J. Hum. Genet.* **37**: 524–532.

Willard HF. (1991) Evolution of alpha satellite. *Curr. Opin. Genet. Dev.* **1**: 509–514.

Willard HF, Waye JS. (1987a) Hierarchical order in chromosome-specific human alpha satellite DNA. *Trends Genet.* **3**: 192–198.

Willard HF, Waye JS. (1987b) Chromosome-specific subsets of human alpha satellite DNA: analysis of sequence divergence within and between chromosomal subsets and evidence for an ancestral pentameric repeat. *J. Mol. Evol.* **25**: 207–214.

Willard HF, Smith KD, Sutherland J. (1983) Isolation and characterisation of a major tandem repeat family from the human X chromosome. *Nucleic Acids Res.* **11**: 2017–2033.

Willard HF, Waye JS, Skolnick MH, Schwartz CE, Powers VE, England SB. (1986) Detection of restriction fragment polymorphisms at the centromeres of human chromosomes by using chromosome-specific alpha satellite DNA probes: implications for development of centromere-based genetic linkage maps. *Proc. Natl Acad. Sci. USA* **83**: 5611–5615.

Willard HF, Greig GM, Powers VE, Waye JS. (1987) Molecular organization and haplotype analysis of centromeric DNA from human chromosome 17: implications for linkage in neurofibromatosis. *Genomics* **1**: 368–373.

Wolfe J, Darling SM, Erickson RP, Craig IW, Buckle VJ, Rigby PWJ, Willard HF, Goodfellow PN. (1985) Isolation and characterisation of an alphoid centromeric repeat family from the human Y chromosome. *J. Mol. Biol.* **182**: 477–485.

Wu JC, Manuelidis L. (1980) Sequence definition and organization of a human repeated DNA. *J. Mol. Biol.* **182**: 363–386.

Yang TP, Hansen SK, Oishi KK, Ryder OA, Hamkalo BA. (1982) Characterisation of a cloned repetitive DNA sequence concentrated on the human X chromosome. *Proc. Natl Acad. Sci. USA* **79**: 6593–6597.

6

Telomeres, subterminal sequences, variation and turnover

Nicola J. Royle

6.1 Telomeres and their functions in eukaryotes

Telomeres were first defined from cytological observations of chromosomes during mitosis. Hermann Muller studied *Drosophila melanogaster* chromosomes after X-ray irradiation and, failing to recover chromosomes without physically recognizable termini, hypothesized that a specialized structure was required to stabilize the natural ends of chromosomes. Barbara McClintock observed in *Zea mays* that the ends of broken chromosomes were more reactive than the natural ends of chromosomes (see Biessmann and Mason, 1993). Broken chromosomes readily fused to produce dicentric chromosomes, which usually formed anaphase bridges during mitosis and further uncontrolled chromosome breakage and fusion events were observed. A cycle of chromosome breakage, fusion and anaphase bridge formation causes instability and is extremely deleterious as the genome undergoes rearrangement, and acentric chromosome fragments that arise during chromosome breakage are lost.

Elucidation of semi-conservative replication prior to mitosis or meiosis, raised questions as to how complete replication of a linear chromosome could be achieved without the loss of terminal sequences during synthesis of the lagging strand (Blackburn and Szostak, 1984) or possibly the leading strand (Lingner *et al.*, 1995). Semi-conservative DNA replication is achieved by DNA polymerases but all known polymerases initiate synthesis from a 3′ hydroxyl group in a 5′ to 3′ direction. Consequently, removal of the RNA primer results in the loss of sequences from the 5′ end of the newly synthesized strand of a linear DNA molecule (*Figure 6.1*). In the absence of any compensating mechanism, the sequence loss leads to a gradual shortening of

Human Genome Evolution, edited by M. Jackson, T. Strachan and G. Dover.

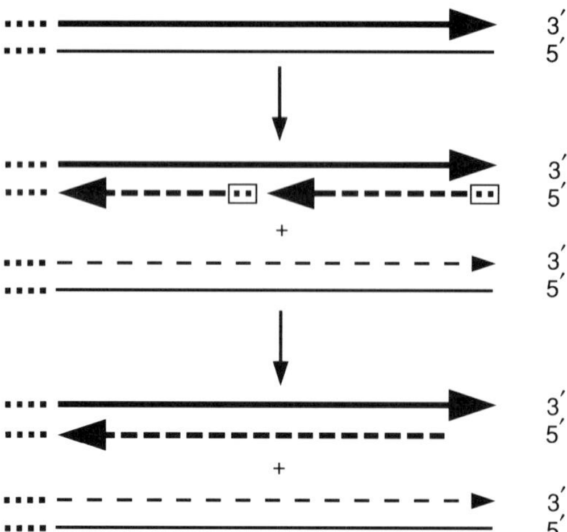

Figure 6.1. Diagram showing incomplete replication of a linear DNA molecule during the synthesis of the lagging strand. Single-stranded gaps are generated after removal of the RNA primers (▪ ▪)for lagging strand synthesis. DNA polymerase and ligase complete the synthesis across internal single-stranded gaps but the terminal single-stranded gap persists, resulting in sequence loss from the terminus. The continuous lines represent template strands and broken lines represent newly synthesised strands of DNA.

the chromosome with each cycle of replication. A variety of mechanisms were proposed that allowed chromosomes to be replicated to the terminus at each cell division (see Blackburn and Szostak, 1984; Zakian, 1989) but it is now clear that sequence loss from the terminus of a eukaryotic chromosome can be tolerated as long as the organism has a mechanism to compensate for the loss.

Telomeres in all eukaryotes are involved in a variety of functions. These include: (i) the ‘capping’ of the chromosome end to prevent inappropriate fusion to other chromosomes and degradation by exonuclease activity; (ii) a role in the nuclear architecture and probably in transient associations between the ends of homologous and non-homologous chromosomes; and (iii) compensation for the sequence loss that arises during replication. Telomeric sequences are conserved in the majority of species and composed of head-to-tail arrays of a short repeat unit (usually 6–8 bp). The sequence of the repeat unit includes a run of G residues with the G-rich strand oriented 5′ to 3′ towards the terminus of the chromosome. In some yeast species the repeat units are irregular in length and in others the repeat unit is longer (23–25 bp) and not G-rich (McEachern and Blackburn, 1995; McEachern and Hicks, 1993). Given the vulnerability of sequences located at the ends of eukaryotic chromosomes, it is perhaps not surprising that an alternative mechanism for the maintenance of chromosome integrity has evolved in some species, such as

D. melanogaster (Biessmann and Mason, 1993; Levis *et al.*, 1993; Mason and Biessmann, 1995) and some other Diptera (Okazaki *et al.*, 1993).

6.2 The structure of human telomeres

The repetitive nature of telomeres in most eukaryotes and the conservation of repeat unit sequence was established following the isolation of telomeres from a number of ciliates, slime moulds and fungi (Zakian, 1989): it was predicted that human telomeres would also be composed of a repetitive array. Moyzis *et al.* (1988) screened a human repetitive DNA sequence library with Chinese hamster repetitive DNA sequences and isolated clones containing arrays of $(TTAGGG)_n$ repeats. They showed by fluorescence *in situ* hybridization (FISH) that TTAGGG repeats were present at the ends of all human chromosome arms. They also demonstrated that the arrays of TTAGGG repeats were sensitive to *Bal*31 exonuclease digestion and were therefore most likely to be located at the termini of human chromosomes. The TTAGGG repeat is highly conserved in telomeres of vertebrates (Meyne *et al.*, 1989) but it is not always limited to terminal locations: in the Chinese hamster, for example, long arrays of TTAGGG repeats have also been detected in the pericentric regions (Meyne *et al.*, 1989, 1990). Arrays of TTAGGG repeats are mainly located at or towards the end of the human chromosomes. However, head-to-head arrays of TTAGGG and variant repeat types have been found at 2q13,

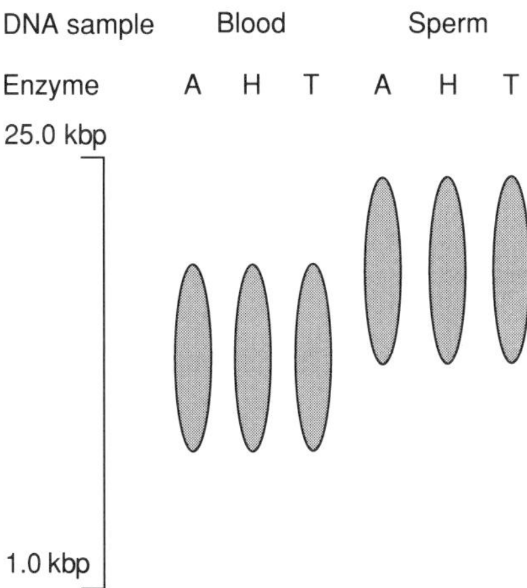

Figure 6.2. Detection of telomeres by hybridization of digested genomic DNA to $(TTAGGG)_n$. Smears of hybridizing products are detected; the length of the smear is largely independent of the restriction enzyme used to digest the DNA but the smear of hybridizing fragments is 5–10 kb longer in germline than in blood DNA. A, *Alu*I; H, *Hae*III; T, *Taq*I.

the presumed fusion point of two ancestral chromosomes that gave rise to human chromosome 2 (Ijdo *et al.*, 1991).

Human telomeres are heterogeneous in length as a result of successive cycles of replication and cell division; consequently, the terminal restriction fragment for any one chromosome end contains a discrete length of non-telomeric DNA adjacent to a variable length of TTAGGG repeats. Such terminal length heterogeneity was first demonstrated by detection of the terminal *Bam*HI restriction fragment of human Xp/Yp using a DNA probe located about 5 kb proximal to the telomere (Cooke *et al.*, 1985). Digestion of human genomic DNA by almost any restriction endonuclease releases terminal restriction fragments from all chromosome ends and the products can be detected as a smear of DNA fragments that hybridize to $(TTAGGG)_n$ (*Figure 6.2*). The average length of the smear of hybridizing DNA fragments varies between individuals but is usually between 5 and 15 kb in somatic tissues and 5–10 kb longer in germline DNA (Allshire *et al.*, 1989; Cross *et al.*, 1990). In somatic tissues, the average length of the telomere repeat array decreases with increasing age (Hastie *et al.*, 1990). The terminal arrays of repeats are not uniform in sequence but contain variant repeat types such as TTGGGG and TGAGGG that are localized in the proximal 3–4 kb of human telomeres (Allshire *et al.*, 1989; Guerrini *et al.*, 1993).

6.3 Average telomere length and secondary structure: variation between species

Although conservation of telomeric sequences has been maintained in the majority of eukaryotes, there is enormous variation in the average length of a telomere repeat array between species. In many lower eukaryotes, telomeres are relatively short homogeneous arrays of repeats; for example, the telomeres of *Tetrahymena* contain approximately 400 bp of TTGGGG repeats (Greider, 1991). Proteins that bind specifically to telomeres have been identified in some lower eukaryotic species and some are probably involved in protecting (capping) the termini of the chromosomes. In *Saccharomyces cerevisiae*, the average telomere length is also short (about 300 bp) and the repeats are organized into non-nucleosomal condensed chromatin structures called telosomes, which are known to affect gene expression and the timing of replication (Gilson *et al.*, 1993; Sandell and Zakain, 1992).

Vertebrate telomere repeat arrays are much longer than those in lower eukaryotes but there is still considerable variation between the average telomere length in different species. For example, telomeres in human somatic tissues vary between 5 and 15 kb, telomeres in the mouse *Mus spretus* are a similar size to human telomeres, telomeres in the mouse *M. musculus* are 5–10 times longer than those in *M. spretus* or in humans (Kipling and Cooke, 1990; Starling *et al.*, 1990). Yet it seems that telomeres in all three species serve the same functions. In rats, chickens and humans, the bulk of the telomere repeat arrays are organized into nucleosomes (Makarov *et al.*, 1993; Muyldermans *et al.*, 1994) although in rats telomeric nucleosomes have a shorter internucleosomal linker length than in the rest of the rat genome. Recently, a human telomere-associated protein, telomere repeat binding factor (TRF), has been identified and cloned (Chong *et al.*, 1996). Immunofluorescence labelling experiments, which demonstrate that TRF co-localizes with telomeric sequences, suggest that human telomeres do contain a nucleoprotein complex which may protectively cap the chromosome ends. In contrast to the observations in *S. cerevisiae* (described above), preliminary investigation of the level of expression from a gene sandwiched between a telomere repeat array and over 750 kb of the human X chromosome α-satellite DNA showed that its expression was unaffected by these repetitive DNAs (Bayne *et al.*, 1994).

6.4 Maintenance of human telomeres by telomerase

It is vital that chromosome integrity is maintained in the pluripotent germ cells and that the risk of coding sequence loss from chromosomes is kept to a minimum. Telomere repeat arrays in human sperm DNA are longer than in somatic tissues. There is no age-related reduction in the average telomere length; instead, the average telomere length in sperm DNA may actually increase with age (Allsopp *et al.*, 1992; Hastie *et al.*, 1990). In fact, the loss of repeats from the termini of chromosomes as a result of incomplete replication is compensated for in both the male and female germlines: TTAGGG repeats are added *de novo* by the enzyme telomerase. Telomerase was first identified in *Tetrahymena* (Greider and Blackburn, 1989) as a specialized reverse transcriptase, containing protein and RNA components. The 159 base RNA molecule encodes within its length the template region (5′-CAACCCCAA-3′), which is complementary to the *Tetrahymena* telomere repeat TTGGGG. The G-rich strand of the *Tetrahymena* telomere anneals to the three 3′ bases of the telomerase RNA template region and the remaining six bases act as a template for the synthesis of a new telomere repeat (Autexier and Greider, 1994). Translocation of telomerase to the end of the newly synthesised repeat enables the enzyme to add another repeat and hence to elongate the telomere processively. *Tetrahymena* telomerase also has a 3′–5′ nucleolytic cleavage activity so it can remove mismatched bases between the annealed DNA and RNA template; therefore, telomerase probably plays a role in telomere repair as well as synthesis (Collins and Greider, 1993). Characterization of the genes that encode the RNA and protein components of telomerase is under way in a number of unicellular eukaryotes such as *T. thermophila* and *S. cerevisiae* (Cohn and Blackburn, 1995; Collins *et al.*, 1995; Singer and Gottschling, 1994).

Human telomerase activity was first detected in protein extracts from an immortalized human cervical carcinoma cell line (Morin, 1989) and recently the gene that encodes the RNA component (hTR) of human telomerase has been cloned. It encodes an RNA molecule of about 450 nucleotides that is not polyadenylated and contains a template region of 11 nucleotides (5′-CUAAC-CCUAAC-3′; Feng *et al.*, 1995). There is only 65% similarity between the human and mouse RNA components of telomerase and little or no similarity to the telomerase RNA components of lower eukaryotes (Blasco *et al.*, 1995). With the use of a sensitive PCR-based assay, telomerase activity has been detected in adult human testes and ovaries but not in most normal human somatic tissues (Kim *et al.*, 1994). It is not known when telomerase is inactivated during the development of the human embryo but telomerase activity has been detected in fetal testes and ovaries. This could mean either that telomerase is inactivated in fetal somatic cells after the germ cells have been established or that telomerase is reactivated soon after the germ cells are defined. Interestingly, analysis of telomerase activity in a mutant *M. musculus* strain that lacks primary spermatogonia but has structurally normal testes shows that the telomerase activity in mouse testicular DNA arises from the germ cells and not the surrounding somatic cells (Prowse and Greider, 1995).

The pattern of telomerase expression in various tissues differs between humans and two species of mice, *M. spretus* and *M. musculus*. Telomerase is active in many somatic tissues in adult mice, for example in the liver, kidney and spleen but not in the brain. Liver tissue has a high turnover of cells and the mouse may gain some advantage by maintaining telomere length in this tissue. In humans, the lack of somatic telomerase activity may reflect an additional level of control on cellular proliferation that in turn gives additional protection against tumorigenesis that is absent in shorter-lived species such as mice (Prowse and Greider, 1995). Although attractive, this hypothesis may be too simplistic; telomeres of *M. musculus* are much longer than human telomeres and yet the mouse telomerase appears to be non-processive, adding only a single or a few repeats on to a template *in vitro* (Prowse *et al.*, 1993). This is in contrast to the human telomerase which is processive and adds many repeats during assays *in vitro*. The different length of template regions of mouse mTR and human hTR (8 and 11 nucleotides, respectively) may contribute to the difference in processivity of their telomerases (Blasco *et al.*, 1995; Feng *et al.*, 1995).

6.5 The telomere hypothesis of cellular senescence and immortalization

In all somatic tissues that have been tested, the average length of human telomere repeat arrays decreases with increasing age of the donor. These investigations have included tissues such as colonic epithelium and haematopoietic cells that undergo numerous cell divisions throughout the life of the individual (Vaziri *et al.*, 1994). In general, DNA is lost from human telomeres at a rate of about 20–40 bp per year in somatic tissues (Hastie *et al.*, 1990; Lindsey *et al.*, 1991; Vaziri *et al.*, 1993).

Normal human fibroblasts have a finite lifespan in culture; after a number of generations they grow more slowly, show chromosome rearrangements (including the formation of dicentric chromosomes as a result of end fusions) and generally exhibit a senescent phenotype. The loss of telomere repeats in cultured human fibroblasts correlates with the number of cell divisions (Allsopp *et al.*, 1992; Harley *et al.*, 1990). These observations have led to speculation about the role of telomere loss in cellular senescence and ageing. Interestingly, fibroblasts from patients with Hutchinson–Gilford progeria, a syndrome of premature ageing, have shorter telomeres and a reduced replicative capacity in culture than fibroblasts from age-matched normal individuals (Allsopp *et al.*, 1992). It has also been noted that Down syndrome (DS) patients have a significantly higher rate of sequence loss from their telomeres (133 bp per year) compared with normal individuals (41 bp per year), which correlates with the premature immunosenescence of lymphocytes observed in DS patients (Vaziri *et al.*, 1993). The control of telomere length and the rate of sequence loss from human telomeres is poorly understood but the high heritability (78%) indicates a strong genetic influence on the length of human telomeres (Slagbloom *et al.*, 1994).

The telomere hypothesis of cellular senescence and immortalization (Allsopp *et al.*, 1992; Greider, 1994) predicts that the presence of one or more short telomeres within a normal cell triggers a cell cycle checkpoint, when the normal cell would exit from the cell cycle and senesce (*Figure 6.3*). However, if cells bypass this checkpoint during oncogenesis they continue to divide and telomere length decreases until a crisis is reached when most telomeres are critically short (Counter *et al.*, 1992; Greider, 1994). Activation or derepression of telomerase is required to bypass this crisis and as a result telomere length is stabilized (Allsopp *et al.*, 1992). However, some recent data indicate that telomerase activation alone is not sufficient for cell immortalization and that telomere length stabilization can be achieved by another mechanism not involving telomerase (Bryan *et al.*, 1995). Telomerase is active in most colonic carcinomas (Kim *et al.*, 1994) and telomere length is stable and usually shorter than in the surrounding normal tissue (Hastie *et al.*, 1990). Furthermore, telomerase activity has been detected in the majority of immortal cultured cell lines (98/100) established from many different tissues and in 90 of 101 biopsies from many types of human tumours, but has not been detected in mortal cell lines or in any normal somatic tissues (Counter *et al.*, 1994; Kim *et al.*, 1994). Interestingly, hTR, the RNA component of human telomerase, is present in some primary cell lines and in some normal tissues but is found at higher levels in the ovary and testis; it has been proposed that hTR is not functional in normal human somatic cells (Feng *et al.*, 1995). Variation in telomere length between tumours suggests that telomerase can be activated at any time during tumour progression but usually after the first checkpoint has been bypassed. If telomerase activation is an important step in

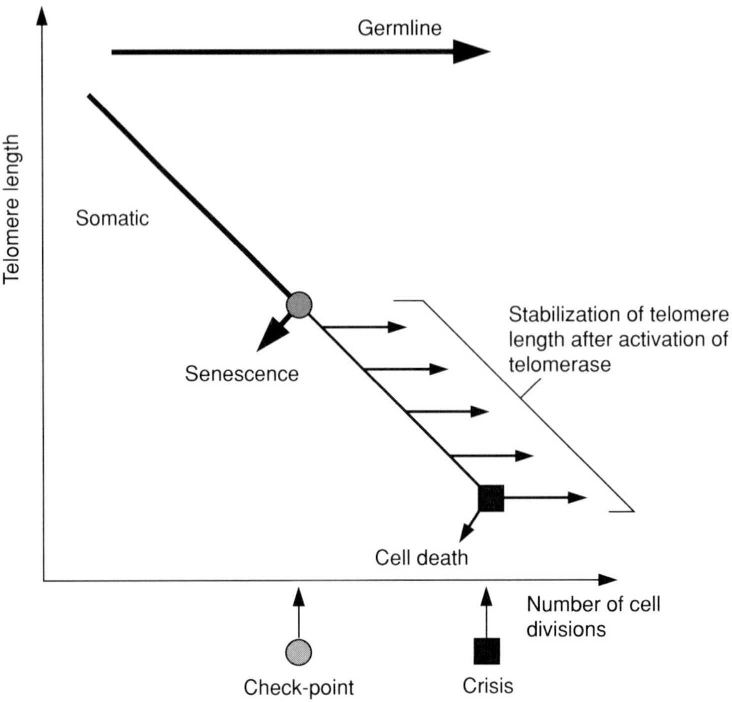

Figure 6.3. Telomere hypothesis of cellular senescence and immortalization. Telomere length is maintained in the germline where telomerase is active. Reduction in telomere length occurs as a result of DNA replication in the absence of telomerase activity in normal somatic tissues. Eventually, the presence of short telomeres serves as a trigger for a cell cycle checkpoint (●). Normal cells exit the cell cycle at this point and senesce. If a cell could bypass this checkpoint after transfection with a tumour virus or during oncogenesis, it would continue to divide until a crisis (■) was reached. However, activation or derepression of telomerase at any point from the checkpoint to crisis results in the stabilization of telomere length. Failure to activate telomerase before or at crisis would result in genome instability and cell death. Activation of telomerase is required for cellular immortalization and it may contribute to oncogenesis.

tumorigenesis, then it could be a target for chemotherapy (Greider, 1994). There is some scepticism as to how effective a telomerase inhibitor would be in arresting tumour growth because the average telomere length at the first cell cycle checkpoint is probably sufficient to allow the completion of many rounds of replication and cell division before a crisis is reached (Kipling, 1995). Recent experiments with telomerase inhibition *in vitro* using antisense hTR allowed 20–26 population doublings prior to crisis (Feng *et al.*, 1995).

What then is the minimum length of a functional telomere in a human cell?

Most of the measurements of telomere length have been carried out on total human DNA and it is difficult to determine the minimum length of a functional telomere from these data. However, telomere chromosome fragmentation experiments (Farr *et al.*, 1991) in telomerase-expressing cultured cells indicated that a repeat array of about 1 kb is sufficient to stabilize a chromosome (Barnett *et al.*, 1993).

6.6 Turnover of telomeric repeats in the human germline

In the human germline, telomerase protects telomeres from 'erosion' through sequence loss during replication. Consequently, events that affect the turnover of repeats at the proximal ends of human telomeres are unlikely to involve telomerase. Indeed, variant repeat types such as TTGGGG and TGAGGG have been identified and grossly localized to the proximal end of human telomeres (Allshire *et al.*, 1989). The variant repeat types could have arisen as base substitutions which did not disrupt the 6 bp periodicity or the run of three G residues within the telomere repeat array.

A PCR strategy (telomere-variant repeat mapping by PCR; TVR-PCR) has been used to determine the distribution of the TTAGGG (T-type), TGAGGG (G-type) and TCAGGG (C-type) repeats at the proximal end of the Xp/Yp pseudoautosomal telomere (Baird *et al.*, 1995). This has shown that the distribution of T-, G- and C-type repeats is extremely variable at the proximal end of the telomere and the extraordinary amount of allelic diversity suggests that there is also a high mutation rate introducing novel telomeres into the population. However, the alleles can be grouped according to the interspersion pattern of the repeat types and according to the association with a particular haplotype in the DNA immediately flanking the telomere (*Figure 6.4*). This suggests that the telomeres have diverged essentially along haploid lineages and it is likely that slippage during replication contributes significantly to the allelic diversity and telomere repeat sequence turnover in germinal cells. There is some indication that exchanges between alleles occur occasionally but these exchanges are unlikely to involve the DNA flanking the Xp/Yp telomere, as there is almost complete linkage disequilbrium across the flanking polymorphisms which would be disrupted by allelic exchanges. The pseudoautosomal region has a specialized function during male meiosis so repeat sequence turnover at this telomere may be atypical, but currently little is known about allelic diversity of autosomal telomeres.

6.7 Generation of new telomeres by chromosome healing

Controlled chromosome breakage and healing by the addition of a new telomere occurs in a number of lower eukaryotes (Biessmann and Mason, 1993). In *Tetrahymena,* chromosome fragmentation at specific sites occurs during macronuclear development and the fragments are stabilized by the *de novo*

Telomeres associated with flanking haplotype A

```
NNGGGGGGGG---G-G---G--GG----------GG-GG--G------------GGG-G--G----G----G--G-GGG-GG-G-G--G--GG-GG-G--------------G--GG--...
NNGGGGGGGG---G-G---G--GG------------GG-GG--G-----------GG-GG--G-G--------G-GGG-GG-G-G--G--------------GGGG-GG--G--G---G---GG...

NNN-CCCNCNCNCNCCNNNNN--C-N-C--CN-CC--NNNNNNNNNNNNNNNNNNNNNNNNNNNN-CCCCCCCCCCCC-----GGGGGGGG-GGGGG-G-G-G-G-G---GG-GGG-GG-G-G...

NNGGGGGG--G----N----N-NN-NN-NNNNNNNN----NNNNNNNNNNNNNNNNNNNNNNNNNNNNNNNNN-----NNNNNNNNNN----------------NNN--NN...
NNGGGGGG--G----N----N-----NNNNNNNNNNNNN--------NNNNNNNNNNNNNNNNNNNNNNNNNNNNNNNNNNNNNNNNNNNNNN----------------------...
NNGGGGGGG--G----N----N-------NNNNNNNNNNNNNNNNNNNNNNNN-NNNNNNNNNNNNNNNNNNNNNNNNNNNNNNNNNNNNNNNNNNNNNNN------------------...
NNGGGGGGG-------N---N----NNNNNNNNNNNNNNNNNNNNNNNN--------------------------------------------------------------------...
NNGGGGGGG------N----N-------NNNNNNNNNN-N---GGGNGGGGNNNNNNNNNNNNNNNNNNNNNNNN---------------N--N-CN-CCCCCCCCCCC------...
```

Telomeres associated with flanking haplotype B

```
G----CCCCCCCCCCCCCCCCCCCCCCCCC----------------------------------------------------------------------------------------...
G---NN-CC-CC-CCCCCCCCCCC----N--C-----N-------------------N------------------------------------------------------------...
G---NN-NNCCCCCCCCCCC--C---------------------N------------N--NN-----NN-------------------------------------------------...
G---NN-NN-NN-CCCCCCCCCCCCCCCCCCNNN-----------N--------------------------------------------------------------------...
G---NN-NN-NN-CCCCCCCCCCCCCCC--NN--------NNNNNNNNNNNNNNNNNNNN-NNN------------------------------------------------------...
G---NN-NN-CCCCCCCCGCGGCGCCGGCGGCCCCCCGGCGGCGNGNCGGCCCCCCCCCCC---------------------------------------------------------...

G-------NNNNNNNNNNNNNNN-NNNNNNNNNNNNNNNNNNNN-NNNN-----------NN----------------------------------------------------...

G-----------------------------------------------------------------------------------------------------------------...
G-----------G--GG-GG-GG-GG-G-GG--G---G------GGG-GGG---------------------------------------------------------------...
```

Telomeres associated with flanking haplotype C

```
G---GGNNNNNNNNNNNNNN--GGGGGGG---GGGGGGG--GG-GG-GG-GG--G-GG-GG-G--G----G-------GG-GG-GG--G-----G--G-G--G--G--G--G-G---...
G---GGNNNNNNNNNNNNN--GGGGGGG--GGGGGGG--GG-GG-GG-GG--G-GG-GG-G----G-------GG-GG-GG--------G--G--G--G-G--G--G---GGGGGGG...
G---GGNNNNNNNNNNNNN--GGGGGGG---GGGGGGG-GGG-GG-GG-GG--G-GG-GG-G--G--G-G-------GG-GG-GG-----G--G--G--G-G--G--G--G-GG--G--G--GGGG...

G---GGNNNNNNNNNNNNNC-GGGGGGG-GG-G-----------NNNN--------------------------------------------------------------...
```

Figure 6.4. Telomere maps. At the proximal end of the Xp/Yp telomere, interspersion patterns of TTAGGG, TGAGGG and TCAGGG repeat types with other repeats of unknown sequence have been determined. Alleles have been grouped according to the haplotype A, B or C (Baird *et al.*, 1995) found in the 850 bp of DNA immediately adjacent to the telomere and divided into subgroups according to similarities at the start of the telomere repeat arrays. TGAGGG, TCAGGG and unknown repeats are coded G, C and N, respectively, while the TTAGGG repeat type has been coded by a dash to aid visualization of the relationship between different telomere codes. Figure kindly supplied by D.M. Baird.

addition of telomeres (Yu and Blackburn, 1991). Furthermore in *Ascaris lumbricoides,* a multicellular organism, chromosome diminution accompanied by the addition of new telomeres in somatic cells results in the eradication of some germline-specific DNA (Muller *et al.*, 1991). The question then arises: does chromosome healing occur in higher eukaryotes and in humans?

The first characterized human chromosome break healed by the addition of TTAGGG repeats was found in a patient with α-thalassaemia (Wilkie *et al.*, 1990). The patient had haemoglobin H disease (HbH), a severe form of α-thalassaemia usually caused by deletion of three out of four copies of the α-globin genes located at the distal end of the short arm of chromosome 16 (16p13.3). However, this patient inherited three normal copies of the α-globin coding sequence (−α/αα); his paternally inherited chromosome 16 carried a deletion of one α globin gene (−α) and his maternally inherited copy of chromosome 16 carried two normal α-globin genes on a chromosome with a terminal deletion of 16p. The breakpoint had occurred 50 kb distal to the α-globin gene cluster causing the loss of the α-globin 5′ hypervariable region (HVR), regulatory sequences for the globin gene locus and all sequences distal to the breakpoint. The patient's mother had one normal copy of chromosome 16 with a normal α-globin locus but her other chromosome 16 carried the terminal deletion that was stably transmitted to her son. The loss of sequences distal to the α-globin locus did not apparently cause additional deleterious effects on the phenotype of the mother or son.

Terminal deletions of 16p that have been healed by the addition of a new telomere have also been identified in a number of other patients that exhibit α-thalassaemia, with or without mental retardation; the severity of the phenotype depends on the size of the terminal deletion (Flint *et al.*, 1994; Lamb *et al.*, 1993). In this region of the genome, at least six breakpoints have been characterized, and were found to have only a few bases of similarity to the TTAGGG repeat. The small amount of sequence complementarity between the breakpoints and the telomere repeat is compatible with the mode of action of *Tetrahymena* telomerase, as three bases in the RNA template region are used for annealing to the DNA and the remaining six bases are used as a template for the synthesis of repeats (Autexier and Greider, 1994; Collins and Greider, 1993; Yu *et al.*, 1990). Human telomerase can seed the addition of telomere repeats *in vitro* on to an oligonucleotide (Morin, 1991) and it has been proposed that stable terminal deletions arise in the human germline as chromosome breaks that are healed by telomerase-mediated addition of a telomere. The minimal sequence requirements for telomerase to add a repeat array also suggests that inappropriate healing of chromosome breaks might occur in the human germline. Currently, the frequency of chromosome healing in the germinal cells is unknown but many such gametes would be inviable because of the loss of a large number of vital genes. Whereas the viable gametes could give rise to patients carrying a small terminal deletion (Ledbetter, 1992); some terminal deletion syndromes are well characterized while others have only recently been recognized as distinct syndromes. The developmental defects of each terminal deletion syndrome are different but the phenotypes may overlap. A recent survey for very small or 'cryptic' terminal rearrangements and deletions among patients with idiopathic mental retardation indicates that 3–6% of such patients may have suffered loss of genes from the end of a chromosome (Flint *et al.*, 1995).

6.8 Sequences adjacent to telomeres of normal human chromosomes

Many strategies have been used to isolate sequences adjacent to human telomeres. Among the most successful was telomere rescue in the yeast *S. cerevisiae* (Brown, 1989; Cross *et al.*, 1989; Riethman *et al.*, 1989), as this allowed the isolation of large fragments of human DNA immediately proximal to a telomere. Analysis of sequences adjacent to human telomeres has revealed classes of repetitive sequences which have been described as subtelomeric repeats (de Lange *et al.*, 1990), telomere-associated repeated sequences (Brown *et al.*, 1990), subtelomeric sequences (Weber *et al.*, 1990; Wells *et al.*, 1990), subterminal sequences (Cross *et al.*, 1990), human subtelomeric repeat (HST) (Cheng *et al.*, 1991) and telomere-adjacent sequences (Royle *et al.*, 1992). The human genome contains at least one family of subterminal repeats present in a few hundred copies; other families may exist but the size, structure and relationship between the subterminal repeat sequence families is not fully understood.

The proximal end of the best characterized family of subterminal repeats is not clearly defined and some copies contain *Alu* elements or truncated portions of L1-like sequences. These subterminal repeats are internally repetitious (Royle, 1995), containing a tandem array of 61 bp repeats that vary in copy number. Distal to the 61 bp repeat array lies an array of 29 bp repeats but again the copy number of these repeats can vary. Distal to the 29 bp repeats lies a discontinuous array of 37 bp repeats (usually five). The arrays of 29 bp and 61 bp repeats have a structure similar to GC-rich minisatellites, but they are not locus-specific as they are found at many chromosome ends. Subterminal repeats can be truncated at any position by the presence of TTAGGG and variant repeats (*Figure 6.5*), which may be an interstitial array or the telomere itself. It has also been shown that some copies of the subterminal repeat sequence family are transcribed into a 6 kb polyadenylated RNA or shorter non-polyadenylated RNA (Cheng *et al.*, 1991). The sequence divergence between different isolates of the subterminal repeat family ranges from 2% to 20% (Cross *et al.*, 1990; Royle *et al.*, 1992), suggesting that the repeat sequence family is ancient; cross-hybridizing DNA fragments that show similar restriction patterns have been detected in the chimpanzee (*Pan troglodytes*) and gorilla (*Gorilla gorilla*) genomes.

Many copies of the subterminal repeat sequence family described above have been isolated from sequences adjacent to telomeres but some have been

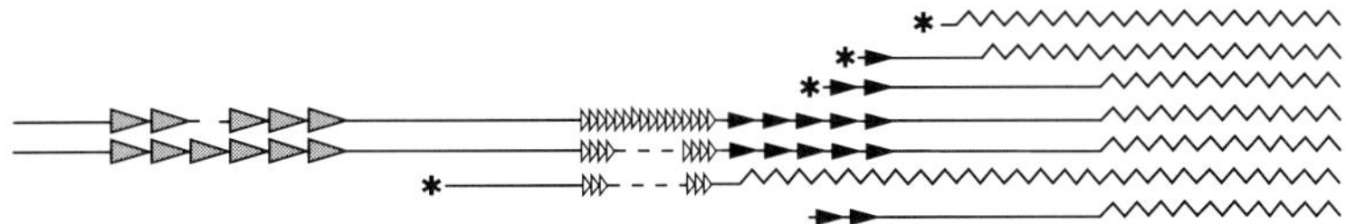

Figure 6.5. Human subterminal repeats. The best characterized family of human subterminal repeats are internally repetitious, containing a variable number of 61 bp repeats (▷) in a tandem array, a variable number of 29 bp repeats (|▷) in a tandem array and five 37 bp repeats (➤) in a discontinuous array. The subterminal repeats can be terminate at almost any position by an array of TTAGGG and variant repeats, (⁄\/\/\/_), which can be either a telomere or an internal array of telomere-like repeats. Different copies of this subterminal repeat sequence family have been aligned to show the variable truncation by the presence of TTAGGG repeats. The proximal limit of each cloned sequence is shown by an asterisk. Non-repetitive sequence is represented by a straight bold line.

isolated from internal loci. FISH analysis of human chromosomes showed that the majority of the subterminal repeats are located close to the ends of human chromosomes. FISH analysis has also revealed that the distribution of subterminal repeats varies between unrelated individuals such that hybridization to neither, one or both homologues of a given chromosome end have been reported (Ijdo *et al.*, 1992). However, normal Mendelian segregation of the different chromosome ends has been observed within families. The variable distribution and copy number of human subterminal repeats might be explained by unequal exchange events between subterminal repeats of homologous and non-homologous chromosome ends. If these exchanges involved a conversion-like process, they would tend to homogenize the subterminal repeat family. However, because the sequence divergence between different copies varies from 2% to 20%, homogenization by terminal conversion-like events may only occur infrequently.

No function has been attributed to human subterminal repeat sequences. They are not required for telomere function, as telomeres lacking adjacent subterminal repeats have been identified. The overall organization of the subterminal repeats and arrays of TTAGGG repeats in the human genome is similar to the organization of terminal sequences in the yeast genome (Louis and Haber, 1990a, b), although, as expected, there is no similarity between the sequences located adjacent to the telomeres in these two species. Instead, the organizational similarity may reflect the processes that operate to homogenize and distribute terminal sequences of eukaryote chromosomes.

6.9 Telomeres and proximal unique sequences

The extent of subterminal repeats interspersed with short arrays of TTAGGG and variant repeats and perhaps some chromosome-specific sequences varies considerably between chromosome ends. The sequence adjacent to the Xp/Yp pseudoautosomal telomere is different from those adjacent to many autosomal telomeres (Brown *et al.*, 1990) because the terminal 850 bp contains four tandemly arranged copies of a 63 bp GC-rich repeat unit, which are unique within the human genome. This minisatellite does not show length variation between alleles but the terminal 850 bp of Xp/Yp has a high frequency of base substitution polymorphisms (one per 65 bp in Caucasians) (Baird *et al.*, 1995). The polymorphisms are in almost complete linkage disequilibrium such that only three common Caucasian haplotypes have been identified. In summary, chromosome-specific sequences are located within the terminal 1 kb of the Xp/Yp telomere but the distance to the nearest gene is unknown.

The end of 7q is also atypical because in FISH analysis of metaphase chromosomes it does not show hybridization to the subterminal repeat family described above, although a very diverged copy has been identified within 2 kb of the start of the 7q telomere (Coleman and Royle, unpublished). The terminal 240 kb contains polymorphic chromosome-specific minisatellite and microsatellite loci (Riethman *et al.*, 1993), and 7q unique sequences are present within 8 kb of the start of the telomere (Brown *et al.*, 1990). However, to date, no coding sequences have been identified in the terminal 240 kb. The ends of other chromosomes have not been studied by sequence analysis, although many have been isolated, and the distance between the most terminal chromosome-specific sequences and the start of the telomere varies from 10 to 200 kb. The proximity of coding sequences to human telomeres is also unknown for most chromosome ends, although recently it has been shown that the distal V_H gene segment (V_H 7/81), which maps to the end of the immunoglobulin V_H locus, is within 10–12 kb of the 14q telomere. The V_H 7/81 gene segment contains an open reading frame but its functional status is uncertain; the terminal 10–12 kb of 14q hybridizes to subterminal repeats (Cook *et al.*, 1994).

6.10 Variation in the start site of a telomere

A long-range physical map of the distal end of 16p has revealed an extraordinary amount of length variation between the end of the α-globin locus and the start of the 16p telomere. Four terminal fragments of different length have been identified where the distance between the distal 5′HVR and the telomere is 170 kb, 245 kb, 350 kb or 430 kb; these have been designated alleles A, D, B and C, respectively (Wilkie *et al.*, 1991; *Figure 6.6*). Allele A is most common and present on about two-thirds of the chromosomes 16 analysed. The terminal 170 kb fragment of allele A and 350 kb of allele B hybridize to different copies (Brown *et al.*, 1990) of the subterminal repeat sequence family described in Section 6.8. The maps of the four different alleles (A, B, C, D) are the same from the ζ2 globin gene to the divergence point located approximately 145 kb distally and four genes are encoded within this region (Higgs *et al.*, 1993), one of which is only about 20 kb from the divergence point. The divergence point between the most common 16pter alleles, A and B, comprises a complex segment of about 4 kb which shows partial loss of homology before the point of complete divergence. This segment of DNA includes a (CA_n) repeat microsatellite, whose length varies between the 16pter alleles,

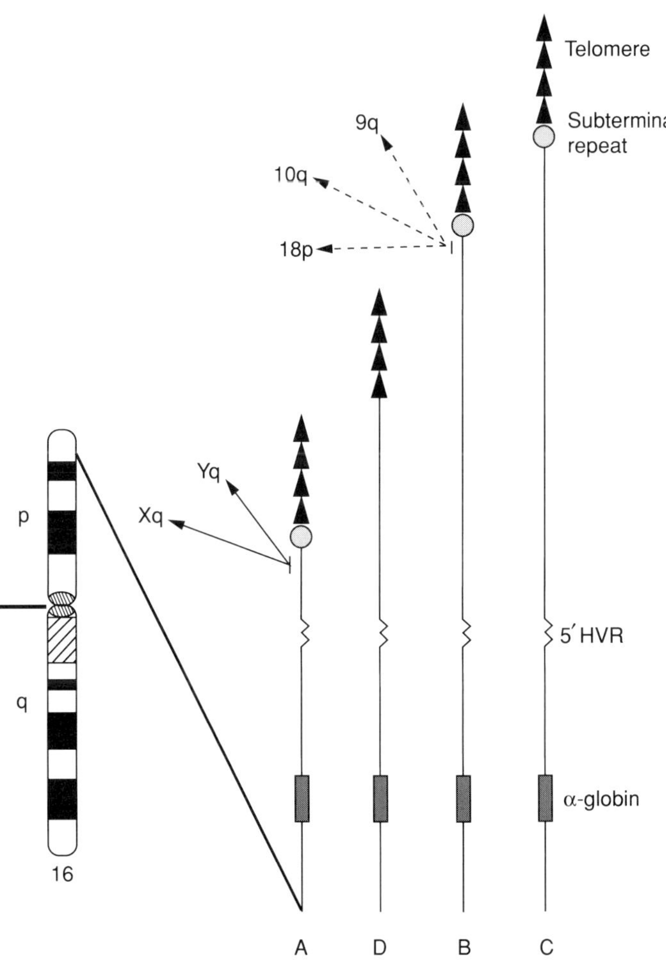

Figure 6.6. Diagram showing the extraordinary length variation adjacent to the 16p telomere. The distance between 5′ HVR (-∿∿∿-), which is distal to the α-globin locus, and the telomere (►►) varies considerably between different 16pter alleles (see text for details) and divergent copies of the subterminal repeat (◯) sequence are associated with the different 16pter alleles. Distal to the 5′ HVR there is a point of sequence divergence between the 16pter alleles; from that point sequences from allele A show more similarity to the ends of Xq and Yq than to allele B and sequences from allele B show similarity to the ends of 9q, 10q and 18p.

with allele A having a very long array (250–350 repeats) of imperfect CA repeats (Wilkie and Higgs, 1992).

Sequences distal to the divergence point of 16pter allele A cross-hybridized to loci at Xqter and Yqter but not to allele B, while sequences distal to the divergence point of the 16pter allele B cross-hybridized to loci on 9qter, 10qter and 18qter. Therefore, sequences distal to the divergence point of 16pter alleles show more similarity to other chromosome ends than to one another. The lack of similarity between the 16pter alleles A and B and their association with different copies of the subterminal repeat suggest that the extraordinary terminal length variation has arisen by exchanges between 16pter and other non-homologous chromosome ends (Higgs *et al.*, 1993). This type of extreme length variation adjacent to the start of the telomere was not observed at the Xp/Yp telomere, which always starts within two or three bases of the same place (Baird *et al.*, 1995). There is preliminary evidence that length polymorphism adjacent to the telomeres of 2q and 7q also occurs but at a low frequency (Macina *et al.*, 1994; Riethman *et al.*, 1993). Given the variable distribution and copy number of subterminal repeat sequences, such length variation is likely to occur at other chromosome ends.

6.11 Unequal terminal exchanges

Genetic map expansion relative to the physical map occurs at the ends of chromosomes (especially in males). Furthermore, recombination nodules and chiasmata tend to localize towards the ends of meiotic bivalents. These data indicate that the ends of human chromosomes are proficient at homologous recombination. There is also cytological evidence that chromosome pairing begins towards the ends of chromosomes and it is presumed that homologue recognition is based on sequence homology. The length variation and reduced homology between alleles of 16pter suggests that homologue recognition must occur more proximally in this chromosome and, as stated in Section 6.9, this suggests that terminal exhanges may occur between non-homologous chromosome ends. Non-homologous terminal exchanges and unequal homologous exchanges may be byproducts of aborted attempts to identify homology during meiosis. Such events may account for the distribution and the copy number of the subterminal repeats and internal arrays of TTAGGG repeats (Royle, 1995). If these exchanges are limited to the regions of subterminal repeats towards chromosome ends, they may be inconsequential for the cell. In yeast, a mutation of the *EST1* gene, which may encode part of the yeast telomerase, leads to sequence loss from telomeres and to senescence of the culture (Lundblad and Szostak, 1989). The sequence loss can be overcome by a *RAD52*-mediated recombination pathway when exchanges between the subterminal repeats of non-homologous chromosome ends compensate for terminal sequence loss in the absense of the *EST1* gene product (Lundblad and Blackburn, 1993).

When terminal exchanges between non-homologous human chromosomes occur proximal to the subterminal repeats, they result in a translocation of genes from one chromosome end to another with deleterious effects on unbalanced carriers (as for all reciprocal translocations). Exchanges between the ends of two chromosomes not associated with an alteration of the conventional banding pattern have been described as 'cryptic' translocations (Ledbetter, 1992). Cryptic translocations may be the underlying cause in a proportion of patients exhibiting the phenotype of a terminal deletion syndrome (Flint *et al.*, 1995; Wilkie, 1993).

6.12 Evolution of terminal sequences in African apes

The subterminal repeats found in the human genome have been detected in chimpanzees and gorillas (Cross *et al.*, 1990), which is further evidence that they are ancient and must have existed in the progenitor to humans, chimpanzee and gorillas. However, they are not necessarily located adjacent to telomeres in all three species. The karyotype similarities between humans, chimpanzees, gorillas and orangutans reflects the phylogeny of these species. Most of the differences consist of inversions of chromosomal segments, a few reciprocal translocations, band insertions and a reduction in chromosome number by the terminal fusion of two chromosomes to generate the human chromosome 2 (Jauch *et al.*, 1992; Weinberg *et al.*, 1990; Yunis and Prakash, 1982). In addition, there are some notable differences in the amount of heterochromatin. Nearly all human and orangutan chromosome arms end in a negatively staining G-band which is usually indicative of euchromatic DNA; over half the chimpanzee chromosome arms and nearly all those in the gorilla terminate with an additional, heterochromatic, positively staining G-band (*Figure 6.7*). Recent analysis has revealed that the chimpanzee genome contains a satellite composed of a 32 bp AT-rich repeat unit, which is located at the ends of 21 chromosome arms and at two interstitial sites in the chimpanzee, and at the ends of nearly all the gorilla chromosomes. The satellite is predominantly subterminal and may be the major component of the additional heterochromatic G-bands in chimpanzees and gorillas but it is absent from the human and orangutan genomes (Royle *et al.*, 1994). The size of the blocks of subterminal satellite is not known for either species but signals from FISH analysis suggest that they are very long and that there is variation in the size of the blocks between homologous chromosome ends. The physical relationship between the subterminal satellite, the human-like sub-terminal repeats and telomeres in chimpanzees and gorillas is not fully understood although it is most likely that the subterminal satellite occurs distal to all other sequences and adjacent to the telomere. The telomere at the end of the Xp/Yp pseudoautosomal region does not start at the same place in the three species of African apes: preliminary evidence

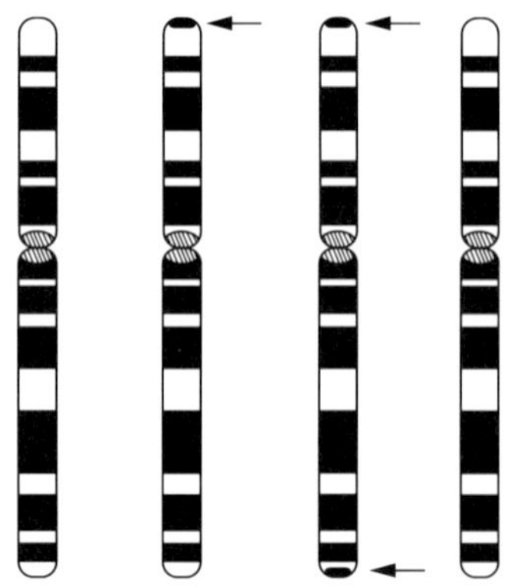

Figure 6.7. Idiogram of a G-banded human chromosome 6 shown adjacent to the homologous chromosome from chimpanzee, gorilla and orangutan. The G-banding patterns are similar, although the chimpanzee and gorilla chromosomes have additional heterochromatic G-bands, indicated by arrows. Additional heterochromatic terminal G-bands are also observed at the ends of many other chimpanzee and most gorilla chromosome arms (Yunis and Prakash, 1982).

suggests that in the chimpanzee the subterminal satellite is located distal to the point where the Xp/Yp telomere starts in humans (Baird, Di and Royle, unpublished).

Terminally located satellites have been described in other species but it is surprising to find such a radical difference (presence or absence) between closely related species (Andrews, 1992; Diamond, 1988; Goodman *et al.*, 1990). These differences give further indication that there has been rapid turnover of terminally located sequences during recent evolution of the African apes. It seems likely that the progenitor to humans, chimpanzees and gorillas contained some subterminal satellite in one or a few terminal locations and that the satellite became massively expanded and dispersed in chimpanzees and gorillas but was lost from the human genome (*Figure 6.8*). How then was the satellite expanded and dispersed so rapidly in chimpanzees and gorillas but lost, apparently completely in humans, since the divergence of these species about 4–7 million years ago? Chromosome breakage followed by telomerase-mediated addition of a telomere at a new site could have contributed to the complete loss of the subterminal satellite from chromosome ends. Unequal exchanges between blocks of subterminal satellite between homologous or non-homologous chromosomes could have increased or decreased the size of the block of repeats. Exchanges between more proximally located subterminal-repeat-like sequences on non-homologous chromosomes could also have resulted in the dispersal or removal of subterminal satellite arrays.

Figure 6.8. Diagram showing the evolutionary relationship between humans, chimpanzees, gorillas and orangutans (based on sequence analysis; see Friday, 1992; Goodman, 1990). A subterminal satellite is present at the ends of most gorilla and nearly half the chimpanzee chromosome arms but absent from the human and orangutan genomes. A progenitor (◎) to the African apes (humans, chimpanzees and gorillas) probably contained some subterminal satellite which became expanded and dispersed in the chimpanzee and gorilla genomes but was lost from the human genome.

6.13 Summary

Telomerase is required for the synthesis of a terminal array of TTAGGG repeats in human cells and therefore it primarily influences the sequence at the termini of human chromosomes. However, in somatic tissues, telomerase is inactive and sequences are lost from the end of chromosomes. Ultimately, this leads to cellular senescence as the replicative capacity of cells is limited in the absence of telomerase activity. Destabilization of the genome as a result of terminal sequence loss and the reactivation of telomerase may both play a role in tumorigenesis.

In germinal cells, telomerase is active and maintains telomere length by the addition of repeats to the terminus of telomeres. Therefore, the turnover of proximally located repeats in the telomere can be influenced by other factors, such as mutations that introduce base substitutions. The accumulation of sequence variation may be limited by selection against mutations that alter the 6 bp periodicity or disrupt the run of three G residues of proximally located repeats in human telomeres. The distribution of variant repeats at the proximal end of the Xp/Yp telomere is very variable and alleles appear to have evolved along haploid lineages. Slippage during replication may have contributed significantly to repeat sequence turnover at the proximal ends of human telomeres, although occasional truncations or deletions within a telomere would presumably result in the renewal of a homogeneous array of TTAGGG repeats by telomerase.

More radical changes to terminal sequence organization may include chromosome breaks that are healed by telomerase, resulting in the sudden truncation of a chromosome and a re-positioning of the telomere. In addition, conversion-like events and unequal exchanges between subterminal repeats of homologous or non-homologous chromosomes may have had an important role in the evolution of families of subterminal repeats and may have contributed to the interspersion of subterminal repeats with short arrays of TTAGGG and variant repeats. These processes are also likely to have had an important role in the loss of subterminal satellite sequences during the recent evolution of humans from chimpanzees and gorillas.

References

Allshire RC, Dempster M, Hastie ND. (1989) Human telomeres contain at least three types of G-rich repeat distributed non randomly. *Nucleic Acids Res.,* **17**: 4611–4627.

Allsopp RC, Vaziri H, Patterson C, Goldstein S, Younglai EV, Futcher AB, Greider CW, Harley CB. (1992) Telomere length predicts replicative capacity of human fibroblasts. *Proc. Natl Acad. Sci. USA* **89**: 10114–10118.

Andrews P. (1992). Evolution and environment in the Hominoidea. *Nature,* **360**: 641–646.

Autexier C, Greider CW. (1994) Functional reconstitution of wild-type and mutant *Tetrahymena* telomerase. *Genes Dev.* **8**: 563–575.

Baird DM, Jeffreys AJ, Royle NJ. (1995) Mechanisms underlying telomere repeat turnover, revealed by hypervariable variant repeat distribution patterns in the human Xp/Yp telomere. *EMBO J.* **14**: 5433–5443.

Barnett MA, Buckle VJ, Evans EP, Porter ACG, Rout D, Smith AG, Brown WRA. (1993) Telomere directed fragmentation of mammalian chromosomes. *Nucleic Acids Res.* **21**: 27–36.

Bayne RAL, Broccoli D, Taggart MH, Thomsom EJ, Farr CJ, Cooke HJ. (1994) Sandwiching of a gene within 12 kb of a functional telomere and alpha satellite does not result in silencing. *Hum. Mol. Genet.* **3**: 539–546.

Biessmann H, Mason JM. (1993) Genetics and molecular biology of telomeres. *Adv. Genet.* **30**: 185–249.

Blackburn EH, Szostak JW. (1984) The molecular structure of centromeres and telomeres. *Annu. Rev. Biochem.* **53**: 163-194.

Blasco MA, Funk W, Villeponteau B, Greider, CW. (1995) Functional characterisation and developmental regulation of mouse telomerase RNA. *Science* **269**: 1267–1270.

Brown WRA. (1989) Molecular cloning of human telomeres in yeast. *Nature* **338**: 774–776.

Brown WRA, MacKinon PJ, Villasante A, Spurr N, Buckle VJ, Dobson MJ. (1990) Structure and polymorphism of human telomere-associated DNA. *Cell* **63**: 119–132.

Bryan TM, Englezou A, Gupta J, Bacchetti S, Reddel RR. (1995) Telomere elongation in immortal human cells without detectable telomerase activity. *EMBO J.* **14**: 4240–4248.

Cheng J-F, Smith CL, Cantor CR. (1991) Structural and transcriptional analysis of a human subtelomeric repeat. *Nucleic Acids Res.* **19**: 149–153.

Chong L, van Steensel B, Broccoli D, Erdjument-Bromage H, Hanish J, Tempst P, de Lange T. (1996) A human telomeric protein. *Science* **270**: 1663–1667.

Cohn M, Blackburn EH. (1995) Telomerase in yeast. *Science* **269**: 396–400.

Collins K, Greider CW. (1993) *Tetrahymena* telomerase catalyzes nucleolytic cleavage and nonprocessive elongation. *Genes Dev.* **7**: 1364–1376.

Collins K, Kobayashi R, Greider CW. (1995) Purification of *Tetrahymena* telomerase and cloning of genes encoding the two protein components of the enzyme. *Cell* **81**: 677–686.

Cook GP, Tomlinson IM, Walter G, Riethman H, Carter NP, Buluwela L, Winter G, Rabbitts TH. (1994) A map of the human immunoglobulin V_H locus completed by analysis of the telomeric region of chromosome 14q. *Nature Genetics* **7**: 162–168.

Cooke HJ, Brown WRA, Rappold GA. (1985) Hypervariable telomeric sequences from the human sex chromosomes are pseudoautosomal. *Nature* **317**: 687–692.

Counter CM, Avilion AA, LeFeuvre CE, Stewart NG, Greider CW, Harley CB, Bacchetti S. (1992) Telomere shortening associated with chromosome instability is arrested in immortal cells which express telomerase activity. *EMBO J.* **11**: 1921–1929.

Counter CM, Hirte HW, Bacchetti S, Harley CB. (1994) Telomerase activity in human ovarian carcinoma. *Proc. Natl Acad. Sci. USA* **91**: 2900–2904.

Cross SH, Allshire RC, McKay SJ, McGill NI, Cooke, HJ. (1989) Cloning of human telomeres by complementation in yeast. *Nature* **338**: 771–774.

Cross SH, Lindsey J, Fantes J, McKay SJ, McGill NI, Cooke HJ. (1990) The structure of a subterminal repeated sequence present on many human chromosomes. *Nucleic Acids Res.* **18**: 6649–6657.

Diamond JM. (1988) DNA-based phylogenies of the three chimpanzees. *Nature* **332**: 865–866.

Farr C, Fantes J, Goodfellow P, Cooke H. (1991) Functional reintroduction of human telomeres into mammalian cells. *Proc. Natl Acad. Sci. USA* **88**: 7006–7010.

Feng J, Funk WD, Wang SS *et al.* (1995) The RNA component of human telomerase. *Science* **269**: 1236–1241.

Flint J, Craddock CF, Villegas A *et al.* (1994) Healing of broken human chromosomes by the addition of telomeric repeats. *Am. J. Hum. Genet.* **55**: 505–512.

Flint J, Wilkie AOM, Buckle VJ, Winter RM, Holland AJ, McDermid HE. (1995) The detection of subtelomeric chromosomal rearrangements in idiopathic mental retardation. *Nature Genetics* **9**: 132–140.

Friday AE. (1992) Human evolution: the evidence from DNA sequencing. In: *The Cambridge Encyclopaedia of Human Evolution* (eds S Jones, R Martin and D Pilbeam). Cambridge University Press, Cambridge, pp. 316–321.

Gilson E, Laroche T, Gasser SM. (1993) Telomeres and the functional architecture of the nucleus. *Trends Cell Biol.* **3**: 128–134.

Goodman M, Tagle DA, Fitch DHA, Bailey W, Czelusniak J, Koop BF, Benson P, Slightom SL. (1990) Primate evolution at the DNA level and a classification of Hominoids. *J. Mol. Evol.* **30**: 260–266.

Greider CW. (1991) Chromosome first aid. *Cell* **67**: 645–647.

Greider CW. (1994) Mammalian telomere dynamics: healing, fragmentation shortening and stabilization. *Curr. Opin. Genet. Dev.* **4**: 203–211.

Greider CW, Blackburn EH. (1989) A telomeric sequence in the RNA of *Tetrahymena* telomerase required for telomere repeat synthesis. *Nature* **337**: 331–337.

Guerrini AM, Camponeschi B, Ascenzioni F, Piccolella E, Donini FL. (1993) Subtelomeric as well as telomeric sequences are lost from chromosomes in proliferating B lymphocytes. *Hum. Mol. Genet.* **2**: 455–460.

Harley CB, Futcher AB, Greider CW. (1990) Telomeres shorten during ageing of human fibroblasts. *Nature* **345**: 458–460.

Hastie ND, Dempster M, Dunlop MG, Thompson AM, Green DK, Allshire RC. (1990) Telomere reduction in human colorectal carcinoma and with ageing. *Nature* **346**: 866–868.

Higgs DR, Wilkie AOM, Vyas P, Vickers MA, Buckle VJ, Harris PC. (1993) Characterization of the telomeric region of human chromosome 16p. *Chromosomes Today* **11**: 35–47.

Ijdo JW, Baldini A, Ward DC, Reeders ST, Wells RA. (1991) Origin of human chromosome 2: an ancestral telomere–telomere fusion. *Proc. Natl Acad. Sci. USA* **88**: 9051–9055.

Ijdo JW, Lindsay EA, Wells RA, Baldini A. (1992) Multiple variants in subtelomeric regions of normal karyotypes. *Genomics* **14**: 1019–1025.

Jauch A, Weinberg J, Stanyon R, Arnold N, Topanulli S, Ishida T, Cremer T. (1992) Reconstruction of genomic rearrangements in great apes and gibbons by chromosome painting. *Proc. Natl Acad. Sci. USA* **89**: 8611–8615.

Kim NW, Piatyszek MA, Prowse KR *et al.* (1994) Specific association of human telomerase activity with immortal cells and cancer. *Science* **266**: 2011–2015.

Kipling D. (1995) Telomerase: immortality enzyme or oncogene? *Nature Genetics* **9**: 104–106.

Kipling D, Cooke HJ. (1990) Hypervariable ultra-long telomeres in mice. *Nature* **347**: 400–402.

Lamb J, Harris PC, Wilkie AOM, Wood WG, Dauwerse JHG, Higgs DR. (1993) *De novo* truncation of chromosome 16p and healing with $(TTAGGG)_n$ in the α-thalassemia/mental retardation syndrome (ATR-16). *Am. J. Hum. Genet.* **52**: 668–676.

de Lange T, Shiue L, Myers RM, Cox DR, Naylor SL, Killery AM, Varmus HE. (1990) Structure and variability of human chromosome ends. *Mol. Cell. Biol.* **10**: 518–527.

Ledbetter DH. (1992) Minireview: cryptic translocations and telomere integrity. *Am. J. Hum. Genet.* **51**: 451–456.

Levis RW, Ganesan R, Houtchens K, Tolar LA, Sheen F-M. (1993) Transposons in place of telomeric repeats at a *Drosophila* telomere. *Cell* **75**: 1083–1093.

Lindsey J, McGill NI, Lindsey LA, Green DK, Cooke HJ. (1991) *In vivo* loss of telomeric repeats with age in humans. *Mut. Res.* **256**: 45–48.

Lingner J, Promisel Cooper J, Cech TR. (1995) Telomerase and DNA end replication: no longer a lagging strand problem? *Science,* **269**: 1533–1534.

Louis EJ, Haber JE. (1990a) Mitotic recombination among subtelomeric Y′ repeats in *Saccharomyces cerevisiae. Genetics* **124**: 547–559.

Louis EJ, Haber JE. (1990b) The subtelomeric Y′ repeat family in *Saccharomyces cerevisiae* : an experimental system for repeated sequence evolution. *Genetics* **124**: 533–545.

Lundblad V, Blackburn EH. (1993) An alternative pathway for yeast telomere maintenance rescues *est1*⁻ senescence. *Cell* **73**: 347–360.

Lundblad V, Szostak JW. (1989) A mutant with a defect in telomere elongation leads to senescence in yeast. *Cell* **57**: 633–643.

Macina RA, Negorev DG, Spais C, Ruthig LA, Hu X-L, Riethman HC. (1994) Sequence organisation of the human chromosome 2q telomere. *Hum. Mol. Genet.* **3**: 1847–1853.

Makarov VL, Lejnine S, Bedoyan J, Langmore JP. (1993) Nucleosomal organization of telomere-specific chromatin in rat. *Cell* **73**: 775–787.

Mason JM, Biessmann H. (1995) The unusual telomeres of *Drosophila. Trends Genet.* **11**: 58–62.

McEachern MJ, Blackburn EH. (1995) Runaway telomere elongation caused by telomerase RNA gene mutations. *Nature* **376**: 403–409.

McEachern MJ, Hicks JB. (1993) Unusually large telomeric repeats in the yeast *Candida albicans. Mol. Cell. Biol.* **13**: 551–560.

Meyne J, Ratliff RL, Moyzis RK. (1989) Conservation of the telomere sequence $(TTAGGG)_n$ among vertebrates. *Proc. Natl Acad. Sci. USA* **86**: 7049–7053.

Meyne J, Baker RJ, Hobart HH *et al.* (1990) Distribution of non-telomeric sites of $(TTAGGG)_n$ telomeric sequence in vertebrate chromosomes. *Chromosoma* **99**: 3–10.

Morin GB. (1989) The human telomere terminal transferase enzyme is a ribonucleoprotein that synthesizes TTAGGG repeats. *Cell* **59**: 521–529.

Morin GB. (1991) Recognition of a chromosome trunction site associated with α-thalassemia by human telomerase. *Nature* **353**: 454–456.

Moyzis RK, Buckingham JM, Cram LS, Dani M, Deaven LL, Jones MD, Meyne J, Ratliff, RL, Wu J-R. (1988) A highly repetitive DNA sequence, $(TTAGGG)_n$, present at the telomeres of human chromosomes. *Proc. Natl Acad. Sci. USA* **85**: 6622–6626.

Muller F, Wicky C, Spicher A, Tobler H. (1991) New telomere formation after developmentally regulated chromosomal breakage during the process of chromosome diminution in *Ascaris lumbricoides. Cell* **67**: 815–822.

Muyldermans S, De Jonge J, Wyns L, Travers AA. (1994) Differential association of linker histones H1 and H5 with telomeric nucleosomes in chicken erythrocytes. *Nucleic Acids Res.* **22**: 5635–5639.

Okazaki S, Tsuchida K, Maekawa H, Ishikawa H, Fujiwara H. (1993) Identification of a pentanucleotide telomeric sequence $(TTAGG)_n$ in the silkworm and in other insects. *Mol. Cell. Biol.* **13**: 1424–1432.

Prowse KR, Greider CW. (1995) Developmental and tissue-specific regulation of mouse telomerase and telomere length. *Proc. Natl Acad. Sci. USA* **92**: 4818–4822.

Prowse KR, Avilion AA, Greider CW. (1993) Identification of a nonprocessive telomerase activity from mouse cells. *Proc. Natl Acad. Sci. USA* **90**: 1493–1497.

Riethman HC, Moyzis RK, Meyne J, Burke DT, Olson MV. (1989) Cloning human telomeric DNA fragments into *Saccharomyces cerevisiae* using a yeast-artificial-chromosome vector. *Proc. Natl Acad. Sci. USA* **86**: 6240–6244.

Riethman HC, Spais C, Buckingham J, Grady D, Moyzis RK. (1993) Physical analysis of the terminal 240 kbp of DNA from human chromosome 7q. *Genomics* **17**: 25–32.

Royle NJ. (1995) The proterminal regions and telomeres of human chromosomes. *Adv. Genet.* **32**: 273–315.

Royle NJ, Hill MC, Jeffreys AJ. (1992) Isolation of telomere junction fragments by anchored polymerase chain reaction. *Proc. R. Soc. Lond. Ser. B.* **247**: 57–61.

Royle NJ, Baird DM, Jeffreys AJ. (1994) A subterminal satellite, located adjacent to telomeres in chimpanzees is absent from the human genome. *Nature Genetics* **6**: 52–56.

Sandell LL, Zakain VA. (1992) Telomeric position effect in yeast. *Trends Cell Biol.* **2**: 10–14.

Singer MS, Gottschling DE. (1994) *TCL1*: template RNA component of *Saccharomyces cerevisiae* telomerase. *Science* **266**: 404–409.

Slagbloom PE, Droog S, Boomsma DI. (1994) Genetic determination of telomere size in humans: a twin study of three age groups. *Am. J. Hum. Genet.* **55**: 876–882.

Starling JA, Maule J, Hastie ND, Allshire RC. (1990) Extensive telomere repeat arrays in mouse are hypervariable. *Nucleic Acids Res.* **18**: 6881–6888.

Vaziri H, Schachter F, Uchida I, Wei L, Zhu X, Effros R, Cohen D, Harley CB. (1993) Loss of telomeric DNA during aging of normal and trisomy 21 human lymphocytes. *Am. J. Hum. Genet.* **52**: 661–667.

Vaziri H, Dragowska W, Allsopp RC, Thomas TE, Harley CB, Lansdorpp PM. (1994) Evidence for a mitotic clock in human hemapoietic stem cells: loss of telomeric DNA with age. *Proc. Natl Acad. Sci. USA* **91**: 9857–9860.

Weber B, Collins C, Robbins C, Magenis RE, Delaney AD, Gray JW, Hayden MR. (1990) Characterization and organization of DNA sequences adjacent to the human telomere associated repeat $(TTAGGG)_n$. *Nucleic Acids Res.* **18**: 3353–3361.

Weinberg J, Jauch A, Stanyon R, Cremer T. (1990) Molecular cytotaxonomy of primates by chromosomal *in situ* suppression hybridisation. *Genomics* **8**: 347–350.

Wells RA, Germino GG, Krishna S, Buckle VJ, Reeders ST. (1990) Telomere-related sequences at interstitial sites in the human genome. *Genomics* **8**: 699–704.

Wilkie AOM. (1993) Detection of cryptic chromosomal abnormalities in unexplained mental retardation: a general strategy using hypervariable subtelomeric DNA polymorphisms. *Am. J. Hum. Genet.* **53**: 688–701.

Wilkie AOM, Higgs DR. (1992) An unusually large $(CA)_n$ repeat in the region of divergence between subtelomeric alleles of human chromosome 16p. *Genomics* **13**: 81–88.

Wilkie AOM, Lamb J, Harris PC, Finney RD, Higgs DR. (1990) A truncated chromosome 16 associated with a thalassaemia is stabilized by addition of telomeric repeat $(TTAGGG)_n$. *Nature* **346**: 868–871.

Wilkie AO, Higgs DR, Rack KA, Buckle VJ, Spurr NK, Fischel-Ghodsian N, Ceccherini I, Brown WRA, Harris PC. (1991) Stable length polymorphism of up to 260 kb at the tip of the short arm of human chromosome 16. *Cell* **64**: 595–606.

Yu G-L, Blackburn EH. (1991) Developmentally programmed healing of chromosomes by telomerase in *Tetrahymena*. *Cell* **67**: 823–832.

Yu G-L, Bradley JD, Attardi LD, Blackburn EH. (1990) *In vivo* alteration of telomere sequences and senescence caused by mutated *Tetrahymena* telomerase RNAs. *Nature* **344**: 126–132.

Yunis JJ, Prakash O. (1982) The origin of man: a chromosomal pictorial legacy. *Science* **215**: 1525–1530.

Zakian VA. (1989) Structure and function of telomeres. *Annu. Rev. Genet.* **23**: 579–604.

7

Tandemly repeated minisatellites: generating human genetic diversity via recombinational mechanisms

John A.L. Armour

7.1 Introduction

7.1.1 Tandemly repeated minisatellites

Highly polymorphic minisatellites, which include loci showing the most extreme diversity in human populations, are part of the complement of repeated DNA in the human genome. In common with most vertebrate genomes, the human genome has a substantial contribution from repeated DNA, including both repeated blocks of DNA of known function, such as ribosomal (r) DNA repeats, telomere repeats and centromere repeats, as well as apparently 'gratuitous' blocks of repeats with no well-defined function.

Among these tandemly repeated blocks there is a wide variety of sequences and scales of repetition. For example, the simplest repeated sequences of all, in which a single base is reiterated, occur frequently; the $(A)_n$ tracts found 3′ to retrotransposons (such as *Alu* elements) are particularly common examples. Dinucleotide repeats which contain, for example, 20 or so $(CA)_n$ repeats are also frequent and have been extremely useful in genetic mapping (Weber, 1990). At the other end of the size scale, some satellite arrays at the centromeres of human chromosomes can extend for up to 5 Mb (Willard, 1991).

Human Genome Evolution, edited by M. Jackson, T. Strachan and G. Dover.

Minisatellites occupy a middle ground both in the size of the repeats and the extent of the array; the original name 'minisatellite' was used to indicate a repeated loci of which the total extent was smaller than the classical satellite DNAs (Jeffreys *et al.*, 1985a). This class of repeated locus is most usefully defined in terms of size, as tandemly repeated blocks of total extent between approximately 0.5 kb and 40 kb. The commonly used alternative name 'VNTR' (variable number of tandem repeats) is less usefully descriptive as it might be correctly used to describe variation in any type of repeat array, including dinucleotide or satellite repeats (Tautz, 1993). Furthermore, 'VNTR' describes only those loci which show *variation* in the number of repeats. It therefore does not include the quite numerous minisatellites which are monomorphic in human populations (Armour *et al.*, 1990).

In many cases minisatellites have been isolated as 'anonymous' clones from genomic libraries but in some cases isolation of genomic DNA at or near a functional gene has shown the presence of a minisatellite, such as those at the α-globin (Higgs *et al.*, 1981; Jarman *et al.*, 1986), insulin (Bell *et al.*, 1982), H-*ras* (Capon *et al.*, 1983) and glucose phosphate isomerase (Faik *et al.*, 1994) loci. Nearly all minisatellites described to date lie outside known coding sequence, but there are interesting exceptions. At the *MUC1* locus, for example, a 60 bp repeat that encodes a 20 amino acid repeat in the protein is reiterated a variable number of times. Not only is the DNA variable in length but the repeats are translated so that the corresponding protein product shows parallel variation in size (Swallow *et al.*, 1987).

It is, however, the frequent variability of minisatellites which has attracted the interest of researchers. Variation in the number of repeats in a block gives rise to concomitant variation in the length of the locus, which can in principle be assayed either by simple Southern blot hybridization or (if the allele is not

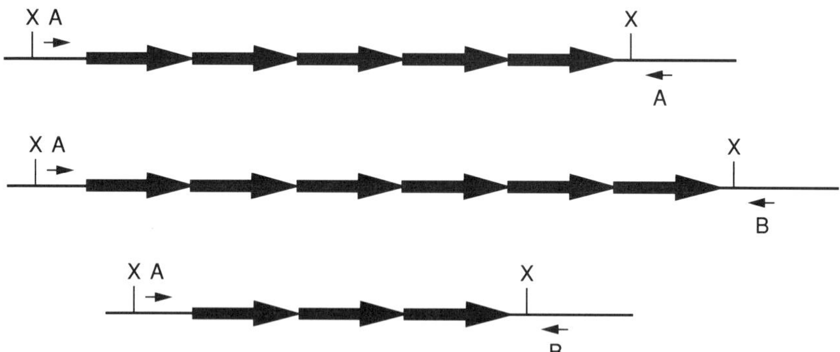

Figure 7.1. Variation in length caused by polymorphism in repeat unit number at a minisatellite. Variation in the numbers of repeat units between alleles, and consequent variation in total allele length, can be assayed simply by Southern blot hybridization after digestion with an enzyme (such as X), which does not cut in the repeat array, or by PCR between flanking primers A and B.

too long) by PCR amplification (*Figure 7.1*). In principle, variation in the number of repeats at a single locus could produce an infinite number of alleles although in practice allele sizes must be constrained if the locus is not to constitute a large fraction of the genome. At the most variable minisatellite loci found in humans there is a large number of different allele lengths; for example, at the hypervariable minisatellite *D1S8*, alleles commonly contain between 120 and 1000 or more repeats of a 29 bp sequence, giving alleles in the range of 3.3 to 29 kb (Wong *et al.*, 1987; A.J. Jeffreys, personal communication).

In addition to variation in the number of repeats in a block, further variation in allelic structure comes from subtle variations in repeat unit sequence. Different interspersion patterns of these repeats along a repeat array gives the potential for multiple different alleles of identical length. At this level of analysis, the true variability of some loci can be astonishing – at the *D1S8* locus there are probably in excess of 10^8 different alleles worldwide (see Section 7.2.3).

7.1.2 Applications of minisatellite loci

The extreme variation shown by some of these loci makes them exceptionally useful in a variety of genetic analyses. If a locus has many alleles, each of which is present only at low frequency in the population, then a typical individual will have inherited two different alleles – in other words, will be heterozygous. The ability to discriminate the two different chromosomes in an individual is of great value in many studies. Although their preferential subtelomeric location limits their use in linkage analysis, minisatellites can be very useful reporters of chromosomal events such as allele losses from tumours (Fearon *et al.*, 1987) and chromosomal deletions (Flint *et al.*, 1995). The general informativeness of the loci is also of great value in establishing paternity and in individual identification in forensic analysis; the lower the allele frequency, the more discriminating the system.

Multi-locus DNA fingerprinting makes combinatorial use of many loci, each of which is hypervariable, in a single, high-resolution test (Jeffreys *et al.*, 1985b, 1985c, 1986). The main use of this technology is in establishing family relationships, and its main advantage is that high statistical confidence can be derived from a single test. In *Figure 7.2*, all the fragments present in the child which are not present in the mother must (if there have been no mutations) have come from the true father – whoever he is. In *Figure 7.2*, there are multiple non-maternal fragments (asterisks) which have no counterpart in Mr X, who can, therefore, be excluded from paternity.

7.1.3 The number and distribution of minisatellites in the human genome

How many minisatellites are there in the human genome? If only variable

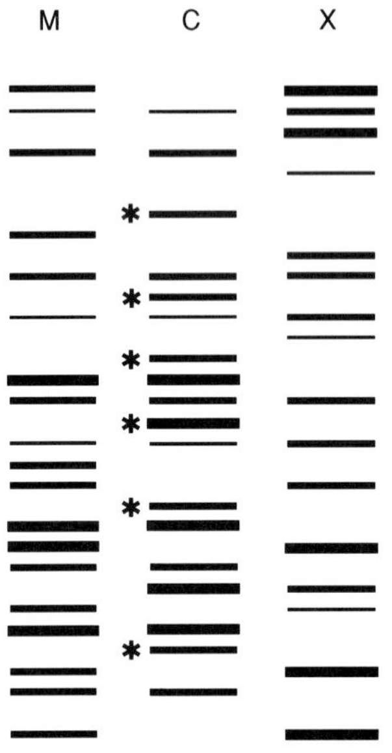

Figure 7.2. Schematic representation of a multi-locus DNA fingerprint, in this case testing for paternity of the child C. In the absence of recent mutation (see *Figure 7.3*) all the fragments present in the child must derive from the mother (M) or the true father. Having identified those fragments inherited by the child from the mother, the remaining non-maternal fragments are compared with the alleged father X. In this case, multiple non-maternal bands (asterisks) are missing from the alleged father X, excluding him as the true father.

minisatellites are counted, and then only those with a heterozygosity greater than 0.7, the frequency of appearance of such loci from clones selected at random from lambda libraries would suggest a total complement of around 1500 loci in a haploid human genome (Braman *et al.*, 1985). This figure is probably relatively accurate but makes the assumption that the genomic libraries screened are representative and in particular that there is no bias against minisatellites. We know, however, from attempts to isolate minisatellites from human and other mammalian genomes that many minisatellite loci are poorly represented in genomic libraries – and lambda libraries in particular – so that these figures could well be underestimates (Armour *et al.*, 1990; Kelly *et al.*, 1989).

An alternative approach to estimating the total complement of minisatellites might be to choose a large well-characterized region of DNA, such as the human leukocyte antigen HLA region of chromosome 6, and extrapolate from the number of minisatellites found. This would be a valid approach if minisatellites were scattered evenly across the genome but it is very clear that minisatellites are not randomly distributed, being more common in the subtelomeric regions of human chromosomes (Armour *et al.*, 1989a; Royle *et al.*, 1988). Thus, there can be very high densities of minisatellites in the regions near chromosome ends but very few in other regions (such as HLA). If 1500 minisatellite loci were distributed within the DNA contained in the terminal 2 Mb of each of the 41 non-acrocentric chromosome ends, then in these regions there would be on average one minisatellite approximately every 50 kb. This high density of minisatellite loci in subtelomeric regions correlates with another feature of biological interest: the terminal regions of human chromosomes are known to have increased frequencies of recombination, an effect which is often most pronounced specifically in the male germline (see Section 7.3.3).

In what follows I will discuss research into the mechanisms by which some minisatellites have generated and maintain such high levels of variability and will relate this to their possible role(s) in the genome. Throughout this discussion it is worth bearing in mind that most research has concentrated on the variable minisatellites and even then mostly on the most variable and unstable loci known. Our picture of how minisatellites behave – and the general themes of this review – may therefore suffer from such a bias; minisatellites which are less spectacularly unstable may not conform to all the properties inferred from the study of the most hypervariable loci.

7.2 Generating diversity: the mutation process

7.2.1 Background

At some loci, most famously the HLA system, existing variation can be explained in terms of selective pressures on functions such as resistance to infection (see Chapter 3). By contrast, it is not possible to explain the extensive variation seen at many minisatellite loci in these terms. First, with the uncommon exceptions noted above, minisatellites are not coding sequences. Furthermore, while evidence is growing that in some cases gene function may *constrain* the behaviour of a minisatellite (see Section 7.3.2), there is no evidence that functional selection is *driving* the generation of diversity. The diversity seen in modern populations is the result of very high rates of mutation in the germline; in other words, there is so much variation because mutation is acting to produce new variants at a high frequency.

These mutations, like the variation they generate, affect the numbers of tandem repeats. At some loci, the observed population diversity is so high that we can predict rates of spontaneous germline mutations high enough to be observed directly, simply by looking at transmission patterns from parents to offspring and scoring new alleles of anomalous size (*Figure 7.3*) – evolution at these loci is occurring so quickly that we can occasionally show it happening.

Mutation rates to new length alleles above about 0.2% per allele per generation (i.e. per gamete) can be estimated by direct observation in pedigrees (Armour *et al.*, 1989b; Jeffreys *et al.*, 1988; Vergnaud *et al.*, 1991). The germline instability varies between loci; at some loci no new alleles have been observed despite extensive surveys, while at (for example) the *D2S90* (*CEB1*) locus, which currently holds the genome record for mutation in normal individuals, male germline mutation rates as high as 15% per sperm have been measured.

Like the rates of mutation, there is considerable variation between loci in the processes and patterns of mutation. The simplest differences to be observed are in the sex bias of the mutation process, with some loci such as *D2S90* (*CEB1*) and *D16S309* (*MS205*) undergoing mutations almost exclu-

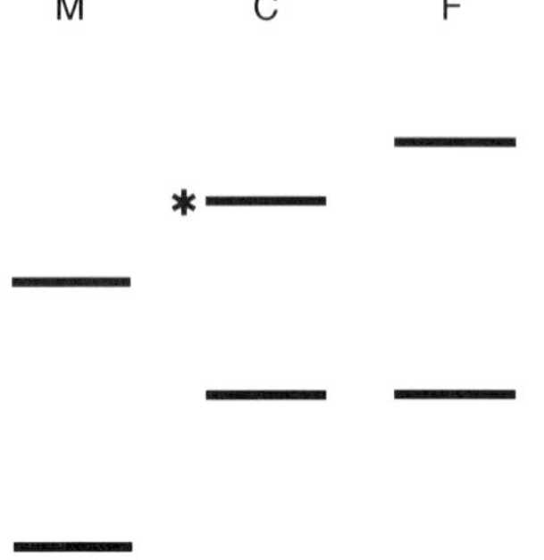

Figure 7.3. Schematic representation of a germline mutation event at a minisatellite. The lower allele in the child (C) has been inherited unchanged from the father (F) but the upper allele (asterisk) has undergone a change in length upon transmission from the mother (M).

sively in the male germline. Other loci, like *D1S8* (*MS32*) and *D7S21* (*MS31*) show mutations in both male and female germlines, but detailed analysis shows that different processes may be operating in the different sexes.

One simple consideration in evaluating evidence for mutation in pedigrees is that a mismatch between offspring and (in particular) the father is not necessarily due to mutation; the possibility of incorrect attribution of parentage needs to be excluded. In practice, the criterion is simple: incorrect paternity will usually lead to anomalies detected at many different loci whereas in a case of germline mutation only one locus is involved and other highly informative loci are consistent with correct parentage.

Analysis of hypervariable minisatellites in non-human primates shows that those loci which are long and highly variable in humans are frequently short (two or three repeats) and monomorphic in other primates (Gray and Jeffreys, 1991). This is a reflection of their very high rates of evolution: the extreme allelic variation seen in humans at these loci has occurred in a very short time. There is nothing necessarily 'special' about humans here: there are presumably other loci that are highly polymorphic in chimpanzees but short and genetically uninteresting in humans and it is simply because we have initially checked for variability in *humans* that we study a particular set of loci.

7.2.2 Mutational mechanisms: the possibilities

So how might tandemly repeated arrays undergo length changes during germline passage? A number of mechanisms might be proposed to account for such events (*Figure 7.4*); this discussion will not consider *de novo* synthesis of repeat units which, although important for telomere repeats, seems implausible for minisatellites. The mechanisms illustrated can be simply divided into those which use one allele only as a template and those which use both alleles in the generation of the new allele.

One simple way to get a new number of tandem repeats is for unequal exchange to occur between sister chromatids (*Figure 7.4a*). Remember that this is not a true recombination since sister chromatids are simply duplicates of the same allele. This is therefore an *inter*molecular but *intra*-allelic mechan-

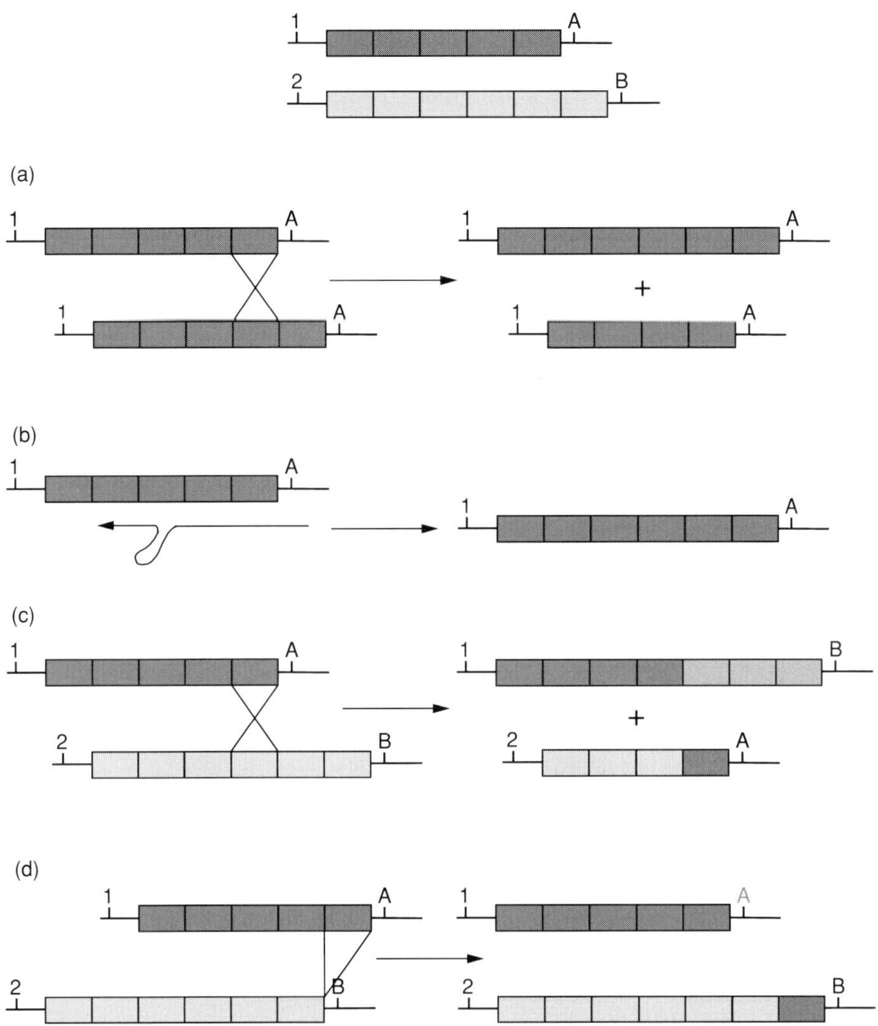

Figure 7.4. Possible mechanisms of length change mutation at minisatellites; the progenitor alleles in this individual are shown at the top. (a) Unequal sister chromatid exchange (USCE) between misaligned copies of one allele. (b) Replication slippage, in which the newly synthesised copy of the allele loses registration within the array to give a copy of (in this case) increased size. (c) Simple unequal recombination between alleles, with exchange of flanking markers. (d) Unequal gene conversion, with exchange of repeats from the donor into the recipient background without rearrangement of the recipient haplotype.

ism. By contrast, slipped-strand mispairing during replication (*Figure 7.4b*) is an entirely intramolecular event.

Two simple points are worth making about 'replication slippage' mechanisms. The first is that this kind of slipped-strand mispairing followed by new synthesis could well be associated with replication but does not have to be. It is possible to have the same intermediates arising, for example, after a nick has been introduced into one strand of DNA within the repeat block; the leading 3′ end could then 'peel off' and reanneal at the wrong template repeat and the resulting gap could be filled in by repair synthesis. The second point is that replication slippage mechanisms are plausible (and popular) explanations for the behaviour of simple repeats (Richards and Sutherland, 1994). However, it remains the case that (in humans at least) there is no direct evidence to establish these mechanisms as the definitive mutational pathway at simple repeats [and exclude, for example, unequal sister chromatid exchange (USCE)]. It is also abundantly clear that they cannot account for many mutations at minisatellites (see Section 7.2.4).

The possibility that minisatellites might be involved in promoting recombination led to initial speculation that unequal recombination between alleles, in which a crossover occurs between misaligned alleles (*Figure 7.4c*), might be a common mechanism for length-change mutation. This simple unequal recombination model would predict a characteristic structural change accompanying the mutation event: markers flanking the minisatellite will be exchanged. Both initial analysis (Wolff *et al.*, 1988, 1989) and subsequent work (see Section 7.2.4) have shown that simple unequal recombination is not a predominant mechanism for length-change mutation in minisatellites: to date, not a single example of this mechanism has been published.

Recombinational mechanisms, however, do play a major role in minisatellite mutation; as described below, a number of mutation events have been observed which involve the exchange of repeat units between alleles but without the exchange of flanking markers. Thus a local 'patch' of repeats is transferred from one allele to the other without disrupting the overall flanking haplotype (*Figure 7.4d*), in a process akin to gene conversion. The high frequency of these exchange events at some minisatellite loci may catalyse the study of gene conversion in the human germline as a subject of direct observation.

7.2.3 Investigating repeat turnover and diversity: internal structural analysis

How do we study the turnover of repeat units within tandem arrays? If the arrays are composed of perfect repeats, as at many dinucleotide and other simple repeat loci, the best we can do is to count the repeats and observe flanking genotypes. However, a further level of analysis is possible at minisatellites because minisatellites frequently show variation in repeat unit sequence. Thus, rather than being composed of perfect repeats of identical

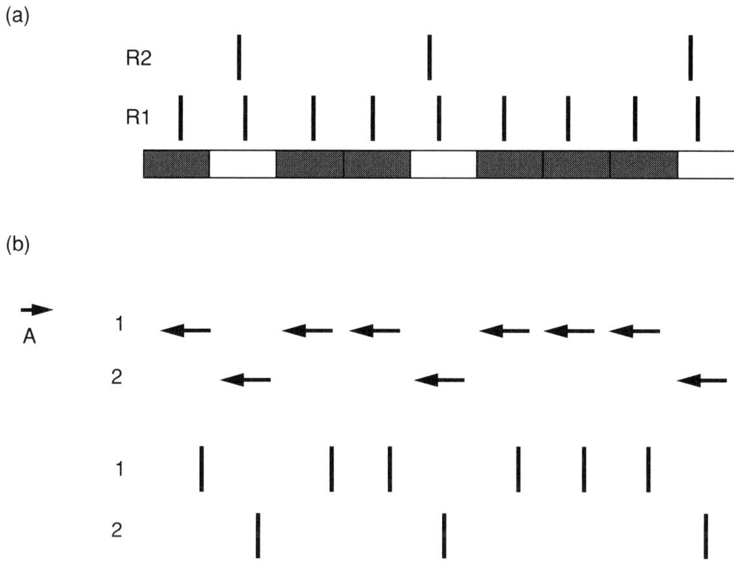

Figure 7.5. Minisatellite variant repeat (MVR) mapping, using (a) restriction enzymes or (b) PCR primers. In (a) all repeat units have a site for enzyme R1, whereas only one repeat unit type has sites for enzyme R2. In (b), PCR primers specific to each repeat unit type are used in conjunction with a fixed primer (A) in the flanking DNA to give products corresponding to the positions of the different repeat types within the array.

units, a minisatellite allele would more usually contain two or more subtly different repeat sequences [minisatellite variant repeats (MVRs)] that differ at perhaps one or two bases. This variation in sequence can then be used to map internal structure within a tandemly repeated allele by plotting the positions of repeat units of different types ('MVR mapping': Armour *et al.*, 1993; Buard and Vergnaud, 1994; Jeffreys *et al.*, 1990, 1991, 1994; Neil and Jeffreys, 1993).

The internal mapping information can be recovered by restriction mapping of the tandem array, if the sequence variation within the repeat creates or destroys a site for a restriction enzyme (*Figure 7.5a*). While this approach has been used to recover useful information on minisatellite structure (Jeffreys *et al.*, 1990), the practical tedium of restriction mapping limits its usefulness. Using PCR primers to sample repeat unit sequence variation (MVR–PCR) has proved a more efficient approach (*Figure 7.5b*; Jeffreys *et al.*, 1991).

MVR analysis allows us to see the astonishing level of diversity at hypervariable minisatellites. Rather than simply typing alleles by length, MVR analysis can distinguish two different alleles which happen to be the same length. At the most variable loci, most apparent examples of homozygotes are not true homozygotes but 'pseudohomozygotes', with two alleles of different internal structure but identical length (Monckton and Jeffreys, 1991).

A similar consideration can be used to estimate the worldwide diversity at such loci. The extreme polymorphism in both allele length and MVR structure would suggest that genuine convergent evolution – in which *de novo* mutation recreates a pre-existing structure – should be very rare; that is, when a mutation creates a new allele, the same structure is almost certain not to exist already. The approximately 5×10^9 diploid people currently populating the earth are the result of the fusion of a total of 10^{10} haploid gametes. At a locus with a *de novo* mutation rate to new allele structures of 1% per gamete (as at *D1S8*), these 10^{10} gametes will have included 1% which bear new alleles, a total of 10^8. Thus, if we include only those alleles newly created in the last generation, the current worldwide diversity at a locus like *D1S8* will encompass a total of some 10^8 different alleles, exceeding by many orders of magnitude the diversity at even highly variable loci of other types.

Mapping the internal structure of minisatellite alleles has been used in two complementary ways to investigate the process of allele length mutation: (i) MVR mapping of alleles selected at random from the population has been informative in demonstrating general patterns of mutation; and (ii) direct analysis of *de novo* mutant alleles, and comparison with progenitor(s), can be used to infer structural changes occurring during mutation.

7.2.4 Polar minisatellite loci

The investigation of allelic variation and mutational change at the first three minisatellites to be analysed by MVR mapping showed that they all shared an unexpected property: mutational change was preferentially happening at one end of the locus (Armour *et al.*, 1993; Jeffreys *et al.*, 1991, 1994; Neil and Jeffreys, 1993). This polarity of mutation and variation was found at three different loci (*D1S8*, *D7S21* and *D16S309*). The polarity could be observed indirectly in that groups of structurally similar alleles often differed by small changes at one end only. These groups of alleles are presumably related by descent and differ by recent, polar mutational changes (*Figure 7.6*). This in turn would suggest that there is a 'time gradient' across the locus; differences at the 'active' end of the locus are due to relatively recent mutations while differences between alleles at the other, less rapidly evolving end of the locus are presumably the result of older mutation events. It is not surprising, then, that differences between populations can be found in the identity and constitution of the major groups of alleles (Armour *et al.*, 1993; Monckton *et al.*, 1994).

MVR analysis of *de novo* mutations ascertained in pedigrees has shown directly that the mutational process at these loci is indeed concentrated at one extremity of the locus (Jeffreys *et al.*, 1994; *Figure 7.7*). Comparisons of MVR structure between a new mutant allele and its progenitor(s) can be used to infer the structural changes giving rise to the new-length allele. In some cases the mechanism is a simple intra-allelic event, like the three-repeat duplication shown in *Figure 7.7a*. There are a number of events, however, which can only be explained as the result of exchange of repeat units between alleles (*Figure*

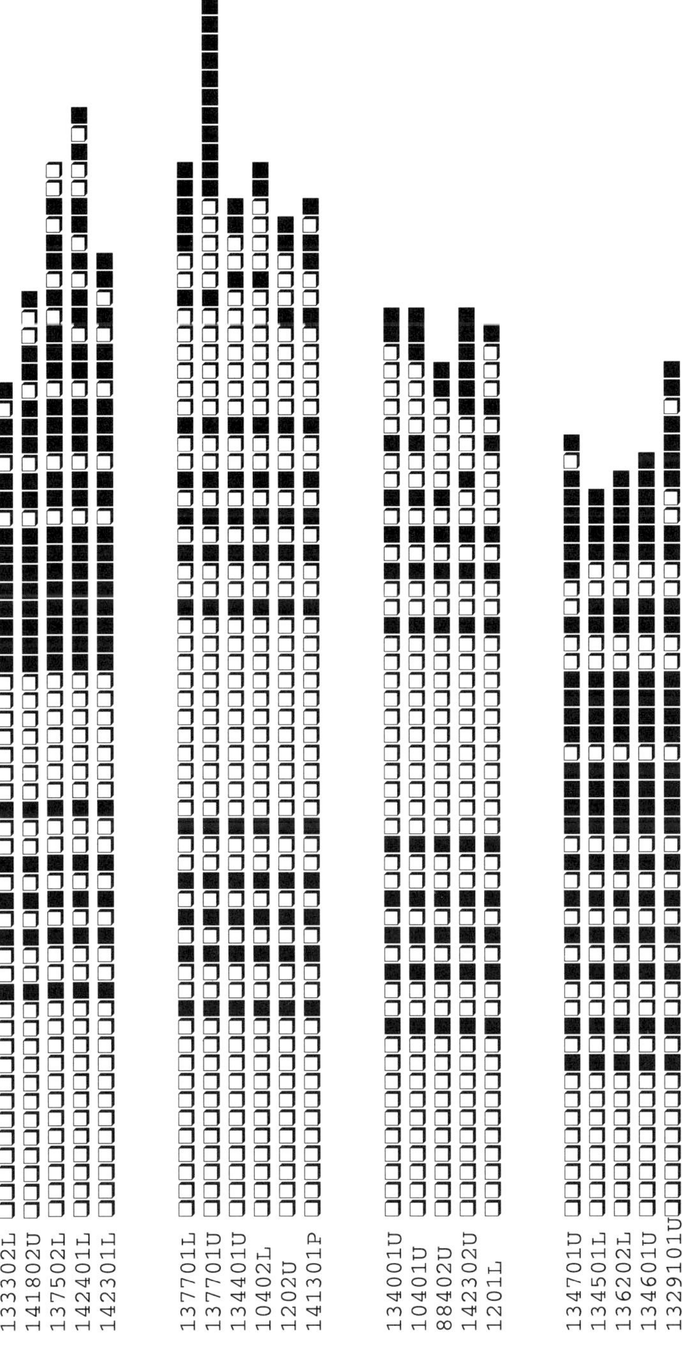

Figure 7.6. Groups of similar, presumably closely related, alleles at minisatellite *D16S309* (*MS205*). Different repeat unit types are shown as black or white boxes (J.A.L. Armour, unpublished).

7.7*b*). None of the mutations of this type reported to date have involved exchange of informative flanking markers and thus the exchange events appear to conform to an event similar to unequal gene conversion, rather than a simple unequal recombination (see Section 7.2.2). Thus, minisatellite mutations frequently involve the transfer of 'patches' of repeat units from one allele into the context of another. A simple way to unify the mutational mechanisms for both intra-allelic and interallelic events at these loci might be to propose that they all involve exchange between minisatellite duplexes: exchanges between sister chromatids lead to intra-allelic mutations; exchanges between alleles lead to interallelic events.

More detailed insights into mutational mechanisms can come from the direct study of germline DNA. In practice, sperm DNA is much easier to obtain in quantity than oocyte DNA and since many loci are preferentially unstable in the male germline, sperm analyses give a very useful, though incomplete, picture. A locus with a germline mutation rate of 1% per gamete would be expected to have a mutational 'load' of 1% in germline DNA: 99% of minisatellite molecules would be of unchanged size, while the remaining 1% would be composed of a variety of different mutant alleles. If small pools of sperm DNA containing, say, 100 genome equivalents per tube are used, then after PCR analysis of such pools any mutant molecule should be represented at a level of about 1% of the products and be reliably detectable.

Small-pool PCR methods, as published for *D1S8* (Jeffreys *et al.*, 1994), have great power in the analysis of mutational processes. Dozens of mutant molecules can be recovered for analysis from the germline of a single person and mutants deriving from each allele of that person can be analysed separately. Analysis of blood and sperm DNA from the same individual clearly shows that mutations occur at high frequency only in the germline, and that somatic events are much less frequent and may result in different patterns of structural change.

The rates and processes of mutation at individual alleles can be assayed by the methods described above, and results from *D1S8* suggest that there is no simple relationship between allele size and mutational instability. While this is clearly different from simple repeats, at which longer tracts are preferentially unstable, this conclusion would be predicted from a consideration of polarity: if only one small part of the locus is active in mutation, then the size of the remaining (inactive) bulk of the allele will be immaterial. Most alleles have similar mutation frequencies but occasional alleles at *D1S8* can be found which have consistently reduced mutation rates. Further work has shown that these unusually stable alleles, which have specifically lost the ability to act as *recipients* of repeat units during exchanges, are associated with a flanking base substitutional variant near the active end of the locus that is not seen in alleles which mutate at a higher rate. It is as if this fortuitous point mutation during human evolution has 'knocked out' the capacity of these alleles to undergo length change mutation (Monckton *et al.*, 1994).

Several of these lines of evidence taken together – polarity of mutation,

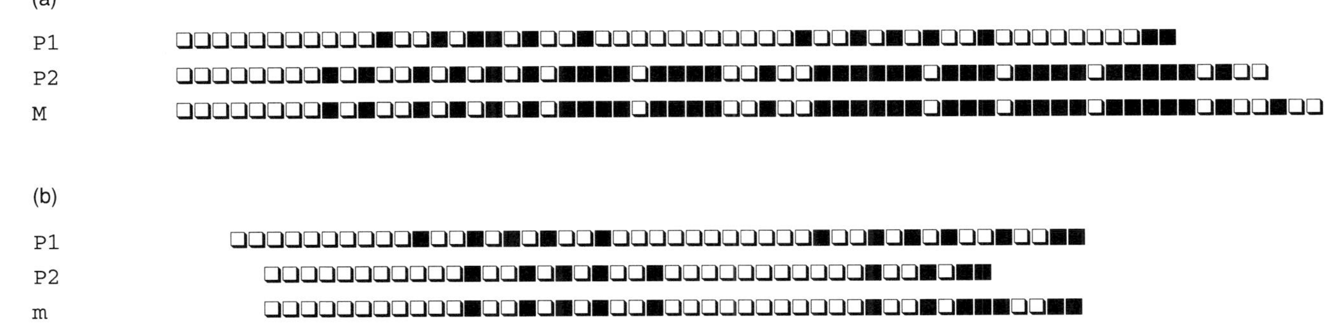

Figure 7.7. MVR analysis of mutation events at the polar minisatellite *D16S309* (*MS205*), showing (a) intra-allelic mutation and (b) an interallelic exchange. Different repeat unit types are shown as black or white boxes. In (a) the new mutant allele M appears to have arisen by reduplication of the last three repeat units of the parental allele P2. In (b), the new mutant allele m can be produced by exchange of the last five repeats from parental allele P1 on to the end of parental allele P2. Note that in both cases the rearrangement of repeat units occurs at one extremity of the locus, here shown on the right.

length independence of instability, and association of unusually stable alleles with a rare variant in the flanking DNA – suggest that instability at these minisatellite loci is not an intrinsic property of the tandemly repeated DNA but is instead in some way directed by elements outside the array at the 'active' end. The involvement of gene conversion-like exchanges has suggested that the initiation of mutation might be achieved by introducing, for example, a double-strand break at the 'active' end of the array (Jeffreys *et al.*, 1994).

7.2.5 Non-polar minisatellite loci

Having established polarity of mutation and variation at the first three different minisatellite loci studied, it seemed that a general principle had been established. However, the large and diverse human genome constantly springs surprises, particularly to those who think that they have discovered general principles.

The *D2S90* minisatellite, detected by probe CEB1, is the most unstable minisatellite locus reported to date. Overall, more than 7% of all parent–offspring transmissions result in a change in the size of the tandem array. This extremely high mutation rate is made all the more remarkable by the fact that nearly all the mutations observed happen in the male germline, thus making the mutation rate to alleles of new length in the male germline approximately 15% per sperm (Vergnaud *et al.*, 1991). This extraordinary locus further prompts the speculation that there may be still more unstable (but as yet undiscovered) minisatellites, at which mutation happens in most transmissions rather than in a minority.

Since *D2S90* mutates at such a high rate, many examples of germline mutation were available for Buard and Vergnaud (1994) to investigate the structural changes by MVR mapping. They showed that the locations at which mutations occurred were spread throughout the locus rather than being concentrated at one end. It is clear, then, that *D2S90* is operating according to different rules than the ones outlined earlier. Looking in detail at the subclasses of mutations at *D2S90*, however, shows that the (relatively few) examples of interallelic exchanges observed all clustered at one end of the locus. This prompts the speculation that *D2S90* may be subject to polar mutational mechanisms, such as predominate in the loci discussed above, but that it also has more frequent non-polar changes.

More recent MVR analysis of variation and mutation at *D5S110* (*MS621*) has also failed to provide any evidence for mutational polarity, suggesting that this locus, like *D2S90*, may be evolving predominantly by processes distributed throughout the locus rather than clustered at one end (Armour *et al.*, 1996).

7.3 Unfinished business

Despite our detailed knowledge of variation and mutation at human minisatellites, some important issues remain unresolved. Below I will discuss briefly some (largely speculative) lines of thought on the nature of the mutational events, the relationship (if any) between minisatellites and nearby genes, and the possible involvement of minisatellites as signals or 'hot-spots' for meiotic recombination.

7.3.1 The machinery of mutation

At polar minisatellites, mutation must be initiated from one end of the tandemly repeated locus. As outlined above, many lines of evidence suggest that elements outside the tandem array are responsible for this initiation and work is under way to define the flanking element(s) and any proteins binding specifically to it. One approach to defining the flanking initiator might be simply to look for sequence similarities in the DNA immediately flanking the active ends of the three known polar minisatellites. No common motif has so far been identified, suggesting either that the common factor is not a simple 'consensus sequence' or that these three minisatellites involve different factors. At non-polar minisatellites, introduction of such an initiating change could occur at any repeat unit rather than the flanking DNA.

Introduction of a double-stranded break into the tandem array would suggest that straightforward mechanistic steps, including enlargement of the double-stranded break and strand invasion from the donor duplex to bridge the gap, subsequently complete the mutation event (Jeffreys *et al.*, 1994). However, a ready-made double-stranded break is not the only way of doing this; a single-stranded nick, for example, could serve as the initiation event. With a gap introduced into the recipient allele, subsequent events in the mutation process depend on the tandemly repeated nature of the substrates. Similar gaps introduced into a non-repeated region could lead to bridging of the gap by a donor segment and gene conversion (*Figure 7.8a*). However, no change in length would have occurred. By contrast, invasion of a gapped tandem array by another, 'bridging', repeated array could give rise to changes in length (*Figure 7.8b*).

These are speculative inferences from the mutational events observed but analogies with meiotic recombination mechanisms are suggested by the germline specificity of high-frequency mutation and the involvement of interallelic exchanges. Both these circumstances suggest that the mutations may be meiotic but there is no direct evidence for this. If they are indeed simply aberrant examples of gene conversion events, of a type which are also happening in non-repeated loci in the human germline, then the study of high-frequency minisatellite mutations may be an interesting and efficient way to study the basic biochemistry of recombination and gene conversion in humans.

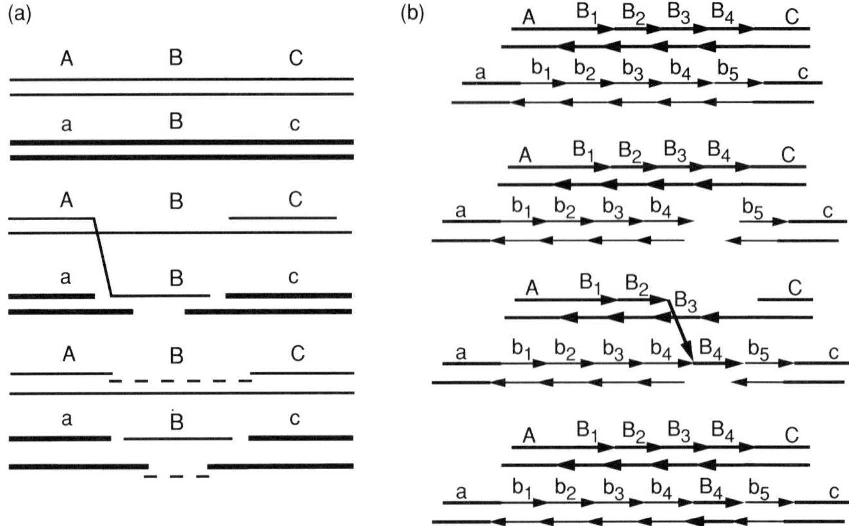

Figure 7.8. Bridging a gap by strand invasion to produce a gene conversion. Where the locus is not tandemly repeated (a) there is no resultant change in length, whereas the same mechanism operating at a tandemly repeated locus can change the number of repeats (b). See text for details.

7.3.2 Minisatellites and genes

Since the discovery of high-frequency germline mutations at hypervariable minisatellites, it has been an underlying assumption that the evolution of these loci follows a selectively neutral pattern (Jeffreys *et al.*, 1988). Indeed, the relationship between mutation rate and heterozygosity would be predicted by a neutral model and it is clear from the sheer diversity at these loci that allelic evolution cannot be severely constrained by selection.

Having said that, it is extremely difficult to be sure that there is *no* selective component to the evolution of these loci, and the examples of the triplet repeat disease loci (see Chapter 1) illustrate how a tandemly repeated locus can evolve neutrally but only up to a point. Variation is selectively neutral as long as the number of repeats is small enough; above a certain size, however, selection acts to curtail the allele size distribution.

Could this also be true of some minisatellites? Two lines of evidence suggest that genes may impose selective constraints upon minisatellite diversity. First, recent studies have shown allelic associations between minisatellites and human disorders related to nearby genes and these associations have been backed up by evidence *in vitro* for modulation of gene expression by the minisatellite. Specifically, predisposition to cancer has been associated with rare alleles at the H-*ras* minisatellite (Krontiris *et al.*, 1993) and the enhancer activity of this minisatellite is different for different alleles (Green and

Krontiris, 1993). Similarly, the locus for inherited predisposition to type I diabetes which maps to the short arm of chromosome 11 (*IDDM2*) appears to be the minisatellite upstream of the insulin gene, rather than the insulin gene itself (Bennett *et al.*, 1995); this minisatellite has been shown to modulate transcription of the insulin gene (Kennedy *et al.*, 1995).

The second line of evidence is less direct but may find more general application. In the examples above, a minisatellite found to be next to a known gene has been analysed in relation to an inherited disorder or predisposition associated with the locus. Most minisatellites, however, and nearly all at which variation and mutation have been studied in detail, have been isolated as anonymous genomic fragments, without any prior knowledge of nearby expressed sequences. It is a curious circumstance that whereas most highly unstable minisatellites have wide-ranging allele size ranges (e.g. 2–22 kb at *D1S7*), some other extremely variable and unstable loci have apparently limited size ranges. Thus, for example, *D16S309* (*MS205*) has a germline mutation rate of about 0.4%, but all alleles so far encountered are less than 5 kb long (Armour *et al.*, 1993).

One simple explanation for this is that at highly unstable loci there will be a large 'random-walk' element to the evolutionary history of allele size distributions and some loci may have remained relatively small purely by chance. However, at some loci there appears to be a small but definite bias in the mutation process towards increases in allele size (Jeffreys *et al.*, 1994), and thus in the long term one would expect not a purely random walk but an overall growth in average allele size. Thus, it remains possible that as yet unknown genes in the vicinity may be exerting selective pressure against some alleles at these minisatellites. Like triplet repeat disease loci, therefore, the evolution of some minisatellite loci may be neutral in the common size range but selection against unusually long alleles may ultimately limit the observed size range. Alternatively, as at the insulin and H-*ras* genes, it may not be simply allele length but allele sequence which is the subject of selective pressures. These hypotheses may be tested by examining allele-specific expression of genes near minisatellites.

7.3.3 Minisatellites and recombination 'hot-spots'

Minisatellites are known to cluster at high concentrations in the subtelomeric regions of human chromosomes, regions which are known to have high rates of meiotic recombination per unit physical distance (especially in the male germline). Minisatellite mutation also frequently involves gene conversion-like exchanges between alleles – although 'full-blown' recombinations (with exchange of flanking markers) have not yet been documented during length-change mutation. This circumstantial evidence suggests that recombinational activity in subtelomeric regions and in minisatellite mutation may be related.

But did the recombination or the minisatellites come first? In our current state of knowledge, both basic possibilities remain open. It could be that minisatellites act as 'hot-spots' to *cause* the high rates of recombination. Alternatively, there may already be high rates of recombination in subtelomeric regions for reasons unconnected with the minisatellites but the recombination helps minisatellites to evolve, and indeed to come into being in the first place (Jarman and Wells, 1989). How do we distinguish these possibilities?

The simplest distinction to make experimentally concerns the spatial placement of crossovers in meiotic recombination; if minisatellites act to promote recombination and gene conversion in some specific way, then we would predict a spatial association between minisatellites and meiotic breakpoints in these regions. Putting to one side the possible complication that a crossover does not need to resolve near where it was initiated, it may therefore be feasible to see whether minisatellites act as recombination hot-spots in human meiosis by fine-structure mapping of crossover location.

However, to reach a situation analogous to the well-defined, discrete recombination hot-spots such as those described in yeast, this analysis will require a much higher level of physical resolution than human genetics is currently used to. Although a number of recent reports have narrowed down regions responsible for high levels of meiotic recombination to relatively small intervals (Hubert *et al.*, 1994; Robinson and Lalande, 1995), some of these relatively small intervals are still large in absolute terms (e.g. 280 kb; Hubert *et al.*, 1994). For such intervals, perhaps 'warm-patch' would be a better description than 'hot-spot'. It will be of great interest to see if somewhere within these large 'warm-patches', smaller, discrete hot-spots can be defined. If they exist, hot-spots like this will be of great value in dissecting the basic components and processes of recombination in the human germline; whether they will be minisatellites remains to be seen.

Even if minisatellites are not hot-spots for homologous recombination in human meiosis, it remains possible that they may function in the correct mutual identification and pairing of homologous chromosomes. During this process, some way must be found to test for homology between chromosomes before pairing and recombination can be allowed to proceed to completion. Sets of chromosome-specific tandem repeat arrays placed in the subtelomeric regions might be efficient substrates for this homology testing, and the successful completion of a strand invasion between homologues might be an appropriate signal to the cell that pairing has been correctly achieved.

Acknowledgements

I am grateful to Professor Alec Jeffreys for his support and encouragement and to the Wellcome Trust for the funding of my research. I would also like to thank Dr Nicola Royle and Professor Alec Jeffreys for their comments on this manuscript.

References

Armour JAL, Wong Z, Wilson V, Royle NJ, Jeffreys AJ. (1989a) Sequences flanking the repeat arrays of human minisatellites: association with tandem and dispersed repeat elements. *Nucleic Acids Res.* **17**: 4925–4935.

Armour JAL, Patel I, Thein SL, Fey M, Jeffreys AJ. (1989b) Analysis of somatic mutations at human minisatellite loci in tumours and cell lines. *Genomics* **4**: 328–334.

Armour JAL, Povey S, Jeremiah S, Jeffreys AJ. (1990) Systematic cloning of human minisatellites from ordered array Charomid libraries. *Genomics* **8**: 501–512.

Armour JAL, Harris PC, Jeffreys AJ. (1993) Allelic diversity at minisatellite MS205 (D16S309): evidence for polarized variability. *Hum. Mol. Genet.* **2**: 1137–1145.

Armour JAL, Gosier M, Jeffreys AJ. (1996) Distribution of tandem repeat polymorphism within minisatellite MS261 (D5S110). *Ann. Hum. Genet.* **60**: 11–20.

Bell GI, Selby MJ, Rutter WJ. (1982) The highly polymorphic region near the human insulin gene is composed of simple tandemly repeating sequences. *Nature* **295**: 31–35.

Bennett ST, Lucassen AM, Gough SCL *et al.* (1995) Susceptibility to human type 1 diabetes at *IDDM2* is determined by tandem repeat variation at the insulin gene minisatellite locus. *Nature Genetics* **9**: 284–292.

Braman J, Barker D, Schumm J, Knowlton R, Donis-Keller H. (1985) Characterization of very highly polymorphic RFLP probes. *Cytogenet. Cell Genet.* **40**: 589.

Buard J, Vergnaud G. (1994) Complex recombination events at the hypermutable minisatellite CEB1 (D2S90). *EMBO J.* **13**: 3203–3210.

Capon DJ, Chen EY, Levinson AD, Seeburg PH, Goeddel DV. (1983) Complete nucleotide sequence of the T24 human bladder carcinoma oncogene and its normal homologue. *Nature* **302**: 33–37.

Faik P, Walker JIH, Morgan M. (1994) Identification of a novel tandemly repeated sequence present in an intron of the glucose phosphate isomerase (GPI) gene in mouse and man. *Genomics* **21**: 122–127.

Fearon ER, Hamilton SR, Vogelstein B. (1987) Clonal analysis of human colorectal tumours. *Science* **238**: 193–197.

Flint J, Wilkie AOM, Buckle VJ, Winter RM, Holland AJ, McDermid HE. (1995) The detection of subtelomeric chromosomal rearrangements in idiopathic mental retardation. *Nature Genetics* **9**: 132–140.

Gray IC, Jeffreys AJ. (1991) Evolutionary transience of hypervariable minisatellites in man and the primates. *Proc. R. Soc. Lond. Ser. B.* **243**: 241–253.

Green M, Krontiris TG. (1993) Allelic variation of reporter gene activation by the HRAS1 minisatellite. *Genomics* **17**: 429–434.

Higgs DR, Goodbourn SEY, Wainscoat JS, Clegg JB, Weatherall DJ. (1981) Highly variable regions of DNA flank the human globin genes. *Nucleic Acids Res.* **9**: 4213–4224.

Hubert R, MacDonald M, Gusella J, Arnheim N. (1994) High resolution localization of recombination hot spots using sperm typing. *Nature Genetics* **7**: 420–424.

Jarman AP, Wells RA. (1989) Hypervariable minisatellites: recombinators or innocent bystanders? *Trends Genet.* **5**: 367–371.

Jarman AP, Nicholls RD, Weatherall DJ, Clegg JB, Higgs DR. (1986) Molecular characterization of a hypervariable region downstream of the human α-globin gene cluster. *EMBO J.* **5**: 1857–1863.

Jeffreys AJ, Wilson V, Thein SL. (1985a) Hypervariable 'minisatellite' regions in human DNA. *Nature* **314**: 67–73.

Jeffreys AJ, Wilson V, Thein SL. (1985b) Individual-specific 'fingerprints' of human DNA. *Nature* **316**: 76–79.

Jeffreys AJ, Brookfield JFY, Semeonoff R. (1985c) Positive identification of an immigration test-case using human DNA fingerprints. *Nature* **317**: 818–819.

Jeffreys AJ, Wilson V, Thein SL, Weatherall DJ, Ponder BAJ. (1986) DNA 'fingerprints' and segregation analysis of multiple markers in human pedigrees. *Am. J. Hum. Genet.* **39**: 11–24.

Jeffreys AJ, Royle NJ, Wilson V, Wong Z. (1988) Spontaneous mutation rates to new length alleles at tandem-repetitive hypervariable loci in human DNA. *Nature* **332**: 278–281.

Jeffreys AJ, Neumann R, Wilson V. (1990) Repeat unit sequence variation in minisatellites: a novel source of DNA polymorphism for studying variation and mutation by single molecule analysis. *Cell* **60**: 473–485.

Jeffreys AJ, MacLeod A, Tamaki K, Neil DL, Monckton DG. (1991) Minisatellite repeat coding as a digital approach to DNA typing. *Nature* **354**: 204–209.

Jeffreys AJ, Tamaki K, MacLeod A, Monckton DG, Neil DL, Armour JAL. (1994) Complex gene conversion events in germline mutation at human minisatellites. *Nature Genetics* **6**: 136–145.

Kelly R, Bulfield G, Collick A, Gibbs M, Jeffreys AJ. (1989) Characterization of a highly unstable mouse minisatellite locus: evidence for somatic mutation during early development. *Genomics* **5**: 844–856.

Kennedy GC, German MS, Rutter WJ. (1995) The minisatellite in the diabetes susceptibility locus *IDDM2* regulates insulin transcription. *Nature Genetics* **9**: 293–298.

Krontiris TG, Devlin B, Karp DD, Robert NJ, Risch N. (1993) An association between the risk of cancer and mutations in the HRAS1 minisatellite locus. *New Engl. J. Med.* **329**: 517–523.

Monckton DG, Jeffreys AJ. (1991) Minisatellite 'isoallele' discrimination in pseudohomozygotes by single molecule PCR and variant repeat mapping. *Genomics* **11**: 465–467.

Monckton DG, Neumann R, Guram T, Fretwell N, Tamaki K, MacLeod A, Jeffreys AJ. (1994) Minisatellite mutation rate variation associated with a flanking DNA sequence polymorphism. *Nature Genetics* **8**: 162–170.

Neil DL, Jeffreys AJ. (1993) Digital DNA typing at a second hypervariable locus by minisatellite variant repeat mapping. *Hum. Mol. Genet.* **2**: 1129–1135.

Richards RI, Sutherland GR. (1994) Simple repeat DNA is not replicated simply. *Nature Genetics* **6**: 114–116.

Robinson WP, Lalande M. (1995) Sex-specific meiotic recombination in the Prader-Willi/Angelman syndrome imprinted region. *Hum. Mol. Genet.* **4**: 801–806.

Royle NJ, Clarkson RE, Wong Z, Jeffreys AJ. (1988) Clustering of hypervariable minisatellites in the preterminal regions of human autosomes. *Genomics* **3**: 352–360.

Swallow DM, Gendler S, Griffiths B, Corney G, Taylor-Papadimitriou J, Bramwell ME. (1987) The human tumour-associated epithelial mucins are coded by an expressed hypervariable gene locus PUM. *Nature* **328**: 82–84.

Tautz D. (1993) Notes on the definition and nomenclature of tandemly repeated DNA sequences. In: *DNA Fingerprinting: State of the Science* (eds SDJ Pena, R Chakraborty, JT Epplen and AJ Jeffreys). Birkhauser Verlag, Basel, pp. 21–28.

Vergnaud G, Mariat D, Apiou F, Aurias A, Lathrop M, Lauthier V. (1991) The use of synthetic tandem repeats to isolate new VNTR loci: cloning of a human hypermutable sequence. *Genomics* **11**: 135–144.

Weber JL. (1990) Informativeness of human (dA-dC)n.(dG-dT)n polymorphisms. *Genomics* 7: 524–530.

Willard HF. (1991) Evolution of alpha satellite. *Curr. Opin. Genet. Dev.* **1**: 509–514.

Wolff RK, Nakamura Y, White R. (1988) Molecular characterization of a spontaneously generated new allele at a VNTR locus: no exchange of flanking DNA sequence. *Genomics* **3**: 347–351.

Wolff RK, Plaetke R, Jeffreys AJ, White R. (1989) Unequal crossingover between homologous chromosomes is not the major mechanism involved in the generation of new alleles at VNTR loci. *Genomics* **5**: 382–394.

Wong Z, Wilson V, Patel I, Povey S, Jeffreys AJ. (1987) Characterization of a panel of highly variable minisatellites cloned from human DNA. *Ann. Hum. Genet.* **51**: 269–288.

8

Microsatellites and other simple sequences in the evolution of the human genome

John M. Hancock

8.1 Introduction

Much of the repetitive DNA in the human genome, rather than being made up of members of the well-characterized sequence families described in other chapters, is made up of tandem repeats of short sequence motifs or sequences that are qualitatively similar. Sequences of this kind which are tandemly repeated have been called microsatellites (Litt and Luty, 1989; Tautz, 1989; Weber and May, 1989) and have become very important in gene mapping. Regions of sequence with concentrations of qualitatively similar short sequence motifs that are not tandemly repeated have been described as 'cryptically simple' (Tautz *et al.*, 1986). A further kind of cryptic simplicity can arise when sequences contain repeats recognizable at the level of patterns of purines and pyrimidines rather than bases (Ricke *et al.*, 1995; Sarkar *et al.*, 1991). In general, these kinds of sequence are known as 'simple sequences' and can be defined as sequence regions that contain a statistically significant over-representation of one or more short motif(s).

This chapter does not attempt to describe the uses of microsatellites in human genetics. Instead, it concentrates on how microsatellites and cryptically simple sequences in the human genome can be detected, the mechanisms giving rise to them, their mode of evolution, and the possible func-

Human Genome Evolution, edited by M. Jackson, T. Strachan and G. Dover.

tional consequences of their incorporation into genomes, particularly coding regions.

8.2 Detection of simple sequence regions

Simple sequence regions within longer sequences can be detected in a number of ways. At the simplest level, search programs such as FASTA can be used to search for specific simple sequences, a method that is widely used but relies on knowing what sequences are being looked for. The locations of simple sequences can also be displayed using dot matrix programs, comparing the sequence in question with an artificial sequence comprising, for instance, arrays of all possible trinucleotide motifs (Dover *et al.*, 1993; Hancock and Dover, 1990). A more sophisticated graphical display method shows long, repeated motifs as peaks in a 'landscape' (DAWG; Clift *et al.*, 1986).

Two kinds of algorithmic approach have been used to identify simple sequence regions. One approach uses measures of the information content of sequences, defining simple sequences as regions containing relatively little information (SMPL; Konopka and Owens, 1990; Miloslavljevich and Jurka, 1993). Konopka and Owens (1990) calculated the Shannon information content of sequences, which in this context measures the deviation of base composition within the sequence from randomness (25% of all four bases). The program SMPL (Miloslavljevich and Jurka, 1993) is related to data compaction algorithms, and determines the minimum quantity of information required to encode a DNA sequence by compressing it using an algorithm that replaces repeated motifs with a reference to their previous occurrence within the sequence. If the number of bits of information saved in this way exceeds a critical value, a sequence is said to be repetitive. By moving a window along sequences, these programs can identify low-information regions within them. However, they do not, strictly speaking, identify regions of clustered motifs. They also ignore some fundamental properties of natural DNA sequences, including variation in overall base composition, biased dinucleotide composition (Nussinov, 1984) and the fact that some coding sequences show different base compositions at first, second and third codon positions (Keese *et al.*, 1989). The second approach, detailed in Section 8.2.1, looks directly for the clustering of short sequence motifs within a DNA sequence (Hancock and Armstrong, 1994; Tautz *et al.*, 1986).

8.2.1 SIMPLE34

SIMPLE34 (Hancock and Armstrong, 1994) is a modification of an original program developed by Tautz and co-workers (Tautz *et al.*, 1986). It provides a measure of the overall level of repetition within a DNA sequence and ascertains whether this level of repetition is greater than would be expected at random. The method has been extended to identify regions that show significantly high levels of repetitiveness and to identify the motifs that are involved in that repetitiveness. It compensates for biased mono- and dinucleotide compositions by comparing the sequence under analysis with quasi-random sequences of the same base and doublet composition as the test sequence. In this way, it can eliminate false positives but at the same time be more sensitive to real fluctuations from random distributions of motifs.

In practice, the program follows a 64 base pair (bp) window that 'slides' along the test sequence. At each position, the program looks for repeats of the 3 and 4 bp motifs located at the centre of the window and awards one point for each repetition of a triplet and four points for every repetition of a tetranucleotide (excluding overlaps between triplets and tetranucleotides). Scores for all the windows on the test sequence are summed. It is clear that scores could be dramatically affected by the nucleotide composition of the test sequence. A highly AT-rich sequence would, purely by virtue of its nucleotide composition, produce higher scores for an AT-rich motif than a sequence which is GC-rich. To control for this, the exercise is also carried out on 10 randomized sequences of the same length and base composition as the test sequence. The score for the test sequence is then divided by the mean value for the randomized sequences to generate a score, the relative simplicity factor (RSF), which indicates how repetitive the test sequence is relative to the random sequences.

As well as counting an overall score for a particular test sequence, the scores reached by individual windows can also be counted. Doing this for both the test and random or quasi-random sequences provides a means of estimating which windows, and therefore which sequence motifs, are primarily responsible for any significantly high level of sequence repetitiveness that might be found. These are termed significantly simple motifs (SSMs; Hancock, 1995b; Hancock and Armstrong, 1994). The program identifies these motifs and their positions. A graphical display of a sliding mean of the scores reached by individual windows provides a graphical indication of the locations of repetitive regions within a sequence. An illustration of this is shown in *Figure 8.1*.

8.3 How simple sequences evolve

Simulation and population studies of microsatellites (Di Rienzo *et al.*, 1994; Valdes *et al.*, 1993) indicate that the frequency distributions of their lengths are consistent with a two-phase model of their evolution. In this model, most of the changes in the size of an array of a motif $(XYZ)_n$ result in new arrays of length $(XYZ)_{n+1}$ or $(XYZ)_{n-1}$, so-called stepwise mutation (Ohta and Kimura, 1973; Wehrhahn, 1975). However, large changes in array size also occur occasionally. This mode of evolution is consistent with slipped-strand mispairing (Levinson and Gutman, 1987a) being the predominant mechanism generating variation in simple sequences. Slipped-strand mispairing results from the mispairing of strands during replication and repair and can result in the gain or loss of motifs from an array, depending both on the way in which the mispair-

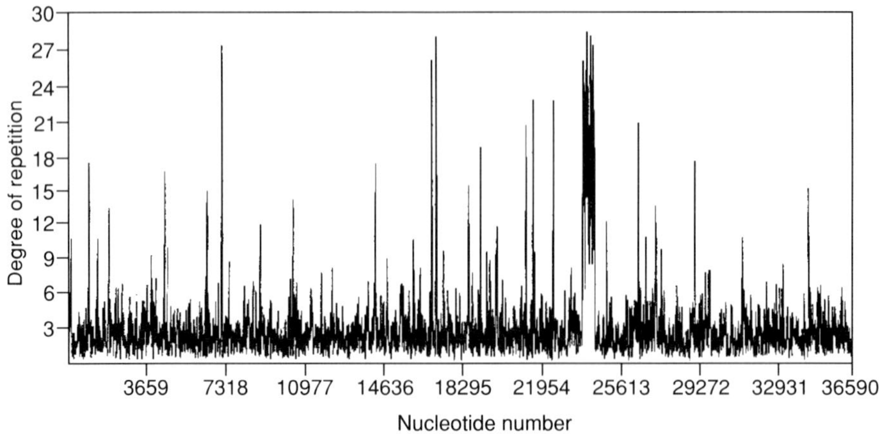

Figure 8.1. A plot of the variation in the level of repetition along the gene encoding the human tissue plasminogen activator (GenBank Accession No. K03021) generated by SIMPLE34 (Hancock and Armstrong, 1994). The vertical axis represents the level of repetition (i.e. points score) measured by the program at the position in the sequence indicated by the horizontal axis. The plot is smoothed by averaging scores obtained for 10 consecutive 64 bp windows (see Section 8.2.1 for more details of how the program analyses sequences for repetition).

ing is repaired and on the strand (nascent or parental) that undergoes mispairing. Classes of repetitive sequence with longer underlying repeats do not appear to evolve predominantly in this way. Rather they appear to undergo extensive unequal crossing-over or unequal sister chromatid exchange events (Chapter 5; Shriver *et al.*, 1993; Stephan and Cho, 1994) while minisatellites appear to mutate predominantly by one or more gene-conversion-like mechanisms (Chapter 7; Jeffreys *et al.*, 1994).

Rates of slippage have been shown to increase with increasing array length in *Escherichia coli* (Levinson and Gutman, 1987b; Murphy *et al.*, 1989; Streisinger and Owen, 1985). Longer arrays also show higher levels of polymorphism as long as they are uninterrupted by point mutations (Weber, 1990). This means that at the evolutionary level slippage tends to feed back on itself, with longer arrays changing more rapidly (Dover and Tautz, 1986).

The replication slippage aspect of slipped-strand mispairing can be modelled *in vitro* by investigating the extension of simple sequence oligonucleotides by DNA polymerases (Behn-Krappa and Doerfler, 1994; Kornberg *et al.*, 1964; Morgan *et al.*, 1974; Wells *et al.*, 1967a,b; Schlötterer and Tautz, 1992). These experiments support the proposition (Levinson and Gutman, 1987a) that replication slippage occurs more readily for short sequence motifs and also suggest that the process shows sequence bias. Experiments using sequences that undergo expansion in triplet expansion diseases (Section 8.8 and Chapter 1) have suggested that aspects of DNA higher-order structure may contribute to the propensity of certain sequences to undergo slipped-strand mispairing (Behn-Krappa and Doerfler, 1994). It should be noted, however, that although population studies support higher rates of mutation for shorter than for longer motifs (Edwards *et al.*, 1992), measurements of mutation rates in cell lines indicate higher rates for tetranucleotide than for dinucleotide repeats (Weber and Wong, 1993). The lack of dependence on array length shown by simple sequences *in vitro* (Schlötterer and Tautz, 1992) suggests that *in vitro* experiments may not represent ideal models for replication slippage *in vivo*.

8.4 Microsatellites in the human genome

The best-characterized simple sequences are microsatellites, which are widespread in the human genome. Arrays of the most commonly used sequence for genetic studies, $(CA/TG)_n$ where n is greater than 12, are estimated to occur approximately 35 000 times (Weber, 1990) or once every 100 kilobases (kb). Arrays where n is greater than 8 are estimated to occur every 30 kb (Stallings *et al.*, 1991). Microsatellite arrays of any sequence over 20 nucleotides long were found once every 6 kb in a 745 kb sample, while only 12% of arrays were longer than 40 nucleotides (Beckmann and Weber, 1992).

Characterization of the frequencies of mono- and dinucleotide microsatellites have shown that $(A/T)_n$ is the most common motif (Stallings, 1992). Microsatellites containing strings of A bases also dominate the frequency distributions of longer motif microsatellites (Beckmann and Weber, 1992). Using the more general definition of simple sequences presented at the beginning of this chapter, Tautz *et al.* (1986) showed a similar bias towards A/T-rich simple sequences in the sequence databases, although they could only survey sequences from broad taxonomic groups because of the relatively small size of the databases at the time. It has been suggested that the preponderance of poly(A/T) microsatellites in the human genome is due to their being seeded by the transposition of *Alu* transposable elements, which may carry with them long poly(A) tails (see Chapter 9; Beckmann and Weber, 1992). However, the lack of association of poly(A/T) microsatellites with *Alu* elements in many cases, together with the predominance of poly(A) microsatellites in *Caenorhabditis elegans* and *Saccharomyces cerevisiae* as well as humans (Hancock, 1995b), suggests that this is not the only explanation for the high frequency of poly(A/T) (see below).

Analysis of mono- and dinucleotide microsatellites has also shown that the dinucleotide $(CG)_n$ is infrequent in the human genome (Stallings, 1992). This dinucleotide contains CpG, the base doublet unmethylated in so-called CpG islands which is methylated and suppressed elsewhere in the genome (Cooper *et al.*, 1983). However, this dinucleotide is also rare in many genomes that do not have methylation (Karlin and Burge, 1995), suggesting that the low frequency of $(CG)_n$ reflects other properties of the sequence, perhaps its ability to form Z-DNA.

Counting trinucleotide microsatellites, Stallings (1994) identified a high frequency of $(CAG)_n$, the motif primarily involved in trinucleotide expansion disorders (Bates and Lehrach, 1994), and low frequencies of $(CGA/TCG)_n$, $(CCA/TGG)_n$ and $(TAC/GTA)_n$. Arrays of $(CAG)_n$ and $(CGG)_n$ have been found in a number of human coding regions (Riggins *et al.*, 1992; Stallings, 1994). However, by comparing frequencies of microsatellites in exons, introns and 5′ and 3′ untranslated regions (UTRs) (but not intergenic regions) the microsatellite $(CAG)_n$ was found to be very rare or absent, in introns, while $(AAT)_n$ was found predominantly in introns.

At the evolutionary level, the best comparisons of microsatellites are available between humans and rodents (i.e. rats and mice), the best characterized non-human mammals. Rat microsatellite arrays appear to be generally longer than human arrays (Beckmann and Weber, 1992). Stallings *et al.* (1991) estimated that approximately 30% of $(CA)_n$ dinucleotide microsatellites in close proximity to genes are conserved in orthologous positions in humans and rodents (which diverged approximately 60–80 million years ago; Nei, 1987). Conservation of microsatellites within coding regions seems to be limited (Stallings, 1994) although this is not the case in the TATA-binding protein TBP, which possesses orthologous $(CAG)_n$ arrays in mouse and human (Hancock, 1993; Tamura *et al.*, 1991) and exhibits length polymorphism in human populations (Gostout *et al.*, 1993; Polymeropoulos *et al.*, 1991). Some human $(CA)_n$ microsatellites have been found to be conserved in other primates (Calvas *et al.*, 1994; Deka *et al.*, 1994).

8.5 SIMPLE34 analysis of human and other genomes

Recently, long sequences have begun to emerge from genome sequencing projects. These have allowed individual species, including humans, to be analysed for simple sequences using the more general definition of sequence simplicity (Hancock, 1995b). It should be stressed that this approach measures clustering of sequence motifs and not necessarily their tandem, microsatellite-like repetition. Because of this, motifs cannot *a priori* be grouped into clusters of circularly permutable motifs and their complements in the following analysis.

Table 8.1 presents the percentage contributions of the 64 different trinucleotide motifs to the simple sequence complement of the human genome (data from Hancock, 1995b, summarizing results obtained from 27 sequences greater than 30 kb long, totalling 1 436 210 bp). *Table 8.2* represents the extent to which each of these classes of sequence shows bias between exons (E), introns (I) and extragenic non-coding regions (EG) of the genome (extragenic regions include 5' and 3' UTRs as well as intergenic regions). The measure of bias (R) for each simple sequence motif is calculated by the formula:

$$R(\text{E, I, EG}) = \frac{\%(\text{E, I or EG})}{(\%\text{E} + \%\text{I} + \%\text{EG})}$$

Table 8.1 Proportions of simple sequence motifs class in human exons, introns and extragenic sequences

Motif	Sequence class			Motif	Sequence class		
	Exons	Introns	Extra-genic		Exons	Introns	Extra-genic
AAA	1.5	15.59	20.18	GAA	0.0	1.40	3.42
AAC	0.0	0.46	0.59	GAC	0.0	0.00	0.00
AAG	0.5	1.10	2.64	GAG	3.0	1.47	2.04
AAT	0.0	1.97	1.86	GAT	0.0	2.03	0.26
ACA	0.0	2.17	1.92	GCA	4.0	0.06	0.00
ACC	0.0	0.00	0.13	GCC	3.5	0.17	0.67
ACG	0.0	0.00	0.00	GCG	0.0	0.04	0.46
ACT	0.0	0.04	0.00	GCT	9.5	0.04	0.03
AGA	0.5	2.34	3.16	GGA	5.0	0.73	1.99
AGC	2.5	0.00	0.13	GGC	0.5	0.22	0.55
AGG	0.0	0.81	1.52	GGG	3.5	0.62	1.17
AGT	0.0	0.04	0.13	GGT	0.0	0.03	0.21
ATA	0.0	5.73	2.76	GTA	0.0	0.07	0.08
ATC	0.0	0.36	0.16	GTC	0.0	0.31	0.00
ATG	0.0	0.56	0.46	GTG	2.5	3.27	2.07
ATT	0.0	2.14	1.73	GTT	1.5	1.02	1.73
CAA	1.0	0.78	0.86	TAA	0.0	1.87	1.29
CAC	0.0	1.30	1.17	TAC	0.0	0.35	0.06
CAG	2.5	0.15	0.46	TAG	0.0	1.19	0.00
CAT	0.0	0.48	0.03	TAT	1.0	5.76	3.21
CCA	1.0	0.28	0.38	TCA	0.0	0.07	0.00
CCC	6.0	0.85	1.56	TCC	2.0	0.94	0.83
CCG	4.5	0.14	0.00	TCG	0.0	0.00	0.00
CCT	14.4	1.33	1.78	TCT	0.0	3.55	2.00
CGA	0.0	0.00	0.00	TGA	0.0	0.92	0.46
CGC	2.0	0.03	0.26	TGC	7.0	0.13	0.20
CGG	0.0	0.07	0.44	TGG	2.5	0.52	0.72
CGT	0.0	0.00	0.00	TGT	0.5	4.63	3.13
CTA	0.0	0.21	0.18	TTA	0.0	1.92	0.98
CTC	4.0	2.28	1.45	TTC	0.0	1.86	1.58
CTG	11.9	0.71	0.70	TTG	0.0	1.08	1.42
CTT	1.0	1.72	1.14	TTT	1.0	20.08	21.65

Simple sequence motifs were identified using the program SIMPLE34 (Hancock and Armstrong, 1994) by comparing the scores attained by individual 64 bp windows with the distribution of scores obtained when analysing random sequences. Simple sequence motifs were identified by scores with less than 10% likelihood of occurring by chance. Proportions are represented as percentages. The triplet motif identified as 'simple' is that at the centre of the window giving the significant high score. Thus, the motif is representative of the motifs identified within the window as being repeats. However, the repeated motif is not necessarily the triplet motif itself (it could, for example, be a more complex repeat based on, or incorporating, the triplet at the centre of the window) (see Section 8.2.1).

where %(E, I or EG) are the percentage contributions of a particular motif in exons, introns or extragenic regions. All three values of R will equal 0.333 if a given motif contributes an equal proportion of the simple sequence motifs in all three types of sequence, zero if it is absent from a region, and 1 if it is solely concentrated in one type of region.

The data generally support conclusions drawn from microsatellites alone (see Section 8.4). In non-coding regions of the human genome, $(A/T)_n$ was by far the predominant simple sequence class, contributing approximately 40% of simple sequence motifs in both introns and extragenic regions. Six other

Table 8.2 Bias of simple sequence motif frequencies between human exons, introns and extragenic sequences

Motif	R			Motif	R		
	Exons	Introns	Extra-genic		Exons	Introns	Extra-genic
AAA	0.040	0.418	0.542	GAA	0.000	0.290	**0.710**
AAC	0.000	0.440	0.560	GAC	–	–	–
AAG	0.117	0.260	0.622	GAG	0.460	0.226	0.314
AAT	0.000	0.515	0.485	GAT	0.000	**0.886**	0.114
ACA	0.000	0.530	0.470	GCA	**0.986**	0.014	0.000
ACC	*0.000*	*0.000*	***1.000***	GCC	**0.806**	0.039	0.155
ACG	–	–	–	GCG	0.000	0.084	**0.916**
ACT	*0.000*	***1.000***	*0.000*	GCT	**0.992**	0.004	0.003
AGA	0.083	0.390	0.527	GGA	0.647	0.095	0.259
AGC	**0.950**	0.000	0.050	GGC	0.390	0.176	0.434
AGG	0.000	0.349	0.651	GGG	0.661	0.117	0.223
AGT	0.000	0.244	**0.756**	GGT	0.000	0.117	**0.883**
ATA	0.000	**0.675**	0.325	GTA	0.000	0.461	0.539
ATC	0.000	**0.691**	0.309	GTC	0.000	**1.000**	0.000
ATG	0.000	0.551	0.449	GTG	0.318	0.418	0.264
ATT	0.000	0.553	0.447	GTT	0.352	0.241	0.407
CAA	0.377	0.296	0.327	TAA	0.000	0.593	0.407
CAC	0.000	0.526	0.474	TAC	0.000	**0.843**	0.157
CAG	**0.803**	0.050	0.147	TAG	0.000	**1.000**	0.000
CAT	0.000	**0.935**	0.065	TAT	0.100	0.578	0.322
CCA	0.603	0.170	0.227	TCA	*0.000*	***1.000***	0.000
CCC	**0.712**	0.102	0.187	TCC	0.530	0.249	0.221
CCG	**0.970**	0.030	0.000	TCG	–	–	–
CCT	**0.823**	0.076	0.101	TCT	0.000	0.639	0.361
CGA	–	–	–	TGA	0.000	**0.669**	0.331
CGC	**0.873**	0.012	0.115	TGC	**0.956**	0.017	0.027
CGG	0.000	0.137	**0.863**	TGG	**0.668**	0.139	0.193
CGT	–	–	–	TGT	0.060	0.561	0.379
CTA	0.000	0.540	0.460	TTA	0.000	0.662	0.338
CTC	0.516	0.296	0.188	TTC	0.000	0.541	0.459
CTG	**0.894**	0.053	0.052	TTG	0.000	0.432	0.568
CTT	0.258	0.446	0.296	TTT	0.023	0.470	0.507

For each motif, identified using SIMPLE34 (Hancock and Armstrong, 1994; see legend to *Table 8.1* for details of how simple sequences are identified), a value *R* was calculated. This represents the extent to which a particular motif is concentrated within a particular sequence class (see Section 8.5 for details). Motifs that are randomly distributed between the three sequence classes will have *R* values of 0.333 for all three. Values of *R* greater than 0.667 are arbitrarily regarded as significantly high and are printed in bold. Values for motifs that are rare (i.e. occur five times or less) are in italics.

motifs made up more than 2% of simple sequence motifs in both introns and extragenic regions: AGA, ATA, GTG, TAT, TCT and TGT. These group into three families of repeats, corresponding to $(AT)_n$, $(CA/GT)_n$ and $(AG/CT)_n$. The other members of these repeat families are also highly represented. Frequencies in coding regions were very different from those in non-coding regions, the predominant motifs being CCT and CTG (=CAG; see also Riggins *et al.*, 1992).

Of the 64 motifs, 28 showed high levels of bias (defined as an *R* score of greater than 0.667) towards particular sequence regions. Of these, three (ACT, TCA and ACC) are present at such low frequency (less than five examples detected) that their high scores cannot be regarded as significant. Of the other 25 motifs, 12 were biased towards exons, these being the six members of the $(CAG)_n$ group (CAG, AGC, GCA, CTG, GCT and TGC), CCG, CGC and GCC (but not their complements; see below), CCC, CCT and TGG. Motifs biased towards introns were ATC/CAT/GAT/TGA, TAC/TAG, ATA and GTC. Motifs biased towards extragenic regions were CGG/GCG, AGT, GAA and GGT. It is noteworthy that the complementary motif groups CCG/CGC/GCC and CGG/GCG/GGC showed very different biases in their locations, with the former being heavily biased towards exons while two of the latter group (CGG and GCG) were biased towards extragenic regions. Furthermore, it has been suggested (Stallings, 1994) that the trinucleotide GGC is excluded from introns. This also appears to be the case from the data in *Table 8.2* ($R < 0.333$), although the effect seems weaker than the effect on other GC-rich sequences.

Stallings (1994) also noted the preference of $(CAG)_n$ repeats to be located in coding regions. He suggested that this might reflect relative exclusion from introns, as $(CAG)_n$ arrays resemble the consensus splice acceptor site CAGG (Shapiro and Senapathy, 1987). Another pressure that might have led to their propagation in coding sequences is the fact that CAG can encode glutamine, and polyglutamine regions in proteins have been implicated in protein–protein interactions (Hancock, 1993; Mitchell and Tjian, 1989; and see Section 8.7). A minisatellite containing CAGG has recently been shown to have become incorporated into a human interferon-induced gene and to result in alternative splicing of the hnRNA (Turri *et al.*, 1995).

How significant is the difference between the simple sequence compositions of coding and non-coding regions, and what is its evolutionary background? A way to test the similarity between the compositions of different regions is to carry out correlation analysis of the frequencies of the different motifs found in simple sequences in the different sequence regions. If this is done, coding and non-coding regions are seen to have different simple sequence compositions (Hancock, 1995b). Introns and extragenic sequences, on the other hand, are highly similar in their simple sequence compositions. The simple sequence compositions of coding and non-coding sequences are also different in two other eukaryotes, the nematode *C. elegans* and brewers' yeast *S. cerevisiae*, although the distinction between the two classes of sequence is weaker in these species. A difference is also observed in two eubacteria, *E. coli* and the leprosy pathogen *Mycobacterium leprae*.

To gain an insight into the pressures that have acted on simple sequences in coding and non-coding regions of the genome, it is interesting to make interspecific comparisons of simple sequences from coding and non-coding sequences. The simple sequence compositions of non-coding sequences in the three eukaryotes mentioned above are very similar, but the simple sequence compositions of coding sequences show no correlation between species (Hancock, 1995b). From these results, it seems that the compositions of simple sequences in non-coding sequences may reflect some underlying processes common to all eukaryotic genomes. This might be bias in the slipped-strand mispairing that gives rise to simple sequences, or common patterns of selection on the sequences resulting from effects such as their DNA conformation or base composition. To resolve the relative contributions of mechanistic and selective factors, the observed frequencies of simple sequence motifs were subjected to analysis of variance to look at the contributions of motif length (e.g. mononucleotide or dinucleotide), motif composition (AT-richness and other base combinations), and the structural complexity of the motifs (e.g. whether they contain alternations of purines and pyrimidines; Hancock, 1995b).

These analyses showed that two factors made major contributions to the observed frequencies of simple sequence motifs: repeat length and AT/GC base composition. Other factors made little or no contribution. Both of these factors would be expected at first sight to affect the frequency of strand mispairing; repeat length is a contributory factor because mispairing by one base should be easier than mispairing by, say, five (see Levinson and Gutman, 1987a) and base composition is involved because AT-rich sequences should denature more easily, allowing mispairing more readily.

The proportion of the variation in simple sequence motif frequency accounted for by these factors showed considerable variability and some consistent patterns. Frequency patterns in non-coding sequences were much more readily explained on the basis of simple properties of the sequence motifs concerned: 66% of the variance can be explained on this basis in humans and similar, although smaller, proportions in the other eukaryotes. In coding regions, much lower proportions of simple sequence frequencies could be explained on the basis of the properties of sequence motifs, with humans showing the lowest proportion, 23%.

The proportion of the variation in simple sequence frequency that can be explained by properties of the sequences of motifs can, to a first approximation, be taken as a lowest estimate of the contribution of mechanistic biases to simple sequence composition. It therefore seems that such biases play a much bigger role in determining simple sequence composition in non-coding than in coding sequences. However, even in human non-coding sequences, as much as 34% of the variance may reflect the action of selection. As might be expected, this contribution seems to be much higher in coding sequences – up to 77%.

A consequence of the stronger action of selection on coding sequences than on non-coding sequences is a lower density of simple sequence motifs in coding than in non-coding sequences. This can be measured in terms of the number of SSMs (see Section 8.2.1) found per kb of sequence. In humans, coding sequences contain around 3.2 SSMs per kb (or one every 300 bp), while non-coding sequences contain around 12.2 SSMs per kb (one every 80 bp). Other species show a similar difference in density of SSMs: 1.3 versus 4.1 in *C. elegans* and 2.3 versus 11.0 in *S. cerevisiae*. However, SSM densities in coding and non-coding regions of eubacteria are much more similar.

8.6 Simple sequences and genome evolution

One of the unsolved mysteries of evolutionary biology is the wide fluctuation shown by the genome size (i.e. haploid DNA content) of organisms. This can be very different between members of even quite restricted taxonomic groups; for example, in Urodele amphibians, genome sizes vary by a factor of more than five from 15.1 to 82.5 pg (see Pagel and Johnstone, 1992). While some variation in genome sizes is clearly due to polyploidy, particularly in plants, the widely varying C values among eukaryotic species (from 0.009 pg in *S. cerevisiae* to 700 pg in the protist *Amoeba dubia*; Cavalier-Smith, 1985) cannot be readily explained.

Genome size correlates strongly with nuclear and cell volume, and it has been suggested that selection acts on genome size by acting on these properties (Cavalier-Smith, 1978). However, most of the extra DNA incorporated into genomes when their size increases is non-coding DNA; the coding genome of organisms varies in size much less than the total. The function of most non-coding DNA is obscure, although it may have structural functions or play a role in the regulation of transcription and/or replication. Evolutionary processes acting to increase or decrease genome size are therefore most likely to result in increases in what has been called 'junk' DNA (Ohno, 1972). It is therefore interesting to ask whether large genomes contain proportionately more simple sequences than small genomes.

This can be done by again looking at SSM densities and, as shown above, the concentration of simple sequence does indeed differ between species. Of the five species considered by Hancock (1995b), the human genome had the highest overall density of simple sequence motifs (11.7 SSMs per kb if sequences whose coding composition is unknown are ignored). Yeast and *C. elegans* showed intermediate densities (5.1 and 3.3 SSMs per kb, respectively), while the two eubacteria had the lowest (0.3 SSMs per kb for *M. leprae* and 0.1 SSMs per kb for *E. coli*). This was also reflected in the mean RSF values of genomic fragments, which were 1.39 ± 0.10 for humans, 1.28 ± 0.07 for *C. elegans*, 1.22 ± 0.05 for yeast, 1.14 ± 0.02 for *E. coli* and 1.10 ± 0.02 for *M. leprae*. All of these means are significantly greater than the expected score for random sequences (1.00). This suggests that slipped-strand mispairing has contributed to the evolution of even the smallest of these genomes, including *S. cerevisiae* which has a very small genome for a eukaryote, and *M. leprae* which has one of the smallest eubacterial genomes (see Herdman, 1985).

To test specifically for a correlation between the increase in genome size and the level of repetition in genomes, the mean RSFs of genomic sequences from the above five species were plotted against the logarithm of genome size. This produced the intriguing result that there is an almost exact linear relationship between the two (*Figure 8.2*). This suggests that, whatever the evolutionary forces may have been that have given rise to fluctuations in genome size, they have also resulted in the incorporation of increasing amounts of simple sequences as genome sizes have increased.

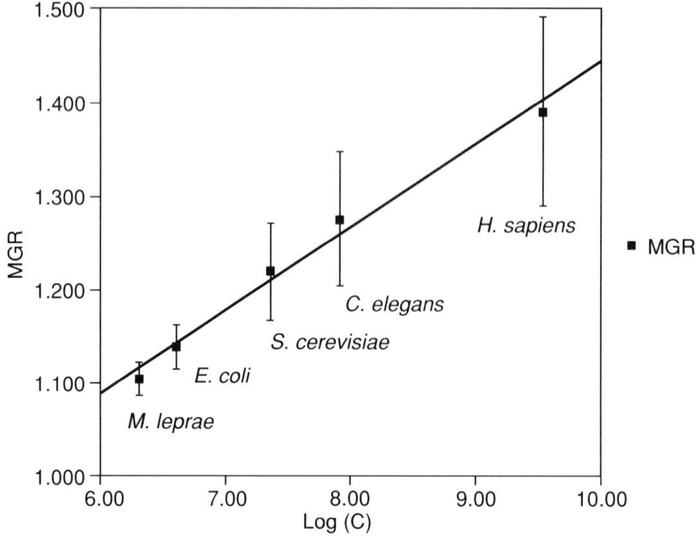

Figure 8.2. A scatter plot of the mean level of repetition (mean genomic relative simplicity factor, MGR) against the logarithm of haploid genome size (C) measured in base pairs (horizontal axis). The MGR was calculated by averaging RSF values (the measure of overall sequence repetitiveness generated by SIMPLE34; Hancock and Armstrong, 1994) for long (30 kb or more) genomic DNA sequences from five species represented in the sequence database by five or more such sequences. The species and numbers analysed were: *M. leprae* (19); *E. coli* (6); *S. cerevisiae* (7); *C. elegans* (67); and *H. sapiens* (27). Error bars show the standard deviations of the data for each species. Details of the sequences analysed can be found in Hancock (1996).

Although the non-coding fractions of genomes appear to be primarily responsible for their variation in size, some variation is also seen in genes. Early analyses of large subunit ribosomal RNAs (Hancock and Dover, 1988) showed that some organisms, primarily metazoans, had highly repetitive RNAs while others had non-repetitive ones. A similar pattern was seen in the TATA-binding proteins (TBP), which contain $(CAG)_n$ arrays in vertebrates but are less repetitive or not at all so in other species (Hancock, 1993). However, the best database of the phylogenetic distribution of sequence simplicity in a functional gene is the small subunit ribosomal RNA. Of these, about 3% show high levels of sequence repetition, while up to another 35% show some evidence of the action of slipped-strand mispairing during their recent evolutionary history (Hancock, 1995a). Again, vertebrates, and in particular mammals, show evidence of slipped-strand mispairing, as do a number of protist groups.

A simple test of whether the phylogenetic pattern of sequence simplicity seen in genomes as a whole is also reflected in their component genes is to plot RSF values of genomic sequences against those for small subunit ribosomal RNAs. This plot is again linear, suggesting that during genome size increase not only non-coding but also coding regions tend to incorporate simple sequences. This is supported by the observation of a correlation between simple sequence motif densities in coding and non-coding regions (Hancock, 1995b).

8.7 Consequences of simple sequence incorporation for genome function: molecular co-evolution

Little is known about the function of non-coding regions of the genome, so it is difficult to speculate on the possible consequences of incorporating novel simple sequences into them. A number of features of the sequences might make them subject to selective forces, including their potential to act as protein binding sites (proteins binding all of the possible short-motif microsatellites have been detected in *in vitro* experiments; Richards *et al.*, 1993), their ability to form specific DNA structures [such as the ability of $(CG)_n$ to form Z-DNA; Section 8.4], and their ability to nucleate nucleosome formation (a property of CAG repeats that may be involved in the aetiology of myotonic dystrophy; Wang *et al.*, 1994).

It is easier to consider the potential consequences of the incorporation of simple sequences into coding regions. Just as most amino acid substitutions are likely to be deleterious (Ohta and Kimura, 1971), so most slippage events within a protein-coding region are likely to disrupt protein structure and have a similar effect. The available evidence on the location of insertions and deletions within protein structures suggests that they tend to occur on protein surfaces, in regions with reverse turn and coil conformations (Pascarella and Argos, 1992). The naive explanation of this observation is that these locations represent positions in which insertions can take place without disrupting protein structure and therefore function.

It is, however, possible to imagine scenarios where proteins could accommodate slippage-derived variation, even in functionally important regions. To do this, we need to invoke the concept of molecular co-evolution (Dover and Flavell, 1984). In molecular co-evolution, changes in one part of an interacting system are compensated for by subsequent changes in another part during evolution. This allows function to be preserved in the absence of long-term sequence conservation. Molecular co-evolution is only possible if individual changes in the elements of the interacting system have little or no deleterious effects on the function of the system; that is, the system shows functional redundancy. A small number of mutant states can then be accommodated by the system at any time without disrupting its function. Such processes can either take place in multigene families, where some unhomogenized, suboptimal variants can coexist with functional sequences (Dover and Flavell, 1984), or in single, interacting proteins if the interactions involve a number of redundant sites (Hancock, 1993).

The most likely examples of molecular co-evolution involve co-evolution between interacting regions on the surfaces of proteins. An example of this appears to be TBP, which has incorporated a polyglutamine region of increasing length during vertebrate evolution. It has been suggested that this reflects the acquisition of new functions in the transcription regulatory machinery over time (Hancock, 1993). This proposal is supported by analyses of interactions of the transcription factor Sp1 and the *Drosophila* developmental regulatory protein antennapedia with TBP. These have shown that the strength of their interactions is directly related to the sizes of glutamine patches on the surfaces of these proteins (Emili *et al.*, 1994). Similarly, the size of the glutamine patch on the surface of the androgen receptor is directly related to its ability to transactivate expression of the gene for growth hormone (Kazemi-Esfarjani *et al.*, 1995)

Other scenarios involving molecular co-evolution can also be imagined. For example, a situation might arise in which the only function of a repetitive region was to maintain the relative spacing of two other regions of a protein. In this case, variations within the repetitive region would be tolerated as long as the essential spacing was not disrupted, and over time the region might change in length as a result of co-evolutionary changes. Such a situation appears to have arisen in the *Drosophila* neurogenic gene *mastermind*, in which simple sequence regions maintain spacing between charged clusters on the protein's surface (Newfeld *et al.*, 1994).

Another possible co-evolutionary scenario would involve a repetitive region that made up part of a protein's core but which interacted with another repetitive region to form an element of structure. As long as variation within these regions was not so great that it prevented proper folding of the protein, a little variation of this kind, and consequent co-evolutionary change, might be possible. No example of this latter kind of evolution has been described in a protein, but a similar process has been described in large subunit ribosomal RNAs, where complementary strands of some secondary structural stems are suggested to have enlarged by 'compensatory slippage' (Hancock and Dover, 1990).

An interesting example of a region within a protein-coding gene that appears to have undergone expansion by slipped-strand mispairing and which has clear phenotypic effects is found in the *period* gene of Drosophila. *Period* is a clock gene involved in the determination of a number of features of *Drosophila* behaviour and development, including the interval from egg laying to hatching, the length of the lovesong cycle, and circadian rhythms (Kyriacou and Hall, 1994). The gene contains a repetitive region that encodes repeats of the amino acid doublet Thr-Gly. The exact structural and functional roles of this repeated region are not known but it has been shown that deleting it affects both the species specificity of the male lovesong, by reducing its length, and results in temperature sensitivity of circadian rhythms (Kyriacou and Hall, 1986). This region has evolved rapidly within the genus *Drosophila*, mostly by changes in length, but intriguingly some conserved regions immediately flanking the repeat appear to have co-evolved with the repeat unit

(Peixoto *et al.*, 1992). This may suggest some kind of structural interaction between the two regions.

The involvement of developmental proteins in such apparently co-evolutionary processes (Dover, 1992) is of particular interest because a high proportion of such genes contain significantly repetitive regions (Hancock, unpublished). Many of these are either transcription factors or cell surface molecules; both of these classes of molecule are ideally suited to co-evolutionary change of the kind outlined above. Processes of this kind may therefore have played a significant role in the evolution of novelty in developmental systems. An indication of this is given by the *Drosophila* developmental gene *hunchback*, which has evolved rapidly by slipped-strand mispairing within simple sequence regions and shows detectably different patterns of secondary expression between different *Drosophila* species (Treier *et al.*, 1989). Another example of co-evolution by transcription factors is found in the ribosomal DNA transcription machinery, in which co-evolutionary changes in TBP-associated factors result in interspecific incompatibility (Rudloff *et al.*, 1994).

The clearest examples of molecular co-evolution outlined above are taken from non-human species, particularly *Drosophila*. However, as we have seen, the human genome is rich in simple sequences, suggesting that similar mechanisms may have come into play during human evolution. There is an indication of this in the long polyglutamine tracts found in the human TBP protein. Other possible examples of genes undergoing molecular co-evolution are the genes involved in trinucleotide expansion disorders.

8.8 Simple sequence evolution and triplet expansion disorders

Simple sequence regions have recently been implicated in the aetiology of a novel class of genetic diseases, the triplet expansion disorders (reviewed in Bates and Lehrach, 1994; see Chapter 1). These are disorders, generally of the nervous system, which appear to result from catastrophic expansion of simple sequence arrays, predominantly of $(CAG)_n$. Of eight diseases of this kind identified to date, five (dentatorubral pallidoluysian atrophy, spinal and bulbar muscular atrophy, Huntington's disease, Machado-Joseph disease, and spinocerebellar ataxia) appear to result from expansion of repeats within coding regions, two (fragile X syndrome and FraxE mental retardation) from expansion of $(CGG/CCG)_n$ repeats in 5′ UTRs of genes and one (myotonic dystrophy) from expansion of a $(CTG)_n$ repeat in a 3′ UTR.

These diseases are evolutionarily interesting for a number of reasons. First, the expansion that takes place to cause the diseases is dependent on the array reaching a certain threshold size, which varies depending on the location of the array. A study of the Huntington's disease array in primates showed that humans have a much larger array than other primates (Rubinsztein *et al.*, 1994) and the puffer fish homologue of this gene is also known to have a very short repetitive region (Baxendale *et al.*, 1995). Similar patterns of evolution have been observed in other genes causing trinucleotide expansion disorders (Djian *et al.*, 1996; Eichler *et al.*, 1995; Rubinsztein *et al.*, 1995; Hancock and Chana, unpublished). This suggests that humans may be uniquely susceptible to diseases of this kind. As the array in the androgen receptor gene responsible for spinal and bulbar muscular atrophy is not genetically unstable in transgenic mice (Bingham *et al.*, 1995), instability of this kind may depend on proofreading deficiencies in the human replication machinery. These deficiencies may also be responsible for the high levels of polymorphism in human microsatellites and the apparent rapid accumulation of simple sequences in human coding regions.

A second interesting issue is the role of interrupting point mutations in the evolution of these arrays. Interrupted microsatellites generally show a lower level of polymorphism than uninterrupted ones (Weber, 1990), although mixed arrays of alternating purines and pyrimidines can also show length polymorphism (Sarkar *et al.*, 1991). It has been suggested that one of the critical steps in the development of triplet expansion disorders is the elimination of interrupting point mutations from arrays (Chung *et al.*, 1993; Kunst and Warren, 1994; Richards and Sutherland, 1993), perhaps by gene conversion or unequal crossing-over. It will be interesting to see whether array interruption is a feature of these arrays in other species.

Finally, as many of these diseases involve expansion of polyglutamine-encoding regions of proteins, their enlargement in humans may have been driven by co-evolutionary processes similar to those that appear to have taken place in other $(CAG)_n$-containing coding sequences (Section 8.6). It is intriguing that these genes are associated with the nervous system and that the evolutionary expansion of $(CAG)_n$ arrays within these genes parallels the rapid increase in brain complexity that took place during primate evolution leading to humans. While this association with the nervous system may reflect a common mechanism of pathogenesis rather than a common feature of gene function, it raises the possibility that triplet expansion disorders may be both a side-effect of the human genetic background that has led to the increase in size of simple sequence arrays in humans, and the price we pay for the level of complexity we have evolved. Indeed, slipped-strand mispairing may have played an important part in making humans what they are today.

References

Bates G, Lehrach H. (1994) Trinucleotide repeat expansions and human genetic disease. *BioEssays* **16**: 277–284.

Baxendale S, Abdulla S, Elgar G *et al.* (1995) Comparative sequence analysis of the human and pufferfish Huntington's disease genes. *Nature Genetics* **10**: 67–76.

Beckmann JS, Weber JL. (1992) Survey of human and rat microsatellites. *Genomics* **12**: 627–631.

Behn-Krappa A, Doerfler W. (1994) Enzymatic amplification of synthetic oligodeoxyribonucleotides – implications for triplet repeat expansions in the human genome. *Hum. Mutat.* **3**: 19–24.

Bingham PM, Scott MO, Wing S, McPhaul MJ, Wilson EM, Gerbern JY, Merry DE, Fischbeck KH. (1995) Stability of an expanded trinucleotide repeat in the androgen receptor gene in transgenic mice. *Nature Genetics* **9**: 191–196.

Calvas P, Blancher A, Salvignol I, Socha WW, Ruffie J. (1994) Length polymorphism of a microsatellite in human and non-human primates. *C.R. Acad. Sci. Ser. III* **317**: 755–763.

Cavalier-Smith T. (1978) Nuclear volume control by nucleoskeletal DNA, selection for cell volume and cell growth rate, and the solution of the DNA C-value paradox. *J. Cell. Sci.* **34**: 247–278.

Cavalier-Smith T. (1985) Eukaryotic gene numbers, non-coding DNA, and the solution of the C-value paradox. In: *The Evolution of Genome Size* (ed. T Cavalier-Smith). John Wiley & Sons, New York, pp. 69–103.

Chung M-y, Ranum LPW, Duvick LA, Servadio A, Zoghbi HY, Orr HT. (1993) Evidence for a mechanism predisposing to intergenerational CAG repeat instability in spinocerebellar ataxia type I. *Nature Genetics* **5**: 254–258.

Clift B, Haussler D, McConnell R, Schneider TD, Stormo GD. (1986) Sequence landscapes. *Nucleic Acids Res.* **14**: 141–158.

Cooper DN, Taggart MH, Bird AP. (1983) Unmethylated domains in vertebrate DNA. *Nucleic Acids Res.* **11**: 647–658.

Deka R, Shriver MD, Yu LM, Jin L, Aston CE, Chakraborty R, Ferrell RE. (1994) Conservation of human-chromosome-13 polymorphic microsatellite $(CA)_n$ repeats in chimpanzees. *Genomics* **22**: 226–230.

Di Rienzo A, Peterson AC, Garza JC, Valdes AM, Slatkin M, Freimer NB. (1994) Mutational processes of simple-sequence repeat loci in human populations. *Proc. Natl Acad. Sci. USA* **91**: 3166–3170.

Djian P, Hancock JM, Chana HS. (1996) Codon repeats in genes associated with human diseases: fewer repeats in the genes of nonhuman primates and nucleotide substitutions concentrated at the sites of reiteration. *Proc. Natl Acad. Sci. USA* **93**: 417–421.

Dover GA. (1992) Observing development through evolutionary eyes: a practical approach. *BioEssays* **14**: 281–287.

Dover GA, Flavell RB. (1984) Molecular co-evolution: DNA divergence and the maintenance of function. *Cell* **38**: 622–623.

Dover GA, Tautz D. (1986) Conservation and divergence in multigene families: alternatives to selection and drift. *Phil. Trans. R. Soc. London Ser. B* **312**: 275–289.

Dover GA, Ruiz Linares A, Bowen T, Hancock JM. (1993) The detection and quantification of concerted evolution and molecular drive. *Meth. Enzymol.* **224**: 525–541.

Edwards A, Hammond HA, Jin L, Caskey CT, Chakraborty R. (1992) Genetic variation at five trimeric and tetrameric tandem repeat loci in four human population groups. *Genomics* **12**: 241–253.

Eichler EE, Kunst CB, Lugenbeel KA, Ryder OA, Davison D, Warren ST, Nelson DL. (1995) Evolution of the cryptic FMR1 CGG repeat. *Nature Genetics* **11**: 301–307.

Emili A, Greenblatt J, Ingles CJ. (1994) Species-specific interaction of the glutamine-rich activation domains of Sp1 with the TATA box-binding protein. *Mol. Cell. Biol.* **14**: 1582–1593.

Gostout B, Liu Q, Sommer SS. (1993) 'Cryptic' repeating triplets of purines and pyrimidines (cRRY(i)) are frequent and polymorphic: analysis of coding cRRY(i) in the proopiomelanocortin (POMC) and TATA-binding protein (TBP) genes. *Am. J. Hum. Genet.* **52**: 1182–1190.

Hancock JM. (1993) Evolution of sequence repetition and gene duplications in the TATA-binding protein TBP (TFIID). *Nucleic Acids Res.* **21**: 2823–2830.

Hancock JM. (1995a) The contribution of DNA slippage to eukaryotic nuclear 18S rRNA evolution. *J. Mol. Evol.* **40**: 629–639.

Hancock JM. (1995b) The contribution of slippage-like processes to genome evolution. *J. Mol. Evol.* **41**: 1038–1047.

Hancock JM, Armstrong JS. (1994) SIMPLE34: an improved and enhanced implementation for VAX and Sun computers of the SIMPLE algorithm for analysis of clustered repetitive motifs in nucleotide sequences. *Comput. Appl. Biosci.* **10**: 67–70.

Hancock JM, Dover GA. (1988) Molecular co-evolution among cryptically simple expansion segments of eukaryotic 26S/28S rRNAs. *Mol. Biol. Evol.* **5**: 377–391.

Hancock JM, Dover GA. (1990) 'Compensatory slippage' in the evolution of ribosomal RNA genes. *Nucleic Acids Res.* **18**: 5949–5954.

Herdman M. (1985) The evolution of bacterial genomes. In: *The Evolution of Genome Size* (ed. T Cavalier-Smith). John Wiley & Sons, New York, pp. 37–68.

Jeffreys AJ, Tamaki K, MacLeod A, Monckton DG, Neil DL, Armour JAL. (1994) Complex gene conversion events in germline mutation at human microsatellites. *Nature Genetics* **6**: 136–145.

Karlin S, Burge C. (1995) Dinucleotide relative abundance extremes: a genomic signature. *Trends Genet.* **11**: 283–290.

Kazemi-Esfarjani P, Trifiro MA, Pinsky L. (1995) Evidence for a repressive function of the long polyglutamine tract in the human androgen receptor: possible pathogenetic relevance for the $(CAG)_n$-expanded neuronopathies. *Hum. Mol. Genet.* **4**: 523–527.

Keese P, Mackenzie A, Gibbs A. (1989) Nucleotide sequence of the genome of an Australian isolate of turnip yellow mosaic virus. *Virology* **172**: 536–546.

Konopka AK, Owens J. (1990) Complexity charts can be used to map functional domains in DNA. *Gene Anal. Technol. Appl.* **7**: 35–38.

Kornberg A, Bertsch LL, Jackson JF, Khorana HG. (1964) Enzymatic synthesis of deoxyribonucleic acid. XVI. Oligonucleotides as templates and the mechanisms of their replication. *Proc. Natl Acad. Sci. USA* **51**: 315–323.

Kunst CB, Warren ST. (1994) Cryptic and polar variation of the fragile X repeat could result in predisposing normal alleles. *Cell* **77**: 853–861.

Kyriacou CP, Hall JC. (1986) Interspecific genetic control of courtship song production and reception in *Drosophila*. *Science* **232**: 494–497.

Kyriacou CP, Hall JC. (1994) Genetic and molecular analysis of *Drosophila* behaviour. *Adv. Genet.* **31**: 139–186.

Levinson G, Gutman GA. (1987a) Slipped-strand mispairing: a major mechanism for DNA sequence evolution. *Mol. Biol. Evol.* **4**: 203–221.

Levinson G, Gutman GA. (1987b) High frequencies of short frameshifts in poly-CA/TG tandem repeats borne by bacteriophage M13 in *Escherichia coli* K-12. *Nucleic Acids Res.* **15**: 5223–5338.

Litt M, Luty JA. (1989) A hypervariable microsatellite revealed by *in vitro* amplification of a dinucleotide repeat within the cardiac muscle actin gene. *Am. J. Hum. Genet.* **44**: 397–401.

Miloslavljevich A, Jurka J. (1993) Discovering simple DNA sequences by the algorithmic significance method. *Comput. Appl. Biosci.* **9**: 407–411.

Mitchell PJ, Tjian R. (1989) Transcriptional regulation in mammalian cells by sequence-specific DNA binding proteins. *Science* **245**: 371–378.

Morgan AR, Coulter MB, Flintoff WF, Paetkau VH. (1974) Enzymatic synthesis of deoxyribonucleic acids with repeating sequences. A new repeating trinucleotide deoxyribonucleic acid, d(T-C-C)*n*-d(G-G-A)*n*. *Biochemistry* **13**: 1596–1603.

Murphy GL, Connell TD, Barritt DS, Koomey M, Cannon JG. (1989) Phase variation of gonococcal protein II: regulation of gene expression by slipped-strand mispairing of a repetitive DNA sequence. *Cell* **56**: 539–547.

Nei M. (1987) *Molecular Evolutionary Genetics*. Columbia University Press, New York, p.10.

Newfeld SJ, Tachida H, Yedvobnick B. (1994) Drive-selection equilibrium: homopolymer evolution in the *Drosophila* gene mastermind. *J. Mol. Evol.* **38**: 637–641.

Nussinov R. (1984) Strong doublet preferences in nucleotide sequences and DNA geometry. *J. Mol. Evol.* **20**: 111–119.

Ohno S. (1972) So much 'junk' in our genome. *Brookhaven Symp. Biol.* **23**: 366–370.

Ohta T, Kimura M. (1971) On the constancy of the evolutionary rate of cistrons. *J. Mol. Evol.* **1**: 18–25.

Ohta T, Kimura M. (1973) The model of mutation appropriate to estimate the number of electrophoretically detectable alleles in a genetic population. *Genet. Res.* **22**: 201–204.

Pagel M, Johnstone RA. (1992) Variation across species in the size of the nuclear genome supports the junk-DNA explanation for the C-value paradox. *Proc. R. Soc. London. Ser. B* **249**: 119–124.

Pascarella S, Argos P. (1992) Analysis of insertions/deletions in protein structures. *J. Mol. Biol.* **224**: 461–471.

Peixoto AA, Costa R, Wheeler JC, Hall JC, Kyriacou CP. (1992) Evolution of the threonine-glycine repeat region of the period gene in the *melanogaster* species subgroup of *Drosophila. J. Mol. Evol.* **35**: 411–419.

Polymeropoulos MH, Rath DS, Xiao H, Merril CR. (1991) Trinucleotide repeat polymorphism at the human transcription factor IID gene. *Nucleic Acids Res.* **19**: 4307.

Richards RI, Sutherland GR. (1993) Simple repeat DNA is not replicated simply. *Nature Genetics* **6**: 114–116.

Richards RI, Holman K, Yu S, Sutherland GR. (1993) Fragile X syndrome unstable element, p$(CCG)_n$, and other simple tandem repeat sequences are binding sites for specific nuclear proteins. *Hum. Mol. Genet.* **2**:1429–1435.

Ricke DO, Liu Q, Gostout B, Sommer SS. (1995) Nonrandom patterns of simple and cryptic triplet repeats in coding and noncoding sequences. *Genomics* **26**: 510–520.

Riggins GJ, Lokey LK, Chastain JL, Leiner HA, Sherman SL, Wilkinson KD, Warren ST. (1992) Human genes containing polymorphic trinucleotide repeats. *Nature Genetics* **2**: 186–191.

Rubinsztein DC, Amos W, Leggo J *et al.* (1994) Mutational bias provides a model for the evolution of Huntington's disease and predicts a general increase in disease prevalence. *Nature Genetics* 7: 525–531.

Rubinsztein DC, Leggo J, Coetzee GA, Irvine RA, Buckley M, Ferguson-Smith MA. (1995) Sequence variation and size ranges of CAG repeats in the Machado-Joseph disease, spinocerebellar ataxia type 1 and androgen receptor genes. *Hum. Mol. Genet.* **4**: 1585–1590.

Rudloff U, Eberhard D, Tora L, Stunnenberg H, Grummt I. (1994) TBP-associated factors interact with DNA and govern species specificity of RNA polymerase I transcription. *EMBO J.* **13**: 2611–2616.

Sarkar G, Paynton C, Sommer SS. (1991) Segments containing alternating purine and pyrimidine dinucleotides: patterns of polymorphism in humans and prevalence throughout phylogeny. *Nucleic Acids Res.* **19**: 631–636.

Schlötterer C, Tautz D. (1992) Slippage synthesis of simple sequence DNA. *Nucleic Acids Res.* **20**: 211–215.

Shapiro MB, Senapathy P. (1987) RNA splice junctions of different classes of eukaryotes: sequence statistics and functional implications in gene expression. *Nucleic Acids Res.* **17**: 7155–7174.

Shriver MD, Jin L, Chakraborty R, Boerwinkle E. (1993) VNTR allele frequency distributions under the stepwise mutation model: a computer simulation approach. *Genetics* **134**: 983–993.

Stallings RL. (1992) CpG suppression in vertebrate genomes does not account for the rarity of $(CpG)_n$ microsatellite repeats. *Genomics* **17**: 890–891.

Stallings RL. (1994) Distribution of trinucleotide microsatellites in different categories of mammalian genomic sequence: implications for human genetic diseases. *Genomics* **21**: 116–121.

Stallings RL, Ford AF, Nelson D, Torney DC, Hildebrand CE, Moyzis RK. (1991) Evolution and distribution of $(GT)_n$ repetitive sequences in mammalian genomes. *Genomics* **10**: 807–815.

Stephan W, Cho S. (1994) Possible role of natural selection in the formation of tandem-repetitive noncoding DNA. *Genetics* **136**: 333–341.

Streisinger G, Owen J. (1985) Mechanisms of spontaneous and induced frameshift mutation in bacteriophage T4. *Genetics* **109**: 633–659.

Tamura T, Sumita K, Fujino I, Aoyama A, Horikoshi M, Hoffmann A, Roeder RG, Muramatsu M, Mikoshiba K. (1991) Striking homology of the 'variable' N-terminal as well as the 'conserved core' domains of the mouse and human TATA-factors (TFIID). *Nucleic Acids Res.* **19**: 3861–3865.

Tautz D. (1989) Hypervariability of simple sequences as a general source for polymorphic DNA markers. *Nucleic Acids Res.* **17**: 6463–6471.

Tautz D, Trick M, Dover GA. (1986) Cryptic simplicity in DNA is a major source of genetic variation. *Nature* **322**: 652–656

Treier M, Pfeifle C, Tautz D. (1989) Comparison of the gap segmentation gene *hunchback* between *Drosophila melanogaster* and *Drosophila virilis* reveals novel modes of evolutionary change. *EMBO J.* **8**: 1517–1525.

Turri M-G, Cuin KA, Porter ACG. (1995) Characterisation of a novel minisatellite that provides multiple splice donor sites in an interferon-induced transcript. *Nucleic Acids Res.* **23**: 1854–1861.

Valdes AM, Slatkin M, Freimer NB. (1993) Allele frequencies at microsatellite loci: the stepwise mutation model revisited. *Genetics* **133**: 737–749.

Wang Y-H, Amirhaeri S, Kang S, Wells RD, Griffith JD. (1994) Preferential nucleosome assembly at DNA triplet repeats from the myotonic dystrophy gene. *Science* **265**: 669–671.

Weber JL. (1990) Informativeness of human $(dC\text{-}dA)_n.(dG\text{-}dT)_n$ polymorphisms. *Genomics* **7**: 524–530.

Weber JL, May PE. (1989) Abundant class of human DNA polymorphisms which can be typed using the polymerase chain reaction. *Am. J. Hum. Genet.* **44**: 388–396.

Weber JL, Wong C. (1993) Mutation of human short tandem repeats. *Hum. Mol. Genet.* **2**: 1123–1128.

Wehrhahn C. (1975) The evolution of selectively similar electrophoretically detectable alleles in finite natural populations. *Genetics* **80**: 375–394.

Wells RD, Buchi H, Kossel E, Ohtsuka E, Khorana HG. (1967a) Studies on polynucleotides. LXX. Synthetic deoxyribopolynucleotides as templates for the DNA polymerase of *Escherichia coli*: DNA-like polymers containing repeating tetranucleotide sequences. *J. Mol. Biol.* **27**: 265–272.

Wells RD, Jacobs TM, Narang SA, Khorana HG. (1967b) Studies on polynucleotides. LXIX. Synthetic deoxyribopolynucleotides as templates for the DNA polymerase of *Escherichia coli:* DNA-like polymers containing repeating trinucleotide sequences. *J. Mol. Biol.* **27**: 237–263.

9

Evolution of *Alu* retroposons

Roy J. Britten

9.1 Introduction

9.1.1 Alu *sequences: structure and properties*

All vertebrate genomes contain a significant proportion of interspersed repetitive DNA which is, or has been, mobile. In humans, the most abundant repeat of this kind is the *Alu* family of short interspersed repetitive elements (SINEs). Estimates of the abundance of this family range from 500 000 to 1 000 000 elements per human haploid genome. Sequence data from humans and other species further suggest that *Alu* elements are specific to primates, having been derived from a sequence related to the 7SL RNA component of the mammalian signal recognition particle (Ullu and Tschudi, 1984) which subsequently spread through the genome over the past 50–65 million years. The presence of a *Pol*III promoter defining the 5′ end of the *Alu* element, a poly(A) tail, short direct repeats flanking intact elements, and empirical evidence of new insertion events all suggest that *Alu* is a mobile element which moves via an RNA intermediate – a process termed retroposition. While sophisticated models for this process exist (e.g. Schmid and Maraia, 1992), the precise mechanism of retroposition is unknown.

An undamaged recently inserted *Alu* sequence is normally 281 nucleotides (nt) long followed by a tail of 20 or 30 adenosine (A) bases. Structural repetition within the element implies that it consists of two related units, the current organization being the result of an ancient duplication: thus, two poly(A) stretches are present within the element. The GC content is high and includes 23 or more CpG dinucleotides. Such unchanged recent inserts are quite rare: perhaps only one is found per 500 or more typical *Alu* inserts. A typical human *Alu* insert (as distinct from a recent insert) will have been present in the human lineage for anything up to 60 million years. Base substitutions will

Human Genome Evolution, edited by M. Jackson, T. Strachan and G. Dover.

have occurred and many of the CpG dinucleotides will have been substituted. It may not be full-length due to 3′ or 5′ region deletions and often can be quite short. It might be half-length, consisting of either a left or right monomer (Quentin, 1992). The typical *Alu* insert is often subject to insertion of foreign sequences or deletions. These do not occur at random within the *Alu* sequence but there is no model of the underlying mechanism(s). (For references and reviews, see Britten, 1994a; Britten *et al.*, 1988; Deininger and Batzer, 1993; Jurka and Milosavljevic, 1991; Jurka and Smith, 1988; Okada, 1991; Quentin, 1994; Schmid and Maraia, 1992.)

At first glance, the pattern of *Alu* insertion is more or less random but there is evidence of clustering that is not expected statistically. The average spacing is about 4 kb but they sometimes occur less than 1 kb apart (see Riccio and Rossolini, 1993; Stoppa-Lyonnet *et al.*, 1990). Furthermore, *Alu* inserts are often densely clustered around certain genes but rarely occur within coding regions (see Riccio and Rossolini, 1993). In all of the cases where *Alu* sequences that occur in coding sequences have been examined, they cause deleterious mutations including, for example, aberrant splicing leading to inclusion of intronic sequences in a transcript (Makalowski *et al.*, 1994). At a chromosomal level, there is also convincing evidence that *Alu* elements are distributed non-randomly (Chen and Manuelidis, 1989; Soriano *et al.* 1983), in particular they are more prevalent in Giemsa-light (R) human metaphase chromosome bands (Korenberg and Rykowski, 1988).

The functional importance of *Alu* elements has been a subject of debate since their discovery and is beyond the scope of this review. However, new evidence that has appeared in the past year is worth mentioning. There are now six known cases in which an *Alu* sequence has inserted into the control region of a gene, remained there, and been modified over many millions of years. In each case, specific short sequences within the *Alu* element bind regulatory proteins and assist in the regulation of transcription (Britten, 1996). There is suggestive evidence that many more such cases may exist. For example, large numbers of *Alu* sequences retain retinoic acid receptor binding sites (Vansant and Reynolds, 1995) or oestrogen receptor binding sites (Norris *et al.*, 1995). Thus, the view that *Alu* sequences are purely parasitic or selfish is no longer tenable. The vast majority of the million or so *Alu* sequences may be without function but their insertion and presence as sources of mutation and rearrangement is important to the human genome (Makalowski *et al.*, 1994; Maraia, 1995).

9.1.2 **Alu** *sequences: basic patterns of evolution*

Initially, it was presumed that most full-length *Alu* elements possessed the potential for retroposition. The dynamics of *Alu* evolution could therefore be analysed using models developed for other eukaryotic transposable elements (Charlesworth and Langley, 1989). For instance, it was envisaged that variation between *Alu* elements in their ability to retropose (due to factors such as genomic location and sequence variation between elements at the *Pol*III promoter site), coupled to selection against the deleterious effects of insertion, could establish a dynamic equilibrium. However, sequence analyses of *Alu* elements, together with the identification of *Alu* inserts where a recent insertion date can be proved or inferred (Section 9.2.1), have rendered this view of *Alu* evolution untenable. First, it is clear that elements which are retropositionally active (i.e. are a source of copies) at any one time represent a very small, specific subset of all *Alu* elements. Thus, elements which are present in the human genome, but absent from other primates, share diagnostic DNA sequence changes not present in any other *Alu* elements. This is wholly consistent with another important observation: *Alu* elements can be divided into families (Section 9.3.1) where diagnostic sequence changes enable one family to be distinguished from another in a sequential manner (Sections 9.3.2 and 9.3.3). These observations have led to the realization that most *Alu* elements are retropositionally inactive, with only one or a small group of 'source' *Alu* inserts being responsible for new insertional events. These are also referred to as master genes (e.g. Deininger *et al.*, 1992). Thus, the *Alu* sequences present in any one *Alu* family owe their sequence similarity to the fact that they are derived from specific source genes retropositionally active during defined periods of human evolution.

Given this pattern of evolution, it is clear that the DNA sequence of present-day *Alu* elements is established by two principal processes: evolution of the source genes that give rise to the inserts, and subsequent mutation of the inserts *in situ*. An important step in the analysis of *Alu* evolution is, therefore, the ability to discriminate between sequence changes which have occurred in the source genes prior to insertion and those changes which have occurred after the insertion of a new, retropositionally inactive, *Alu* element. Thus, a suitable starting point for a discussion of *Alu* evolution is not an ancient progenitor *Alu* sequence but the consensus sequence of modern *Alu* elements, which provides the best means of discriminating between base changes which have occurred before and after insertion.

9.2 Modern *Alu* consensus and *Alu* sequence divergence data

9.2.1 The modern consensus

A number of *Alu* sequences have been identified as recent inserts due to the fact that they are polymorphic in the human population or due to interspecific differences among the apes. In addition, short oligonucleotide probes, incorporating base changes specific to new inserts, have been used to identify other human-specific *Alu* elements which must have inserted since humans diverged from other hominoids (Bailey and Shen, 1993; Batzer and Deininger, 1991; Batzer *et al.*, 1990, 1994, 1995; Hammer, 1994; Leeflang *et al.*, 1992; Matera

```
1     GGCcgGGcgc gGTGGCTCAc gCCTGTAATC CCAGCACTTT GGGAGGCcgA  50
51    GGcgGGcgGA TCacgAGgtC AGGAGaTcgA GACCAtCCcg gctAAaAcgG  100
101   tGAAACCCcg TCTCTACtAA AAATAcaAAA AATTAGCcgG GcgTaGTGGc  150
151   gggcgCCTGT AgTCCCAGCT ACTtgGGAGG CTGAGGcagG AGaATggcgT  200
201   GAaCCcgGGA GGcgGAGctT GCaGTGAGCc gaGATcccgC CAcTGCACTC  250
251   CAGCCTgGGc gACAGAGcgA gACtccgTCT c  281
```

Figure 9.1. The modern *Alu* consensus sequence. The nucleotides shown in bold uppercase are the positions that are conserved before insertion (CONSBI), having less than 14% average divergence from this consensus sequence. The nucleotides in lower case are the known diagnostic positions, as well as a few positions showing greater than 14% divergence but not otherwise recognized. The CpG dinucleotides are in lower case and underlined. The sequence is the consensus of a group of recent *Alu* inserts and represent a very small proportion of all *Alu* sequences (for references, see text and Britten, 1994a).

et al., 1990a, b; Ziętkiewicz *et al.*, 1994). These modern *Alu* inserts have been aligned and the consensus sequence is shown in *Figure 9.1*. Because these modern sequences have recently been derived from source genes, the consensus sequence can be considered to be the product of more than 50 million years of *Alu* source gene evolution. Therefore, by comparing the DNA sequence of other *Alu* sequences (which are not active) with this consensus it is possible to gain insight into *Alu* sequence evolution. These comparisons are described below.

9.2.2 *Divergence of* Alu *sequences from the modern consensus*

Figure 9.2 shows the results for a set of 1579 almost full-length *Alu* sequences compared with the modern consensus, with the divergence from the modern consensus (in 2% intervals) as the horizontal axis and the number of base positions showing that percentage divergence as the vertical axis. The largest number of positions (mode) differ from the modern consensus in only 8% of all 1579 *Alu* inserts studied. The interpretation of this result, central to the analysis of *Alu* sequences which follows, is that the nucleotides at these positions have not changed in the source genes but that the 8% divergence represents mutations which have occurred in *Alu* elements after insertion. These positions, which have not changed in source genes, are defined as being conserved before insertion (CONSBI). A total of 195 of these can be identified within the 281 bp element and they play a special role in the analysis of *Alu* sequence evolution (Britten, 1994a).

The lower curve in *Figure 9.2* is for the same data with the CpG dinucleotides removed. Looking at the difference between the upper and lower curves, the cytosine (C) and guanine (G) bases of the CpGs show a mode divergence of about 36%. That means that, on average, about 72% of the CpGs are mutated in this population of *Alu* sequences, which is consistent

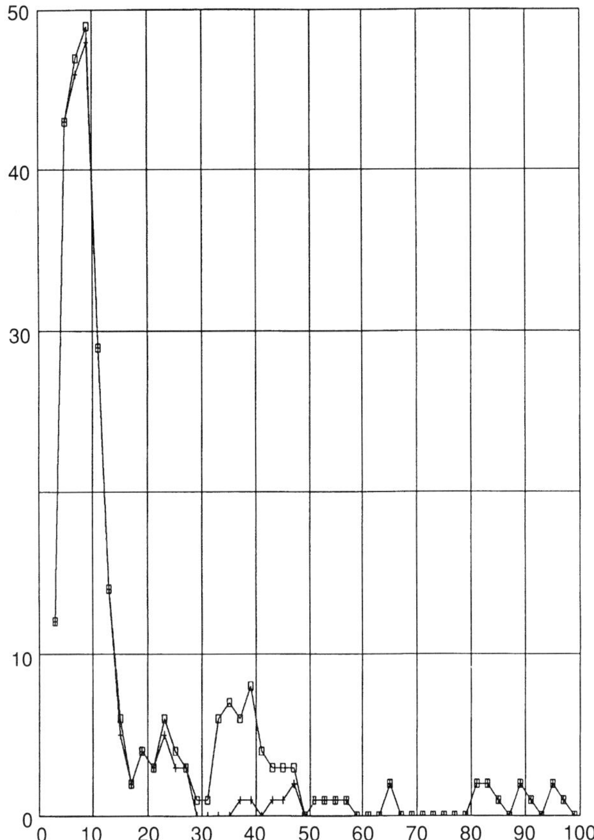

Figure 9.2. Divergence of 1579 *Alu* sequences from the modern consensus and the distribution of average divergence for each position. The horizontal axis shows divergence from the modern consensus (the sequence of which is given in *Figure 9.1*) divided into 2% intervals. The vertical axis shows the number of base positions with the appropriate average percentage divergence. The upper curve is for all 281 positions of the consensus sequence, while the lower curve does not include the CpG positions because they are known to evolve rapidly (with a few exceptions). The peak at the left includes the 195 positions that have been conserved in the source genes and therefore are known at the time of insertion of each of the *Alu* sequences. These positions have mutated *in situ* about 8 ± 5% since their insertion. The positions showing larger differences from the modern consensus include the CpG dinucleotides and those that have been mutated in the source genes; such positions are potentially diagnostic for classes of *Alu* sequences.

with the observation that CpGs evolve about eight times as fast as the average nucleotide substitution (for references, see Britten *et al.*, 1988). There are 56 positions on the left-hand side of the major peak (less than 6% divergence) and only one is a G or C base in the DNA sequence. Comparably, the right-hand side of the peak is made up primarily of Gs and Cs. This is because in these relatively freely drifting positions, C and G bases (not including CpG dinucleotides) change 1.5 times faster than A and T (thymine) bases. This phenomenon is not restricted to these *Alu* sequences (Li and Graur, 1991). It is also observed in primate globin synonymous substitutions where a ratio of about 1.5 to 1 is also observed for changes in G and C bases compared with A and T bases.

9.3 Classes of *Alu* sequences

9.3.1 Investigation of classes

It has been recognized for a decade that some sets of *Alu* sequences are more closely related to each other than to others (e.g. Deininger and Daniels, 1986; Slagel *et al.*, 1987; Willard *et al.*, 1987), suggesting identifiably separate subfamilies. Standard methods for forming trees of relationship gave weak results because divergence, such as that shown in *Figure 9.2,* was present. It was recognized independently by three groups that there were diagnostic positions that could be used to identify classes (Britten *et al.*, 1988; Jurka and Smith, 1988; Quentin, 1988). As a result, there are several systems of nomenclature which have persisted in the literature and the relationship between two of these is presented in *Table 9.1.* A recent attempt to find a unified nomenclature is also included. Identification of the diagnostic nucleotides used to distinguish these groups was accomplished manually and to date there is no algorithm (working from just the *Alu* sequences) that will list the classes or subfamilies and their diagnostic positions. However, collections of diagnostic positions have been constructed (Jurka and Milosavljevic, 1991) and computer procedures using this information have been established that define the subfamily to which *Alu* sequences can be assigned. While this is a considerable accomplishment, there are some details which suggest that future research may lead to modification of the definition of subfamilies and the criteria for membership. It is an unlikely possibility that the present classes and subclasses will be considered in the future to be artificial subdivisions of a continuum of types.

Table 9.1 Alternative nomenclature for classifying *Alu* sequences

Classes	I	II	II	II	III	IV	V
Subfamilies	J	Sx	Sq	Sp	Sc	Sb	Sb1, Sb2
Subfamilies	J	Sx	Sq	Sp	Sc	Y	Ya5, Ya8, Yb8

The first row of classes is from Britten *et al.* (1988) and Britten (1994a). The second row of subfamilies is from Jurka and Smith (1988), as more recently modified. The third row of subfamilies is an agreement among scientists in the field which is an attempt at a consistent system for naming the new recent inserts that have been identified from current studies of human polymorphism and comparisons among apes (Batzer *et al.*, 1996). The sequence in *Figure 9.1* is that of Ya5.

9.3.2 Diagnostic positions are more divergent from the modern consensus

When comparing *Alu* sequences with a modern consensus (*Figure 9.2*), positions with excess divergence from the consensus (defined as greater than 14% divergence, or all positions to the right of the major peak in *Figure 9.2*) can be considered to be diagnostic for different families or groups of *Alu* sequences. Consider three of the positions at the right of *Figure 9.2*, showing 90.3, 94.6 and 96.1% average divergence from the modern consensus. These are positions 145, 96 and 89 of the sequence in *Figure 9.1*, which have mutated very recently in the evolution of source genes. Thus, the new bases A, A and C (present in the modern consensus) are shared by very few other *Alu* sequences, the great majority having G, C and T in these positions. The converse, that nucleotides with less divergence from the modern consensus are changes that in general occurred earlier, is also true. This is because the resulting nucleotides after change (matching the modern consensus) have been present in all the succeeding source genes and therefore in all new inserts up to the present time. Thus, they agree with the modern consensus nucleotides in a larger number of *Alu* inserts.

In principle, each source gene change can be dated by the divergence or lack of divergence from the modern consensus at that position. *Table 9.2* illustrates this using the clearest examples of diagnostic positions that distinguish between *Alu* classes or families. There is, of course, some background 'noise' in this analysis since changes that occurred early in the history of source genes have been present in the inserts for a long time and therefore some will have been mutated *in situ*. In addition, there are cases of reversion and some nucleotides are, or were, in CpG dinucleotides before or after change, increasing the variation. Despite these ambiguities, *Table 9.2* demonstrates that there has been a discernable series of changes in source gene evolution, with *Alu* inserts in class I (family J) being the most ancient retroposed copies of early source genes, *Alu* inserts in classes II and III (families Sx, Sq, Sp and Sc) being more recent insertions and *Alu* inserts in classes IV and V (the Sb and Y families) being the most recently retroposed copies of the *Alu* source genes. In this context, it is noteworthy that class I (subfamily J) members share the closest sequence relationship to the 7SL sequence (Britten *et al.*, 1988; Quentin, 1988). This further supports the conclusion that the earliest class of *Alu* sequences evolved from the 7SL RNA gene and that later classes are the product of additional evolutionary change.

9.3.3 Evolution of source genes and the resulting classes

There is now no reasonable doubt that a pathway of evolution of *Alu* source genes occurred by the addition of new changes to older versions. There have been some reversions and the pathways must have branched. Furthermore, a number of source genes are simultaneously active at present in creating *Alu*

Table 9.2 Divergence of diagnostic positions from the modern consensus

Position	Classes distinguished	Divergence
92	I/II+	0.264
193	I/II+	0.213
203	I/II+	0.225
		(0.234)
162	I/II, II/III	0.536
		(Reversion[a])
76	II/III	0.813
86	II/III	0.811
		(0.812)
196	III/IV	0.852
199	III/IV	0.898
98	III/IV	0.837
93	III/IV	0.837
		(0.856)
145	IV/V	0.903
96	IV/V	0.946
237	IV/V	0.951
89	IV/V	0.961
		(0.940)

The upper 10 diagnostic positions are the most statistically significant (Jurka and Milosavljevic, 1991), and the bottom four are those more recently recognized as diagnostic for the class V (modern) *Alu* insert class. The values for divergence are taken from the analysis shown in *Figure 9.2* of 1579 *Alu* sequences. 'Classes distinguished' refers to the classes of *Alu* sequences shown in *Table 9.1*. I/II+ means that position is distinctive for class I (subfamily J) with respect to all other classes. This table shows that the diagnostic position nucleotides that resulted from early changes in the source genes are similar to those in the modern consensus; that is, class I nucleotides mutated and were conserved in the succeeding source genes (see text for details).
[a] Position 162 was a G base in class I, became an A base in class II and reverted to a G base in classes III and IV.

inserts in the human genome (e.g. Batzer *et al.*, 1995). This has probably been the case throughout *Alu* evolution, with the source genes of the major classes or subfamilies being active for long overlapping periods of time. The full tree of relationships, constructed using modern methods and reporting uncertainties for each of the branches, has not yet been established and forming it probably has some inherent difficulties. It is likely that doubt about the evolutionary significance of the classification of *Alu* inserts will remain until the human genome is sequenced and the relationships of the million or so *Alu* inserts can be analysed.

One question to be resolved is whether classes exist as distinctly separate entities or whether there has been a continuum of change in the diagnostic positions and a continuous production of *Alu* inserts. There are about 40 positions in *Figure 9.2* that are apparently excessively divergent and most are used for diagnostic positions (Jurka and Milosavljevic, 1991). Examination of

Figure 9.2 indicates that there is not a continuous distribution of excessively divergent positions but they appear to be clustered in three regions with low, middle and high divergence. There is a peak at 23%, which agrees with the class I/II distinction shown in *Table 9.2* with an average of 23%. The number of positions in this peak and neighbouring points (14) suggests that much more can be learned about class I by examining them when more *Alu* sequences become available. The three upper averages in *Table 9.2* of 81%, 86% and 94% are all part of the more divergent (more recent) region corresponding to the appearance and evolution of classes III, IV and V. The dozen positions that fall between 36% and 65% in *Figure 9.2* probably correspond to changes leading to the middle-aged subfamilies Sx, Sp and Sq. This intricate story will also benefit from future examination with more *Alu* sequences. Some may turn out to be positions that are unstable in the source genes, having higher rates of mutation. At present, a continuum of source genes that created significant numbers of inserts seems unlikely, as also indicated by the data in Section 9.4.

The rate of evolution of the source genes allows us to make an estimate of the sequence-dependent selection, if any. Twenty-four CpG dinucleotides have been maintained in the source genes while the expectation based on the modern rate of mutation of CpGs is that few would remain. In fact, once the *Alu* sequences are inserted, the CpGs are rapidly lost. However, the rate of change of the remainder of the positions can be estimated from *Figure 9.2*. There are 30 non-CpG positions that have more than 20% divergence and these are presumed to have resulted from changes in the source genes. In addition, the diagnostic positions can provide a measure of changes that have occurred in the source genes and there are about 30 of these, depending on assessment of their statistical significance (Jurka and Milosavljevic, 1991). Using these positions, the author estimates that, since the earliest recognizable *Alu* sequences were being inserted 50–65 million years ago, somewhere between 10 and 15% of the *Alu* sequence has been subject to base substitution. This gives a rate of about 0.2% per million years which is not significantly different from the estimate of the present rate of 'neutral' drift of DNA in higher primates (0.16% every million years). This presents us with a paradox. The CpG dinucleotides imply conservation whereas the diagnostic positions imply free drift near the 'neutral rate'. A possible solution would be if the source genes are located in a CpG island and are not subject to the high mutation rate observed at methylated CpG dinucleotides. Irrespective of this, the author now believes that earlier assertions that the source genes are subject to rigid sequence-dependent selection were premature, despite the fact that the data from CpGs are consistent with this view.

9.4 Divergence of classes and time of insertion of class members

An understanding of both the form and timing of *Alu* evolution can be

Table 9.3. *Alu* classifications and timing of insertion events during human evolution

Class of *Alu* insert	Subfamily name of *Alu* insert	Quantity[a]	Poisson fit[b] *N*	*a*	Event time[c] (million years before present)	Number of *Alu* inserts in human genome[d]
I	J (J)	226	132	17.6	56	65 000
			127	12.7	41	62 000
II	Sx (Sx)	708	415	10.1	32	203 000
			242	16.5	53	119 000
II	Sp (Sp)	247	175	10.0	32	86 000
	Sq (Sq)		60	17.8	57	29 000
III	Sc (Sc)	112	87	8.0	26	43 000
			25	14.1	45	13 000
IV	Sb (Y)	170	32	3.3	11	16 000
			130	7.5	24	64 000
V	Sb1 (Ya5)	23[e]	2–3	(1.7)	(5)	(1000)
	Sb2 (Ya8)					
Totals		1528				750 000[f]

[a] The number of sequences of each class in the set of 1528 nearly full-length *Alu* inserts remaining after removal of duplicates and those including unrelated DNA sequences.
[b] A least-squares fit to the sum of two Poisson distributions was performed, each having the form $N^{\star}\exp(-a)^{\star}a^{x}/x!$ where x is the number of mutations in the 195 CONSBI positions (see text) and a is the average. N is the number of members of the component.
[c] The time since the multiplication event was calculated from a, the average number of mutations at the 195 CONSBI positions, assuming a rate of mutation of 0.16% per million years.
[d] The number of inserts estimated to be present in the human genome was calculated as $N^{\star}750\,000/1528$.
[e] Most of the known *Alu* sequences of this class derived from human genetic studies where they cause variation or were specifically screened on the basis of diagnostic positions. Thus, the number 23 is not comparable to the number of other classes of *Alu* sequences, which were obtained by sequencing gene regions. Only three of the 23 known class V members appear to have been identified by sequencing gene regions and these form a statistical sample equivalent to that of other *Alu* sequences; class V (Sb1) *Alu* sequences are therefore rare. The average number of nucleotide differences from the modern consensus sequence, shown in column 5, is calculated for the total known set of 23, as is the time estimate in column 6. The number estimate in column 7 is based on the three elements discovered by sequencing.
[f] This is an arbitrary figure for the total number of *Alu* inserts in the human genome for which estimates range from 500 000 to 1 000 000.

obtained by separating *Alu* sequences into their respective classes (subfamilies) and calculating the average degree of divergence at the 195 CONSBI positions for each element (Britten, 1994b). Heterogeneity within classes can be investigated by fitting the results from each class to Poisson distributions, and estimates of insertion times can be obtained based on the assumption that *Alu* sequences drifted (at the CONSBI positons) at 0.16% per million years (Bailey *et al.*, 1991; Britten, 1986).

The results of calculations performed on 1500 elements, fitted to Poisson distributions, are summarized in *Table 9.3.* It should, however, be stressed that the statistical uncertainty of these calculations is large because the number of differences from the modern consensus is small – only 10 differences for the major component. Nevertheless, for the two major classes of *Alu* inserts (I and II) a good fit was obtained from a sum of two Poisson distributions, implying the existence of more than one component in the evolution of each major class.

Table 9.3 indicates that source genes for class I were responsible for the earliest insertions of recognizable *Alu* sequences. Indeed, there is a major component of 132 observed class I *Alu* inserts, with an average of about 18 differences from the modern consensus, suggesting an age of about 56 million years. This is consistent with the concept that class I *Alu* sequences were being inserted early in primate evolution. Although class I is the oldest, not all of its members appear to be old. About half (127 cases) appear in a second Poisson component with 12.7 divergences (at the CONSBI positions) from the modern consensus. Examination of individual class I sequences shows that some retain a certain number of CpG dinucleotides, although the majority have lost almost all. It has been suggested that class I can be subdivided into two subfamilies (Jurka and Milosavljevic, 1991), a suggestion which would be consistent with the analysis presented here. These data also suggest that the class I source genes may have been responsible for *Alu* inserts right up to times that overlap the major activity of class II source genes.

Class II contains by far the largest number of *Alu* sequences and the probable components of class II, subfamilies Sx, Sp and Sq, are not distinguishable in this analysis. This suggests that the subfamilies in class II have been inserted principally in overlapping periods of time. The Poisson distributions include less divergent inserts but the number of these is consistent with the expected fluctuation and so it was concluded (Britten, 1994b) that the insertion of class II *Alu* sequences has ceased and probably has not been occurring for tens of millions of years. This is consistent with recent evidence on polymorphism of *Alu* inserts in human genomes and interspecific differences which suggests that only class V *Alu* sequences are currently being inserted (e.g. Batzer *et al.*, 1995, 1996).

The data of *Table 9.3* support the concept that the source genes slowly evolved and were responsible for *Alu* sequence insertion over a long period of time. The rate of insertion appears to have been variable, peaking perhaps 30–40 million years ago, with different source genes possibly being active for overlapping periods.

9.5 Lack of lineages of *Alu* sequences, except the source genes

A further prediction of the source gene mode of *Alu* evolution is that all *Alu* elements should essentially form a single lineage, due to the derivation of all *Alu* inserts from one, or a small number of, slowly evolving source genes. The identification of clearly defined lineages within *Alu* sequences would provide strong evidence for many, retropositionally active, source genes evolving independently. Although initial analyses provided no clear evidence of lineages (Section 9.3.1), this possibility has been investigated in detail by analysing variation at the CONSBI positions in multiple comparisons of randomly chosen *Alu* inserts. If, as the source gene model predicts, *Alu* sequences are derived from a single source, then any pair of *Alu* elements will differ from each other by just the sum of their differences from the source sequence. Accidental identical mutations in both *Alu* inserts could falsely increase identity between them, but few such duplicate mutations would be expected for sequences of modest divergence. However, if lineages do exist, then some *Alu* sequences will differ from each other much less than they do from the source sequence.

To test this hypothesis, an analysis consisting of pairwise comparisons of the known, nearly full-length, *Alu* sequences was performed and the results of the 1 115 000 comparisons have been presented elsewhere (Britten, 1995). They show that *Alu* repeats that are significantly different from the source sequence (at the CONSBI positions) are rarely replicated. When the divergence of pairs of sequences from each other was plotted against the sum of their divergence from the modern consensus, a total of 98% of all sequence comparisons was found to lie along the main diagonal. Thus, the sequence divergence of 98% of all pairwise comparisons is equal to the sum of their divergences from the modern consensus. This pattern is exactly what is predicted from a source gene model of evolution with all *Alu* sequences being derived from a single common source (for the CONSBI positions).

To perform a more rigorous analysis of these comparisons, a Monte Carlo calculation was carried out. For this purpose, 1500 copies of the modern consensus sequence that had been mutated to varying extent in the CONSBI positions were stored. To calculate the expected effect of duplicate mutations properly, the wide range of rates of mutation among the CONSBI positions was modelled by mutating each position so that the average number of substitutions of each kind matched those of the known set of *Alu* sequences. These 1500 mutated copies were compared with each other and the result for the vast majority of comparisons was identical to the above comparison (referred to as the 'real data'). However, there was an interesting difference in the width of the pattern for the real data compared with the Monte Carlo control. This indicated that a very small minority of pairs of actual *Alu* inserts are more similar to each other than is expected from the simple model that they all derive from a single source identical to the modern consensus in the CONSBI positions. The explanation is probably that some families of *Alu* inserts are copies of source genes which differ at a few CONSBI positions. In other words, there is a subset of positions that might be weakly diagnostic. However, there is no large deviation of the real data from the Monte Carlo calculation, which shows that the dominant process is the insertion of copies of source genes sharing the 195 CONSBI nucleotides.

This analysis does not rule out the possibility of copies being made of 'young' inserts, since these sequences have not been subject to enough mutations for their history to be recognized. It also does not rule out rare cases of copying by other mechanisms such as unequal crossover or gene conversion among *Alu* repeats. For instance, there is one reported example of a human *Alu* sequence which has apparently been subject to gene conversion and is now similar to the modern consensus, whereas in apes and monkeys there is an older type of *Alu* insert in the same location (Kass *et al.*, 1995). However, the contribution from such events appears to be very limited. In the above analysis, a total of only 86 pairs (with more than 5% divergence from the consensus) had less than one-third of the expected divergence between the members of the pair. These *Alu* sequences were all compared with each other and an attempt was made to form trees of relationship by the neighbour-joining method. No sets of lineage relationships were found. One group of eight inserts differed from each other by 0–5% divergence (CONSBI positions) and they each differed from the modern consensus by 6% of CONSBI positions. All of them were in the same locations in the 5′ regions of the known human leukocyte antigen (HLA) *DQ1α* gene set (Morzycka-Wroblewska *et al.*, 1993). Each was just 5′ of the promoter region and it is not known if they have a function in that location. Regardless of possible function, however, the salient feature of these *Alu* inserts is that their higher-than-expected sequence identity is due to their common origin as a result of HLA gene duplication events, rather than retroposition. Other groups of *Alu* sequences examined also turned out to be the result of past regional duplications. Thus, an exhaustive search for self-replicating lineages of *Alu* inserts revealed none and it is safe to say that they do not exist for inserts with more than 5% divergence (CONSBI positions) from the consensus.

9.6 Inserted *Alu* sequences drift at most positions but some changes almost never occur

An important, but as yet unexplained, aspect of the evolution of *Alu* inserts is that some types of mutation at certain positions essentially do not occur. This is difficult to reconcile with the concept that the very many inserts are simply free to drift in sequence once they have been inserted. The evidence is presented in *Table 9.4*, which shows the ratio of observed changes to expected changes. The expectation of each type of change for each nucleotide is simply the observed number of differences of that type found in a collection of 1750 nearly full-length *Alu* sequences, each compared with the modern consensus. For example, the average A base anywhere in the modern consensus sequence has a probability of 0.0099 of being a C (result of transversion) in the set of 1750 *Alu* inserts in the human genome, or a 0.0237 chance of being a G base (result of transition).

The most extreme cases of 'forbidden' changes are four that occur at about 18% of the expected rate. They are G to T at position 25, deletion at position 81, T to G at position 115, and A to T at position 116. The probability of all of these occurring by chance is essentially zero and there are many additional cases. *Table 9.4* also shows the results of a Monte Carlo calculation where only three mutations occur at less than half of the expected rate. This contrasts sharply with the 88 mutations (of a total 780 possibilities) that occur at

Table 9.4 Occurrence of specific mutations at particular CONSBI positions of 1750 *Alu* inserts

Occurrence[a]	M.C.[b]	Ratio[c]
4	0	0.1775
1	0	0.1900
2	0	0.2150
6	0	0.2317
7	0	0.2643
5	0	0.2920
3	0	0.3233
7	0	0.3457
19	1	0.3805
8	1	0.4213
21	1	0.4733
28	5	0.5214
28	14	0.5779
34	23	0.6441
38	20	0.7074
52	56	0.7706
61	109	0.8597
99	162	0.9638
158	166	1.0291
52	128	1.1600
38	52	1.2837
38	30	1.4179
16	9	1.5681
12	3	1.7392
16	0	1.9231
8	0	2.1287
2	0	2.3250
3	0	2.5600
3	0	2.7733
1	0	3.0800
1	0	3.6100
4	0	3.8575
1	0	4.2400
2	0	4.7400
0	0	—
2	0	5.8050

[a] The distribution of the ratio of occurrence to expectation for the four possible kinds of mutation at 195 CONSBI positions, totalling 780 entries. To prepare this table, all of the mutations (differences from the modern consensus sequence) that occurred at each position were counted for the 1750 *Alu* sequences in the set. A table was made of the ratio of the fraction for each type of mutation to the expected fraction. The entries in this table were then placed in order and the number in each of the classes of ratio was counted and are listed in column 1. For example, the four minimally occurring cases at the head of the column are positions 25 (G to T), 81 (G deletion), 115 (T to G) and 116 (A to T). The average ratio for these four was 17.75% of the expected occurence, as listed in column 3.
[b] The distribution of ratio of occurrence to average expected occurrence for a Monte Carlo (M.C.) calculation as a measure of the stochastic spread of the distribution.
[c] The average ratio of occurrence to expectation for the set of cases in column 1. For example, on the fourth line are six types of change in specific positions that average approximately 23% of the expected occurrence.

half the expected rate or less in the *Alu* data. There are clearly some specific restrictions on the free drift after insertion of the *Alu* sequences.

Several other observations are germane. The total amount of change to any of the four possibilities at certain positions is markedly low and these positions are clustered together (Britten, 1994a). Finally, there is long-distance correlation of the reduced amount of change at a number of positions. The data of *Table 9.4* show in a particularly sensitive way that mutations occur at lower rates at specific positions in the *Alu* sequence. The correlation of the lack of mutation at any one position with that at another position indicates that a subset of *Alu* inserts have special properties that limit the rate of incorporation of mutations at both positions. It is difficult to visualize how this could happen except by selection and it can be presumed that the binding of proteins to these clusters is important. The implication is that these positions in a subset of *Alu* inserts carry out a selectively important function. Does this selectively important function concern the host primates or the *Alu* sequences themselves? In Section 9.5, it was argued that the selection does not have to do with *Alu* sequence replication, but there is an additional possibility. The *Alu* sequences being examined are those that have survived the risk of unequal crossover leading to deletion. In this context, it is interesting that frequent recombination events affecting the low-density lipoprotein receptor gene and the γ, δ and β globin genes show that some regions of the *Alu* sequence are much more likely to undergo recombination than other regions (Lehrman *et al.*, 1985, 1987). Thus, the process appears to be sequence-dependent and there could be a resulting selective effect that would influence the sequences of the surviving population of *Alu* inserts in the human genome. In this model, the survivors would be those with a smaller probability of undergoing unequal crossover and being deleted. This might very well suppress certain kinds of mutations and lead to correlation in the suppression of changes in widely spaced regions of the *Alu* inserts.

9.7 Summary

Our understanding of *Alu* sequence evolution has altered dramatically since their discovery. In particular, extensive sequence comparisons and the identification of new *Alu* inserts have enabled some unexpected conclusions to be drawn. It is now clear that one or a small number of slowly evolving source genes, have been responsible for the insertion of the 500 000–1 000 000 *Alu* inserts present in the human genome. Furthermore, the identification of nucleotides in the *Alu* source gene CONSBI has further strengthened this conclusion and has enabled selection on some nucleotides within the *Alu* sequence to be inferred. With new *Alu* sequences being deposited in databases on a daily basis, our understanding of *Alu* sequence evolution will continue to be refined. It is hoped, for instance, that active source genes can be identified and that the fundamental question of why *Alu* sequences have become an integral part of the human genome can be addressed.

Alu sequences have apparently been important in primate variation and evolution both in the regulation of expression of genes and through mutations and rearrangements in the genome (Britten, 1996; Makalowski *et al.*, 1994; Maraia, 1995). Many more *Alu* sequences will be found to cause variation and many examples of positively selected changes will be identified. This is likely to become a growing part of future research.

Acknowledgements

This work was supported by NIH grants. I am grateful for a prepublication copy of Batzer *et al.* (1996).

References

Bailey AD, Shen C-KJ. (1993) Sequential insertion of *Alu* family repeats into specific genomic sites of higher primates. *Proc. Natl Acad. Sci. USA* **90:** 7205–7209.

Bailey WJ, Fitch DHA, Tagle DA, Czelusniak J, Slightom JL, Goodman M. (1991) Molecular evolution of the ψη-globin gene locus: gibbon phylogeny and the hominoid slowdown. *Mol. Biol. Evol.* **8:** 155–184.

Batzer MA, Deininger PL. (1991) A human-specific subfamily of *Alu* sequences. *Genomics* **9:** 481–487.

Batzer MA, Kilroy GE, Richard PE, Shaikh TH, Desselle TD, Hoppens CL, Deininger PL. (1990) Structure and variability of recently inserted Alu family members. *Nucleic Acids Res.* **18**: 6793–6798.

Batzer MA, Stoneking M, Alegria-Hartman M. *et al.* (1994) African origin of human-specific polymorphic *Alu* insertions. *Proc. Natl Acad. Sci. USA* **91:** 12288–12292.

Batzer MA, Rubin CM, Hellmann-Blumberg U *et al.* (1995) Dispersion and insertion polymorphism in two small subfamilies of recently amplified human *Alu* repeats. *J. Mol. Biol.* **247:** 418–427.

Batzer MA, Deininger PL, Hellmann-Blumberg U, Jurka J, Labuda D, Rubin CM, Schmid CW, Ziętkiewicz E, Zuckerkandl E. (1996) Standardized nomenclature for Alu repeats. *J. Mol. Evol.* **42:** 3–6.

Britten RJ. (1986) Rates of DNA sequence evolution differ between taxonomic groups. *Science* **231:** 1393–1398.

Britten RJ. (1994a) Evolutionary selection against change in many *Alu* repeat sequences interspersed through primate genomes. *Proc. Natl Acad. Sci. USA* **91:** 5992–5996.

Britten RJ. (1994b) Evidence that most human *Alu* sequences were inserted in a process that ceased about 30 million years ago. *Proc. Natl Acad. Sci. USA* **91:** 6148–6150.

Britten RJ. (1995) Quantitative study of Alu repeated sequences in primate genomes, yielding insight into their sources and evolution. In: *The Impact of Short Interspersed Elements (SINEs) on the Host Genome* (ed. RJ Maraia). R.G. Landes Company, Austin, TX, pp. 223–232.

Britten RJ. (1996) DNA sequence insertion and evolutionary variation in gene regulation. *Proc. Natl Acad. Sci. USA*, in press.

Britten RJ, Baron WF, Stout DB, Davidson EH. (1988) Sources and evolution of human *Alu* repeated sequences. *Proc. Natl Acad. Sci. USA* **85:** 4770–4774.

Charlesworth B, Langley CH. (1989) The population genetics of *Drosophila* transposable elements. *Annu. Rev. Genet.* **23:** 251–287.

Chen TL, Manuelidis L. (1989) SINEs and LINEs cluster in distinct DNA fragments of Giemsa band size. *Chromosoma* **98:** 309–316

Deininger PL, Batzer MA. (1993) Evolution of retroposons. *Evol. Biol.* **27:** 157–196.

Deininger PL, Daniels GR. (1986) The recent evolution of mammalian repetitive DNA elements. *Trends Genet.* **2:** 76–80.

Deininger PL, Batzer MA, Hutchison CA, Edgell MH. (1992) Master genes in mammalian repetitive DNA amplification. *Trends Genet.* **8**: 307–311.

Hammer MF. (1994) A recent insertion of an Alu element on the Y chromosome is a useful marker for human population studies. *Mol. Biol. Evol.* **11:** 749–761.

Jurka J, Milosavljevic A. (1991) Reconstruction and analysis of human Alu genes. *J. Mol. Evol.* **32:** 105–121.

Jurka J, Smith T. (1988) A fundamental division in the *Alu* family of repeated sequences. *Proc. Natl Acad. Sci. USA* **85:** 4775–4778.

Kass DH, Batzer MA, Deininger PL. (1995) Gene conversion as a secondary mechanism of short interspersed element (SINE) evolution. *Mol. Cell. Biol.* **15:** 19–25.

Korenberg JR, Rykowski MC. (1988) Human genome organisation: Alu, lines, and the molecular structure of metaphase chromosome bands. *Cell* **53:** 391–400.

Leeflang EP, Liu W-M, Hashimoto C, Choudary PV, Schmid CW. (1992) Phylogenetic evidence for multiple Alu source genes. *J. Mol. Evol.* **35:** 7–16.

Lehrman MA, Schneider WJ, Südhof TC, Brown MS, Goldstein JL, Russell DW. (1985) Mutation in LDL receptor: Alu–Alu recombination deletes exons encoding transmembrane and cytoplasmic domains. *Science* **227:** 140–146.

Lehrman MA, Goldstein JL, Russell DW, Brown MS. (1987) Duplication of seven exons in LDL receptor gene caused by Alu–Alu recombination in a subject with familial hypercholesterolemia. *Cell* **48:** 827-835.

Li W-H, Graur D. (1991) *Fundamentals of Molecular Evolution.* Sinauer Associates, Sunderland, MA.

Makalowski W, Mitchell GA, Labuda D. (1994) Alu sequences in the coding regions of messenger RNA: a source of protein variability. *Trends Genet.* **10:** 188–193.

Maraia RJ. (ed.) (1995) *The Impact of Short Interspersed Elements (SINEs) on the Host Genome.* R.G. Landes Company, Austin, TX.

Matera AG, Hellmann U, Hintz MF, Schmid CW. (1990a) Recently transposed Alu repeats result from multiple source genes. *Nucleic Acids Res.* **18:** 6019–6023.

Matera AG, Hellmann U, Schmid CW. (1990b) A transpositionally and transcriptionally competent Alu subfamily. *Mol. Cell. Biol.* **10:** 5424–5432.

Morzycka-Wroblewska E, Harwood JI, Smith JR, Kagnoff MF. (1993) Structure and evolution of the promoter regions of the DQA genes. *Immunogenetics* **37:** 364–372.

Norris J, Fan D, Aleman C, Marks JR, Futreal PA, Wiseman RW, Iglehart JD, Deininger PL, McDonnell DP. (1995) Identification of a new subclass of *Alu* DNA repeats which can function as estrogen receptor-dependent transcriptional enhancers. *J. Biol. Chem.* **270:** 22777–22782.

Okada N. (1991) SINEs. *Curr. Opin. Genet. Dev.* **1:** 498-504.

Quentin Y. (1988) The Alu family developed through successive waves of fixation closely connected with primate lineage history. *J. Mol. Evol.* **27:** 194-202.

Quentin Y. (1992) Origin of the Alu family: a family of Alu-like monomers gave birth to the left and right arms of the Alu elements. *Nucleic Acids Res.* **20:** 3397–3401.

Quentin Y. (1994) Emergence of master sequences in families of retroposons derived from 7sl RNA. *Genetica* **93:** 203–215.

Riccio ML, Rossolini GM. (1993) Unusual clustering of Alu repeats within the 5′-flanking region of the human lysozyme gene. *DNA Sequence* **4:** 129–134.

Schmid CW, Maraia R. (1992) Transcriptional regulation and transpositional selection of active SINE sequences. *Curr. Opin. Genet. Dev.* **2:** 874–882.

Slagel V, Flemington E, Traina-Dorge V, Bradshaw H, Deininger P. (1987) Clustering and subfamily relationships of the Alu family in the human genome. *Mol. Biol. Evol.* **4:** 19–29.

Soriano P, Meunier-Rotival M, Bernardi, G. (1983) The distribution of interspersed repeats is nonuniform and conserved in the mouse and human genomes. *Proc. Natl Acad. Sci. USA* **80:** 1816–1820.

Stoppa-Lyonnet D, Carter PE, Meo T, Tosi M. (1990) Clusters of intragenic *Alu* repeats predispose the human C1 inhibitor locus to deleterious rearrangements. *Proc. Natl Acad. Sci. USA* **87:** 1551–1555.
Ullu E, Tschudi C. (1984) *Alu* sequences are processed 7SL RNA genes. *Nature* **312**: 171–172.
Vansant G, Reynolds WF. (1995) The consensus sequence of a major *Alu* subfamily contains a functional retinoic acid response element. *Proc. Natl Acad. Sci. USA* **92:** 8229–8233.
Willard C, Nguyen HT, Schmid CW. (1987) Existence of at least three distinct Alu subfamilies. *J. Mol. Evol.* **26**: 180–186.
Zietkięwicz E, Richer C, Makalowski W, Jurka J, Labuda D. (1994) A young *Alu* subfamily amplified independently in human and African great apes lineages. *Nucleic Acids Res.* **22:** 5608–5612.

10

Human sex chromosome evolution

Nathan A. Ellis

10.1 Introduction

Sex chromosomes are fundamental to the process of sex determination (the developmental process by which males and females are produced) as they carry genetic factors that influence the development of both sexes. Sex chromosomes are formally defined as chromosomes that segregate during meiosis to form different gametes in one sex (the heterogametic sex) but not the other (the homogametic sex). The heterogametic sex can be either male or female: in a species where the males are heterogametic the sex chromosomes are termed X and Y (XX females/XY males), whereas in a species where the females are heterogametic they are termed Z and W (ZZ males/ZW females). Sex chromosomes are often heteromorphic (i.e. they have microscopically distinguishable shapes, banding patterns, or both). They can also be homomorphic (microscopically indistinguishable), and in such cases the conclusion that a species has sex chromosomes at all is based on genetic experiments (*Figure 10.1*). Most mammals, however, have extremely heteromorphic sex chromosomes.

The differentiation between the sexes is initiated during embryonic development. In mammals, gonads arise from cells in the medial mesonephros at the genital ridge to form an undifferentiated gonad that is phenotypically indistinguishable in males and females. The gonads then differentiate into either the ovary (the organ that produces the gametes of the female) or the testis (the organ that produces the gametes of the male). From experiments in rabbits in which undifferentiated or differentiated embryonic gonads were removed, the differentiation of male internal sex organs and external genitalia was shown to depend on hormones produced by the testis (Jost *et al.*, 1973). In the absence of these hormones, female internal sex organs and external genitalia were produced. Hence, mammalian sex determination is fundamen-

Human Genome Evolution, edited by M. Jackson, T. Strachan and G. Dover.

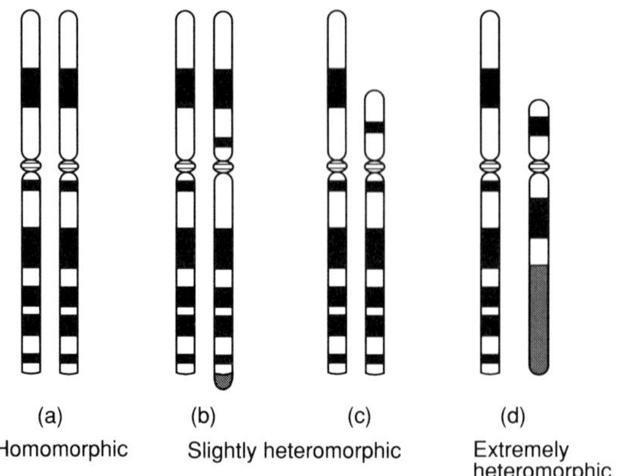

Figure 10.1. Representations of sex chromosomes detected cytogenetically in vertebrates. (a) Homomorphic sex chromosomes; (b and c) slightly heteromorphic sex chromosomes; and (d) extremely heteromorphic sex chromosomes. Homomorphic sex chromosomes have been demonstrated in fish, amphibian and reptilian species (see Bull, 1983, and Ohno, 1967). Slightly heteromorphic chromosomes have been detected in all classes of vertebrates. Shown are the two extremes of slightly heteromorphic chromosomes: (b), in which a small structural difference between the chromosomes is noted, such as an inversion or the acquisition of a small piece of heterochromatin (here both of these are shown on a single chromosome); and (c), in which one chromosome arm is homologous between the two chromosomes but the other arm is entirely different cytogenetically. The chromosomes in (c) are similar to those found in one species of monotreme (Wrigley and Graves, 1988). (d) Extremely heteromorphic chromosomes have been detected in some taxa of reptilia, in birds, and in metatherian and eutherian mammals. Between these sex chromosomes, there are no detectable cytogenetic homologies.

tally a question of gonad determination. In humans and mouse, individuals who have a 45,X chromosome constitution are phenotypic females. In humans, persons with a 47,XXY chromosome constitution are phenotypic males. Therefore, gonad determination in mammals is genotypically controlled, with the Y chromosome carrying a gene that plays a critical role in the development of testes.

The mammalian system of sex determination and sex chromosomes is not utilized in other vertebrates. Birds and snakes have a ZZ/ZW sex chromosome constitution. Interestingly, the avian Z chromosome has a different gene content than the mammalian X chromosome, indicating that different starting material was used in its evolution (Baverstock *et al.*, 1982; Dominguez-Steglich *et al.*, 1990). In certain species of lizards, turtles and alligators, sex determination is controlled by the temperature at which the egg is incubated, indicating that an environmental rather than a genotypic mechanism of sex determination operates (Bull, 1983). In different fish and amphibian species, sex chromosomes can be either XX/XY or ZZ/ZW (Schmid *et al.*, 1989; Sola *et al.*, 1981) and differentiation of the gonads can be influenced by externally administered hormones (Ohno, 1967). However, with few exceptions, notably in wood lemmings (Fredga, 1988) and voles (Just *et al.*, 1995), the mammalian sex determination system is extremely well conserved.

The fact that one system of sex determination and sex chromosomes is used by virtually all mammals, and not by other vertebrate species, suggests that this system must have evolved early in mammalian radiation. Mammals are divided into three major groups: monotremes (Prototheria), marsupials (Metatheria) and so-called placental mammals (Eutheria). The prototherian lineage, in which only three species are extant, diverged from the therian lineage approximately 150–180 million years ago (Hope *et al.*, 1990). The metatherian and eutherian lineages diverged approximately 120–150 million years ago (Hope *et al.*, 1990). Although the sex determination system is highly stable, mammalian sex chromosomes have continued to change in diverse ways. Comparative genetic mapping of sex chromosomes in each of these infraclasses has uncovered conserved and evolving features of mammalian sex chromosomes. This chapter summarizes the current state of this knowledge in the context of human sex chromosome evolution.

10.2 Mechanisms of mammalian sex chromosome evolution

Leaving aside the complex question of why sex should evolve at all (Hamilton *et al.*, 1990; Maynard Smith, 1978), one observes heteromorphic sex chromosomes in many plants and animals. This suggests that such chromosome systems have evolved independently numerous times. Thus, the conditions and forces involved in sex chromosome evolution have arisen more than once and involve properties intrinsic to many genetic systems. While the actual sequence of evolutionary events leading to the present-day heteromorphic sex chromosomes in mammals is unknown, certain hypotheses about their origin are consistent with observation. These hypotheses are presented here in the form of a model that is a modified version of many previous models (Bull, 1983; Charlesworth, 1991; Fisher, 1935; Muller, 1964; Ohno, 1967).

10.2.1 An autosomal origin for sex chromosomes

The mammalian sex chromosomes most probably evolved from an ordinary pair of autosomes (*Figure 10.2a*). As noted above, in vertebrates, the sex chromosomes range in cytogenetic appearance from indistinguishable homomorphic sex chromosomes, to intermediate forms of chromosomes containing both cytologically visible homologous and differential (non-homologous) chromosomal segments, to extremely heteromorphic chromosomes with no sign of homologous segments (Bull, 1983; Christidis, 1989; Schmid *et al.*, 1989; Sola *et al.*, 1981). For heuristic purposes, this range of forms is often viewed as steps along an evolutionary pathway, the intermediates representing transitional forms between the two extremes (*Figure 10.1*). The frequent examples of

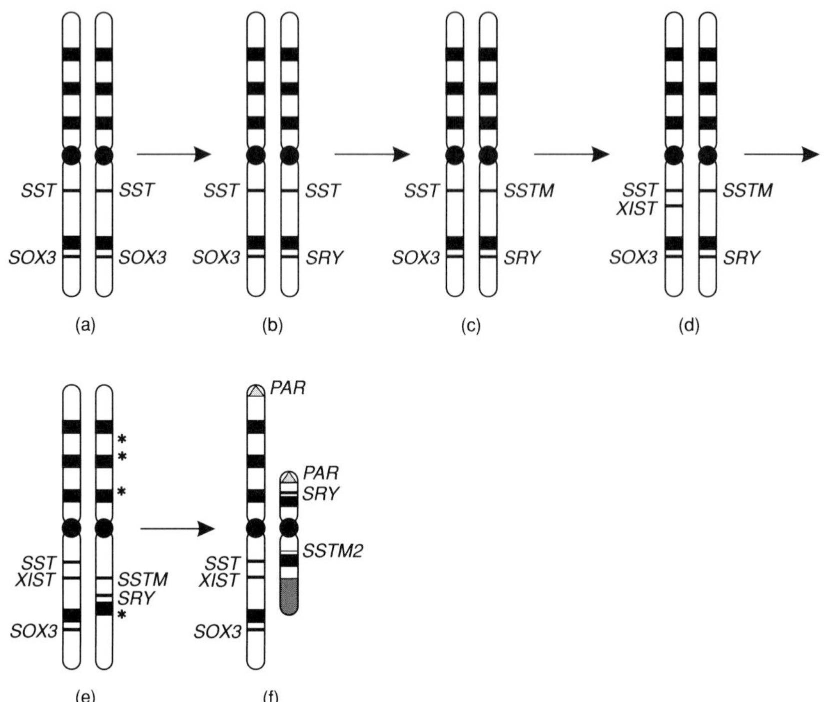

Figure 10.2. Pathway of mammalian sex chromosome evolution: a hypothesis. (a) Before the mammalian system of sex determination evolved, the proto X and Y chromosomes were simple, ordinary autosomes. Sex was determined either by an environmentally controlled system or by a genotypically controlled system but at a locus different from that controlling sex determination today. Two genes are shown, *SOX3* and *SST*, that are examples of potential sex-determining and sexually selected genes, respectively; *SOX3* is a known gene on the mammalian X chromosome (see below) and *SST* (for sexually selected trait) is currently theoretical. Initially, the two alleles of the *SOX3* gene and *SST* genes were functionally identical. (b) Mutation at *SOX3* on one chromosome occurred that made the gene product testis-determining. The gene is now called *SRY* to differentiate it from *SOX3*. The *SRY*-bearing chromosome is by definition a Y chromosome (right), the *SOX3*-bearing chromosome an X chromosome (left). *SRY*/*SOX3* individuals were male and *SOX3*/*SOX3* individuals were female. At this stage, there is no selective force on the sex chromosomes to differentiate. (c) A mutation, *SSTM*, arose at the second locus *SST* on the *SRY*-bearing chromosome that was advantageous to the male carrying it. This chromosome began to spread throughout the population. Recombination between *SRY* and *SSTM* created chromosomes that were not at an advantage, and these recombinants did not spread. (d) Another mutation arose, *XIST*, that suppressed recombination between *SRY* and *SSTM* by condensing the sex chromosomes prematurely during male meiosis. Recombination suppression stabilized the Y chromosome to keep the advantageous arrangement of *SSTM* and *SRY*. (e) Under the

genes in human, mouse, and even marsupials that are present on both the X and Y chromosomes (see below) is consistent with this idea of evolution from a formerly homomorphic pair of chromosomes. These observations alone are not conclusive; nevertheless, an autosomal starting point is the most favourable hypothesis.

10.2.2 Evolution of sex-determining and sexually selected alleles

Imagine that, in an emerging mammalian species, a simple allelic sex-determining mechanism is operating such that the presence of sex-determining allele, *SXD*, determines the male sex (*Figure 10.2b*). The genotype *SXD/sxd* confers maleness with gametes *SXD* and *sxd*, while the genotype *sxd/sxd* produces females with only *sxd* gametes, the genotype *SXD/SXD* being absent. The *SXD*-bearing chromosomes by definition are Y chromosomes and the *sxd*-bearing chromosomes are X chromosomes. In *Figure 10.2*, the *SXD* allele is called *SRY*, because *SRY* is now known to be the major testis-determining factor in mammals, and the *sxd* allele is referred to as *SOX3* as it has been suggested that *SOX3* is the X-linked homologue of *SRY* (see Section 10.4.1). Now suppose that an allele *SSTM* (for sexually selected trait in males) arises by mutation at a linked locus (*SST*) on a Y (*SXD*-bearing) chromosome that confers an advantage in males but not in females (*Figure 10.2c*). The Y chromosome carrying this new allele will spread throughout the population because it confers a fitness advantage to males carrying it. However, the selective force causing the novel Y chromosome to spread will be countered by the force of recombination between the sex chromosomes because recombination will replace the favourable allele *SSTM* with the non-favourable allele *SST* (and place the favourable allele on *sxd*-bearing chromosomes where it has no selective advantage). It is clear from this scenario that any genetic alteration which suppresses recombination between the *SXD* (male-determining) allele and the *SSTM* (advantageous in males) allele, resulting in the inheritance of

---

action of Muller's ratchet (see text), the Y chromosome began to accumulate mutations (depicted by stars) and there was no force to act against rearrangements, for example, inversions (indicated by the new locations for *SRY*, *SSTM* and the lowest chromosome band on the Y chromosome). (f) As the Y chromosome was accumulating mutations, complementary increases in the activity of genes on the X chromosome occurred to compensate for the loss of Y-gene activity. The activity of *XIST* was then adapted to function in X-chromosome inactivation (see text) which acted to equalize the levels of expression of X-linked genes in males and females. As the Y chromosome accumulated mutations, large blocks of genes were deleted, and the Y chromosome acquired large blocks of heterochromatin (cross-hatched region). The old sexually selected gene (*SSTM*) could be lost and new male-advantageous mutations acquired, here represented by *SSTM2*. Such genes could arise on the Y chromosome, for example, by the addition of chromosomal segments to the sex chromosomes, which may have served for a time as pseudoautosomal regions (PAR). Some of these added genes were later incorporated into the Y-specific region and perhaps acquired novel functions.

the favourable combination of alleles essentially as a single locus, would be strongly selected for and would spread through the population towards fixation (*Figure 10.2d*).

What sorts of mutations are representative of *SSTM* alleles? One example is particularly illustrative: brightly coloured male guppies are preferred by females but are more susceptible to predation (Winge, 1932). The 'colour' genes localize to the guppy Y chromosome and recombination between the guppy sex chromosomes, which are homomorphic, is suppressed. Thus, in this example, the *SSTM* alleles are sexually selected traits that are advantageous to males and disadvantageous to females. Another example might be a trait that enhances male fertility and concomitantly decreases female fertility.

Irrespective of the nature of the *SST* alleles present during mammalian evolution, the suppression of recombination on the Y chromosome between favourable combinations of traits advantageous in males is presumed to be the catalyst for a chain of events that ultimately endowed the sex chromosomes with the properties we observe today.

10.2.3 Recombination suppression and degeneration of the Y chromosome

The suppression of recombination outlined above, central to models of sex chromosome evolution, can be accomplished in a number of ways: (i) recombination in the heterogametic sex can be suppressed, such as occurs in the fruitfly *Drosophila melanogaster*; (ii) mutations at *cis*-acting loci that normally promote recombination could be selected for, eliminating their normal recombination function; (iii) the two loci could be relocated close to one another or into a region that recombines infrequently (for example, close to centromeres) by a structural chromosomal alteration like an inversion (inversions have the effect of suppressing recombination in the inverted DNA segment because recombination there is likely to produce abnormal gametes); (iv) sequences that suppress recombination could have been recruited into the region. Below I hypothesize that *XIST*, which is a gene crucial to the mammalian system of dosage compensation known as X-chromosome inactivation, has played a key role in the evolution of recombination suppression in mammals.

In vertebrate species that have only slightly heteromorphic sex chromosomes, the appearance of a small segment of heterochromatin on the Y chromosome or a minor structural difference between the sex chromosomes is quite common (Schmid *et al.*, 1989). How the sex chromosomes go from small differences like these to extreme heteromorphism is unknown. Perhaps it depends on how far apart on the Y chromosome the *SXD* and *SST* loci were when the advantageous mutations arose or on what structural changes occurred on the Y chromosome early in the evolutionary process. Perhaps, as the process of chromosomal differentiation got underway, other alleles that increased the fitness of the male sex arose on the Y chromosome so that recombination suppression was selected for in gradually larger chromosomal regions. These thoughts, although speculative, are instructive because different genetic mechanisms exist which can suppress recombination and one or more could have operated early in the evolutionary process that differentiated the mammalian sex chromosomes.

The next stage in the model of sex chromosome evolution involves the accumulation of mutations on the Y chromosome, primarily by a process known as Muller's ratchet (Charlesworth, 1978; Felsenstein, 1974; Muller, 1964): recombination suppression (selected for to maintain advantageous combinations of alleles) effectively prevents mutations occurring on the Y chromosome from being removed by recombination. Imagine a population where most Y chromosomes contain some deleterious mutations, but where some mutant-free chromosomes also exist. This class of chromosome (like all others) is vulnerable to stochastic loss from the population by random drift. Critically, this loss is permanent, since the mutant-free class cannot be reconstituted by recombination. Further classes of Y chromosomes (with one, two, three, etc., deleterious mutations) would subsequently be open to further rounds of random drift and removal from the population (each one a turn of Muller's ratchet), resulting in a steady increase in the number of mutants at loci in the region of suppressed recombination. Furthermore, the total number of Y chromosomes in a population is only one-third the total number of X chromosomes and one-quarter the total number of autosomes. Consequently, the Y chromosome is particularly vulnerable to this process because population bottlenecks greatly reduce the effective population size of Y chromosomes (Chapter 12). The combination of Muller's ratchet and population bottlenecks is likely to have resulted in a steady accumulation of mutations on the evolving Y chromosome.

The proposal that recombination suppression can cause a loss of fitness, presumably due to the accumulation of recessive mutations, has been verified experimentally in *D. melanogaster* (Rice, 1994). Making use of the fact that male meiosis occurs in the absence of recombination, Rice passaged a given number of wild-type *Drosophila* chromosomes II and III solely through the male sex in small replicate populations. After 34 generations, a test cross was performed and a significant reduction in fitness of heterozygous flies was observed in experimental lines compared with control flies similarly passaged but with recombination allowed.

The suppression of recombination also has a profound effect on the consequences of advantageous mutations. Any advantageous mutation arising in the region of recombination suppression will be driven rapidly to fixation in the population by a 'selective sweep' that will take with it any recessive mutations that have accumulated in the region where recombination is suppressed. A consequence of these mechanisms of drift and selective sweeps is that the Y chromosome carries fewer polymorphic alterations because all the Y chromosomes in a population will have a recent common ancestor. This consequence has been well documented in the evolution of the human Y chromosome (Chapter 12; Dorit *et al.*, 1995; Hammer, 1995; Whitfield *et al.*, 1995).

Thus, at the beginning of sex chromosome evolution we can envisage that different sets of slightly degenerate Y chromosomes would be segregating in the population until one Y chromosome, with a given number of recessive mutations on it, becomes fixed. This chromosome would be subject to subsequent rounds of accumulation of recessive mutations and fixation (Muller's ratchet). This degenerating material could, however, be subject to deletions (leading to heteromorphic sex chromosomes), as these would be selectively neutral. Finally, the Y chromosome would become devoid of most of the genes it formerly contained, retaining only those directly involved in sex determination and those which have become linked due to being advantageous in males. This process of degeneration can be viewed as an evolutionary sink that isolates the Y chromosome permanently and opens the way for the rapid and frequent remoulding of it. The Y chromosome, in its genetic isolation, becomes the subject of numerous random and inconsequential alterations. New chromosomal segments, non-functional pseudogenes, and repetitive sequences can be acquired and lost. For these reasons, the Y chromosome is the most rapidly changing and least conserved chromosomal element in the genome.

10.3 Evolution of X-chromosome inactivation

As outlined above, a natural consequence of the process of Y chromosome degeneration is likely to be reduced population fitness, presumably from haploinsufficiency caused by recessive mutations at many genes. However, the deleterious effects of haploinsufficiency could be alleviated by doubling the expression, in males, of X-linked genes affected by mutations on the Y chromosome. This would create equal amounts of X-linked gene products in males and females, a process known as dosage compensation. In *D. melanogaster*, dosage compensation is achieved in this way. In contrast, mammals have evolved a chromosome-wide inactivation system which has silenced the majority of genes on one X chromosome in females, thus achieving equal levels of expression in males and females by halving the amount of X-linked gene products in females.

This chromosome-wide inactivation system has created functional X-monosomy in females that matches the monosomy present in males. Monosomy of an autosome is not observed in mammals and presumably is deleterious, whereas for some autosomes trisomy is viable. In general, segmental aneuploidy causing an increase in gene dosage is less deleterious than that causing a decrease in gene dosage (Sandler and Hecht, 1973). Because of this, it seems likely that the evolutionary path to silencing one X chromosome in females would include an intermediate stage in which expression of certain genes on the X chromosome is increased in a non-sex-limited fashion as these genes are being lost from the Y chromosome. This process would have the effect: (i) of maintaining the amount of X-linked gene product in males at the level which existed before expression from the Y-linked genes was reduced by mutation and (ii) of increasing the amount of X-linked gene product in females in excess of the level that existed before. Thus, as the expression of more and more genes increased in a non-sex-limited fashion, X chromosome inactivation could evolve, restoring the level of X-linked gene expression in females to the same as that present in males.

The cytological signs of the inactive X chromosome are a condensed heterochromatic triangular structure, the Barr body, which is always localized to the nuclear membrane in interphase cells, and late replication of the inactive X chromosome in S phase (Gartler *et al.*, 1992). X-inactivation is one of the earliest developmental events that regulates the activity of genes and, in cells contributing to the embryo, the maternally or the paternally derived X chromosome are equally likely to be inactivated (Migeon, 1994).

As a result of X-inactivation, the X chromosome has become the most stable and most highly conserved chromosomal element in the genome, a fact given the name Ohno's law from his rigorous documentation of it (Ohno, 1967). Ohno's law holds true for two reasons. First, over and above the usual problems associated with the inheritance of translocations in balanced or unbalanced states, an autosome carrying a segment of X-linked genes meets the special problem of going from monosomy of the formerly X-linked material to disomy of it (excessive gene expression). The second, unexpected reason for the conservation of genes on the X chromosome arises from the fact that males carrying X–autosome translocations are usually infertile (Henderson, 1964; Lifchytz and Lindsley, 1972; Tres, 1975). Infertility probably occurs because the X chromosome normally condenses prematurely in male meiosis, inactivating it and any material attached to it. The X-linked gene for phosphoglycerate kinase (*PGK1*) provides an example of this effect. As a glycolytic enzyme, PGK is a critical product generating energy for sperm. Because the X-linked gene is inactivated, there is an autosomal gene *PGK2* which is expressed in the haploid phase of spermatogenesis. Autosomal *PGK2* is a functional, processed cDNA generated by retrotransposition of the X-linked gene to an autosome (McCarrey and Thomas, 1987; Boer *et al.*, 1987). Autosomal segments attached to the X chromosome as a result of translocation have been shown experimentally to be inactivated (Jaafar *et al.*, 1993). Thus, as a force conserving the X chromosome in mammals, X–autosome translocations lead to infertility because spermatogenesis is adversely affected by haploinsufficiency of the translocated autosomal genes (insufficient gene expression).

10.3.1 Marsupial and extra-embryonic X-inactivation

It has been suggested that the premature condensation of the X chromosome in male meiosis was adapted for X-inactivation in somatic cells and therefore constituted the evolutionary precursor to inactivation there (Lifschytz and Lindsley, 1972). In mouse, the paternal X chromosome is inactivated preferentially in the extra-embryonic membranes (Takagi and Sasaki, 1975), indicating that the X chromosomes coming from the father and from the mother are in fact different as the result of an imprint (see below). In human, X-inactivation is 'leaky' in the extra-embryonic membranes (Migeon *et al.*, 1985). In metatherians, however, the inactivation system is less stable and is tissue-specific; inactivation is also preferentially paternal in metatherians (Cooper *et al.*, 1993; VandeBerg *et al.*, 1987). Thus, inactivation in the eutherian extra-embryonic membranes is similar to metatherian X-inactivation. Because it is unstable, tissue-specific and non-random, the metatherian X-inactivation has been regarded as more primitive than the eutherian system.

From the analysis of XO females and XXY males, the metatherian Y chromosome is known to be sex-determining. However, XO females develop a scrotum and XXY males develop pouches (Cooper, 1995), suggesting that X chromosome dosage plays a role in the development of certain metatherian sexual characteristics and that these sexual characteristics cannot be under the control of the X-inactivation system. Under these circumstances, X-inactivation would be an antagonistic force that would create male pseudohermaphrodites (XX individuals with scrota). Perhaps this gene dosage phenomenon in metatherians acted as a selective constraint against the development of a stable and tissue-non-specific mechanism of X-inactivation. Perhaps in the eutherian lineage, the gene dosage control of sexual characteristics was superseded, allowing the X-inactivation system to become more stable and applied to all tissues.

10.3.2 Counting and the inactivation centre

In X-chromosome inactivation, a counting mechanism must exist that measures the number of X chromosomes to autosomes in a cell, since in certain persons with more than two X chromosomes all but one X chromosome is inactivated (Grumbach *et al.*, 1963). In a zygote with two X chromosomes one is chosen for inactivation, and the cell establishes an inactivation signal that silences nearly the whole X chromosome. From analysis of X–autosome translocations, this signal is initiated at a site on the X chromosome designated the X-inactivation centre (*Xce*) (Rastan, 1994). The inactivation centre is required in *cis* for choice and propagation of the inactivation signal but, after inactivation is accomplished, the inactive state is stable and the inactivation centre is dispensable (Brown and Willard, 1994). From the minimum interval defined by X–autosome translocations in human and mouse (Brockdorff *et al.*, 1991; Brown *et al.*, 1991a), a strong candidate for the *Xce*, referred to as *XIST*, has been identified that has exceptional properties (Brown *et al.*, 1991b).

XIST is the only element on the X chromosome that is transcribed from the inactive X and not transcribed from the active X chromosome (Borsani *et al.*, 1991; Brown *et al.*, 1991b). This observation alone suggests that *XIST* is subject to some form of imprinting (see below). Transcription of the *XIST* gene is first detected in males at meiosis when the X and Y chromosomes prematurely condense (McCarrey and Dilworth, 1992; Richler *et al.*, 1992; Salido *et al.*, 1992). In female embryos, transcriptional activity is first detected just prior to the time of X-inactivation as measured by the X's late replication phenotype (Kay *et al.*, 1993, 1994). The *XIST* mRNA transcript is processed in multiply spliced forms, but none of these forms appear to encode a protein product. *XIST* RNA has been detected in direct association with the Barr body, leading to the hypothesis that *XIST* RNA itself mediates inactivation in *cis* (Brockdorff *et al.*, 1992; Brown *et al.*, 1992). RNA is a diffusible molecule, so how it might remain associated with the inactive X chromosome is unknown.

Recently, a 7 kb deletion including the first exon of the *XIST* transcription unit was introduced into the *XIST* gene of embryonic stem (ES) cells by homologous recombination (Penny *et al.*, 1996). Experiments in cell culture and in chimeric mice showed that the mutation destroyed the ability of the chromosome carrying it to undergo X-inactivation, demonstrating definitively that *XIST* is an integral part of the inactivation system. The data support a model in which the counting mechanism establishes a block of transcription of *XIST* on one X chromosome. On the remaining X chromosome(s), *XIST* expression is activated and operates in *cis* to establish and propagate the silencing signal to X-linked genes. The *XIST* gene is not known to be present on or absent from metatherian X chromosomes, possibly because sequence homology between human or mouse *XIST* and metatherian *XIST* is too low for detection. This problem leaves open the key questions of when and how *XIST* arose.

10.3.3 Imprinting and X-inactivation

The preferential inactivation of the paternal X chromosome in marsupials and in mouse extra-embryonic tissues was initially taken as evidence for an imprint on the paternal X chromosome. Transformation experiments indicated, however, that the imprint was not at the level of the individual X-linked gene (Kratzer *et al.*, 1983). The recent work on *XIST* turns this thinking on its head: active *XIST* is an X-inactivation signal, while an inactive *XIST* blocks this X-inactivation. Studies on the methylation state of *XIST* in sperm, eggs and preimplantation embryos have demonstrated that the promoter region of *XIST* is methylated on the maternal X chromosome and unmethylated on the paternal X chromosome (Ariel *et al.*, 1995; Zuccotti and Monk, 1995). Therefore, *XIST* is not imprinted in spermatogenesis but is imprinted in oogenesis.

These observations suggest the following evolutionary model (*Figure 10.2d–f*): recombination suppression between the sex chromosomes has been selected for to maintain an advantageous haplotypic arrangement of alleles. This suppression only needs to be enforced in male meiosis. It is possible, therefore, that the *XIST* gene originated on the sex chromosomes as a factor that suppresses recombination in male meiosis by creating a signal in its active state that results in the premature condensation of the sex chromosomes. This condensation would limit the access of recombination complexes to the chromatin and coincidentally inactivate the sex chromosomes there. The *XIST* gene could subsequently have evolved additional functions in the inactivation of the X chromosome in the zygote. An imprint on the X chromosome could have subsequently been adapted to establish the inactivity of *XIST* in female gametogenesis. This would result in males receiving an X chromosome from their mothers that remains active, whereas females would receive a maternal X that is preferentially active, and a paternal X chromosome, lacking an imprint, that would be preferentially inactivated. For this model to be complete, the imprint must be removed in the male germ line. To account for observation, it is presumed that in the eutherian lineage, this imprinting mechanism has subsequently been superseded so that X-inactivation is random in the tissues of the embryo proper. This inactivation system has the advantage that new recessive mutations are no longer exposed in females: the recessive mutation can be on either the active or the inactive X chromosome, and selection of cells with the mutation on the inactive X chromosome ensures embryo survival.

10.4 Genes on the Y chromosome

The evolutionary processes outlined in the preceding sections provide plausible explanations for the widely observed conservation of genes on the X chromosome (by dosage compensation) and paucity of genes on the Y chromosome (due to the action of successive rounds of genetic drift, selective sweeps, or both). Viewed in this context, it is apparent that exceptions to the rule of conservation of genes on the X chromosome, or of non-conservation of genes on the Y chromosome, can provide insight into the selective forces at work on the sex chromosomes.

Genes on the Y chromosome can be divided into two groups: those that are Y-linked (strict father-to-son transmission) and those that are pseudoautosomal (i.e. that recombine between the sex chromosomes, see below). The Y-linked genes that have been identified (*Table 10.1*) can be further divided into two groups: those that are present on both sex chromosomes and those that are present exclusively on the Y chromosome. Genes on the Y chromosome have been shown to evolve rapidly, most likely by the process of male-driven evolution (Shimmin *et al.*, 1993; Tucker and Lundrigen, 1993; Whitfield *et al.*, 1993). Consequently, the sequences of orthologous genes can diverge to the point where it is no longer possible to identify a gene in one species using the molecular probe that identifies it in another species. This mechanism could also explain how a gene sequence on the Y chromosome could diverge rapidly from its X-linked homologue to the point where it is no longer homologous. Even with this limitation, it has been possible to identify genes on the

Table 10.1. Cloned Y-linked genes in human and mouse

Human			Mouse			Gene name
Y-linked gene	X-linked homologue	Inactivated	Y-linked gene	X-linked homologue	Inactivated	
SRY	None	N.A.	*Sry*	None	N.A.	Sex-determining region Y
RPS4Y	*RPS4X*	No	None	*Rps4x*	Yes	Ribosomal protein small subunit no. 4
ZFY	*ZFX*	No	*Zfy1, Zfy2*	*Zfx*	Yes	Zinc finger
TSPY	None	N.A.	None	None	N.A.	Testis-specific protein Y
PRKY[a]	*PRKX*	No	N.I.	N.I.		Protein kinase
AMELY[a]	*AMELX*	Partially?	None	*Amel*	Yes	Amelogenin
XGP[b]	*XG*	No	N.I.	N.I.		Xg blood group
DFFRY	*DFFRX*					*Drosophila* fat facets-related
KALP[b]	*KAL*	Yes	N.I.	N.I.		Kalman's syndrome
STSP[b]	*STS*	Partially	*Sts*[c]	*Sts*[c]	No	Steroid sulphatase
ARSDY[b,d]	*ARSD*	No	N.I.	N.I.		Arylsulphatase D
ARSEY[b,d]	*ARSE*	No	N.I.	N.I.		Arylsulphatase E
SMCY	*SMCX*	No	*Smcy*	*Smcx*	No	Selected mouse cDNA, H–Y antigen
RBM	None	N.A.	*Rbm*	None	N.A.	RNA binding motif
DAZ	None	N.A.				Deleted in azoospermia
None	*UBE1*	Partially?	*Ube1y*	*Ube1x*	Yes	Ubiquitin-activating enzyme E1

N.A., not applicable. N.I., not identified. Blanks represent information that is unknown. Genes are listed from distal short arm to distal long arm of the Y chromosome. References for the genes in the table are cited in the text except for *DFFRY* (Kwaja *et al.*, 1994), *ARSD* and *ARSE* (Franco *et al.*, 1995) and *XG* (Ellis *et al.*, 1994).

[a] The Y-linked homologue is expressed at a lower level compared with the X-linked gene.

[b] Evidence exists that the Y-linked homologue is non-functional.

[c] Pseudoautosomal.

[d] Order on the Y chromosome long arm relative to other genes is not known.

ancient Y chromosome and to infer the existence of evolutionary mechanisms that have caused major changes in the structure of the sex chromosomes since the time they differentiated into heteromorphic sex chromosomes.

10.4.1 The testis-determining factor

Mammalian testis-determination is controlled by a single Y-linked gene. This gene was identified by the genetic analysis of phenotypic human males with an XX chromosome constitution. The majority of XX males have received an X chromosome in which a small portion of Y-specific chromatin from the Y short arm (Yp) was transferred to the X short arm (Xp) by an illegitimate recombinational event between the sex chromosomes during male meiosis (Ferguson-Smith, 1966). Different lengths of Y-specific chromatin can be transferred to the X chromosome, thereby defining a nested set of intervals (Vergnaud *et al.*, 1986). The minimum region defined in XX males was 35 kb (Palmer *et al.*, 1989). From that minimum region a single gene was identified, referred to as *SRY* (Sinclair *et al.*, 1990). In mouse, a very similar genetic system (referred to as *Sxr* for sex-reversed) is known in which the testis-determining gene was translocated from the short arm to the pseudoautosomal region on the long arm. The sex-reversed mouse carrying this translocation (XX*Sxr*) also defined a small region in which *Sry* was localized (Gubbay *et al.*, 1990).

Two sorts of genetic observation have demonstrated that *SRY* encodes the single dominantly acting testis-determining factor of the Y chromosome. First, people who inherit a *de novo* mutation in *SRY* develop into phenotypic females with infantile female internal sex organs and with streak gonads (gonads that failed to develop into mature ovaries), an entity known as XY gonadal dysgenesis (Berta *et al.*, 1990; Jager *et al.*, 1990). Similarly, in mouse, a Y chromosome carrying a deletion of *Sry* fails to determine testes. (Mice that carry the deleted Y chromosome develop as fertile females due to a difference in the genetics of the mouse and human X chromosome; Mahadevaiah *et al.*, 1993.) Secondly, a mouse *Sry* transgene has been shown to cause XX embryos to develop as males, indicating that *Sry* expression acts dominantly to produce testis determination (Koopman *et al.*, 1991). Thus, the presence of an active, functional, SRY in the zygote is sufficient to determine testes.

SRY is a member of a family of proteins which contains a peptide motif approximately 80 amino acids long referred to as an HMG box, named after the high mobility group protein HMG1. The HMG box gives SRY the capacity to bind to DNA in a sequence-specific manner (Harley *et al.*, 1992, 1994). When SRY binds to DNA it bends the DNA molecule by as much as 120°. SRY will also bind preferentially to cruciform DNA structures (Ferrari *et al.*, 1992). DNA hybridization experiments using *SRY* as a probe identified a large family of genes, each member having an HMG box domain closely related to *SRY*'s HMG box (Gubbay *et al.*, 1990). These genes have been designated *SOX* genes (for SRY-box-related genes). Many of these genes are expressed specifically during early development, implicating them in the developmental fate of different embryonic cell lineages.

One of these is the X-linked *SOX3* gene (Stevanovic *et al.*, 1993). Of the known *SOX* genes, *SRY* and *SOX3* share the highest homology across the HMG box domain. Outside this domain, nearly all *SOX* genes (*SRY* included) are highly divergent. Both *SOX3* and *SRY* are intronless genes. *SOX3* is expressed in neural crest cells and is implicated in nerve growth or differentiation. In mouse, *Sox3* is expressed in the genital ridge. It has been suggested that *SRY* and *SOX3* may once have been homologous loci and that a mutation arose that transformed one ancestral *SOX3* allele into a testis-determining gene (Foster and Graves, 1994). If this is true, *SRY* and *SOX3* are equivalent to the alleles *SXD* and *sxd*, respectively, in the model of sex chromosome evolution elucidated above (*Figure 10.2*). *SRY* has been detected in males of virtually all eutherian species tested (Sinclair *et al.*, 1990) and it is found in metatherians (Foster *et al.*, 1992). Identification of the *SRY* orthologue on the metatherian Y chromosome initially was difficult because many autosomal *SOX* genes hybridize to the *SRY* probe at the reduced stringency conditions that had to be used, and these genes masked the presence of *SRY*. If proof could be obtained that *SRY* is present on the prototherian Y chromosome, then *SRY* doubtless would have been part of that original genetic system on the ancestral sex chromosomes.

10.4.2 Fertility factors

In human, *SRY*-positive XX males are infertile, and deletions of the long arm of the Y chromosome (Yq) have been associated with azoospermia (Tiepolo and Zuffardi, 1976). The XX*Sxr* mouse is also infertile (Sutcliffe and Burgoyne, 1989). These observations imply the existence of genes on the Y chromosome that act in spermatogenesis. Recently, candidate genes for being fertility factors, referred to as *RBM* (Ma *et al.*, 1993) and *DAZ* (Reijo *et al.*, 1995), have been isolated from Yq (*Table 10.1*). *RBM* represents a gene family composed of approximately 20 members spread from the proximal Yp to the distal part of the euchromatic region on Yq. *DAZ* is a single-copy gene; however, the coding sequences contain seven tandem repeats of a sequence that is duplicated many times on the Y chromosome (referred to as *DYS1*; see Lucotte and Ngo, 1985). Whether either or both of *RBM* and *DAZ* contribute to male fertility is presently controversial, but the facts that both genes are expressed specifically in testis, and that both genes contain a protein motif that potentially binds to RNA (Burd and Dreyfuss, 1994), suggest a common function for *RBM* and *DAZ* in spermatogenesis. *TSPY* is another repeat gene family on the Y chromosome that is expressed specifically in testis (Arnemann *et al.*, 1987). The evolutionary source of *RBM*, *DAZ* and *TSPY* genes is unknown. *RBM* is conserved in eutherians (Ma *et al.*, 1993), but the mouse gene was difficult to identify, again underscoring the problem of identification of orthologues in other species and of potential homologues on the X chromosome to which genes like *RBM* and *DAZ* might be ancestrally related.

10.4.3 Homologous X-linked and Y-linked genes

The majority of known genes on the Y chromosome have X-linked homologues. In human, the functional Y-linked genes (*RPS4Y*, *ZFY*, *AMELY*, *SMCY*; see *Table 10.1*) that are also present on the X chromosome all escape X-inactivation. This observation is consistent with the model for the evolution of X-inactivation presented earlier; if the X- and Y-linked genes perform the same function, then dosage balance between males and females is desirable. For example, in human, there is evidence that the *RPS4Y* gene product is functionally the same as the *RPS4X* gene product (Watanabe *et al.*, 1993). If the gene products perform identical functions, it raises the question: why have Y-linked homologues been retained, given the tendency of the Y chromosome to degenerate? If a gene is not conserved on the Y chromosome in all mammals, then chance retention, rather than selection, is the most likely reason. By this argument, *RPS4Y* and *AMELY* are present on the human Y chromosome purely by chance, since *Rps4* and *Amel* in mouse are X-linked with no Y-homologues present (Chapman *et al.*, 1991; Fisher *et al.*, 1990; Nakahori *et al.*, 1991; Zinn *et al.*, 1991).

ZFY is present on the Y chromosome of all eutherians tested (Page *et al.*, 1987), but is autosomal in metatherians (Sinclair *et al.*, 1988), again suggesting a chance retention event; however, *ZFY* presents an interesting case. *ZFY* encodes a DNA-binding protein with six and a half two-finger motifs that most likely chelate zinc ions, one to each finger (Page *et al.*, 1987). The human *ZFY* and *ZFX* genes are expressed in all tissues examined (Palmer *et al.*, 1990; Schneider-Gadicke *et al.*, 1989), and *ZFX* escapes X-inactivation (Schneider-Gadicke *et al.*, 1989). The *ZFY* and *ZFX* gene products are highly conserved (95% identity), and quite possibly they bind to the same DNA sequence and carry out identical functions. In mouse, there are two *Zfy* genes on the Y chromosome; their expression is testis-specific (Mardon *et al.*, 1989; Nagamine *et al.*, 1989), but the mouse *Zfx* gene is subject to X-inactivation (Adler *et al.*, 1991; Ashworth *et al.*, 1991). Both mouse *Zfy* genes are significantly diverged from the mouse *Zfx* gene and equally so from the human *ZFX* and *ZFY* genes. These observations indicate that the mouse *Zfy* genes have evolved functions different from these other *ZFX* and *ZFY* genes. The mouse *Zfy* may have evolved a testis-specific function that is advantageous in males.

The same could be true of the human *ZFY*, but another explanation is more likely. If the dosage of the *ZFY* gene product is critical during early development, then it would be difficult to bring it under the control of X-inactivation which would, presumably, involve transitory dosage alterations likely to be highly deleterious. Functionally similar, active, Y-linked and X-linked copies would, therefore, be selectively maintained. For the *Zfx* genes in mouse, this difficulty may have been overcome with the duplication of the *Zfy* gene on the Y chromosome as an intermediate stage before the X-linked copy could come under X-inactivation control. Once *Zfx* was X-inactivated, the Y-linked copies were free to take on novel functions.

A process similar to this may explain the puzzling *UBE1* genes (Kay *et al.*, 1991; Mitchell *et al.*, 1991). *UBE1* was most likely present on the sex chromosome ancestral to all mammals because X-linked and Y-linked *UBE1* genes are found in metatherians (Mitchell *et al.*, 1992) and prototherians (Wilcox and Graves, personal communication). Such conservation on the mammalian Y chromosome is rare. It suggests that some selective mechanism must be responsible for its conservation. Although *UBE1* is present on the Y chromosome of rodent, carnivore and artiodactyl species, it is absent from the Y chromosome of Old World monkeys and apes. In mouse, *Ube1x* is subject to X-inactivation (Kay *et al.*, 1991) and is present in several copies on the mouse Y chromosome (Mitchell *et al.*, 1991). *Ube1y* is a candidate for the mouse spermatogenesis gene *Spy* (Sutcliffe and Burgoyne, 1989). These diverse observations can be reconciled as follows. Because mouse *Ube1x* is X-inactivated, the *Ube1y* gene was able to evolve a novel function. In the primate lineage, *UBE1* on the X chromosome is coming under X-inactivation control where it is partially inactivated (see below), so that the Y gene was free to evolve as well. Instead of evolving a novel function, however, mutation and drift resulted in the gene being lost.

Another example of this process is observed for the *SMCY* – recently demonstrated to encode the classic H–Y antigen (Scott *et al.*, 1995; Wang *et al.*, 1995) – and *SMCX* genes. These genes are present on the X and Y chromosomes of many mammalian species, including the red kangaroo, which suggests that the gene has been on the sex chromosomes for the last 150 million years (Agulnik *et al.*, 1994a). But there are several mammalian species known in which a Y-linked copy has not been detected. In human and mouse, *SMCX* genes escape X-inactivation (Agulnik *et al.*, 1994b). Consequently, if the Y-linked copy were deleted, the dosage of *SMCX/SMCY* gene product in males would be half that in females. Perhaps, in the lineages leading to the mammalian species that lack a *SMCY* gene, the *SMCX* gene came under X-inactivation control, releasing *SMCY* from selective pressure and ending in deletion from the Y chromosome.

In summary, genes can come under X-inactivation control independently in different mammalian lineages. Thus, a gene that is conserved on the Y chromosome in one lineage can evolve a novel function in another lineage or can be lost from the Y chromosome in yet another. X-inactivation itself can be partial (Migeon *et al.*, 1982), such that the allele on the inactive X chromosome is expressed but not at the normal level. This state may be transitional – the final adjustment to full inactivation has not yet been made. Similarly, genes on the Y chromosome such as *AMELY* (Salido *et al.*, 1992) and *PKRY* (Klink *et al.*, 1995), both of whose X-linked homologues escape X-inactivation at least to some degree (Whitkop, 1967; G. A. Rappold, personal communication), can be expressed at a lower level than their X-linked homologues. Perhaps these Y-linked genes are also in a transitional state because they must co-evolve with the X-inactivation process: the Y-linked copy still has some functional contribution, but until adjustments in the level of expression of the X-linked copy are made, expression of the Y-linked copy is required and X-inactivation is either not possible or only partial. Once X-inactivation does take over completely, the Y-linked copy can evolve a novel function (on rare occasions) or it can be deleted.

10.5 The pseudoautosomal region

The sex chromosomes must segregate to the opposite poles at meiosis, otherwise the chromosomal basis of sex determination breaks down. As long as the region of recombination suppression and Y-chromosomal degeneration is restricted to one part of the sex chromosomes, there could remain a segment in which the sex chromosomes maintained sequence homology, engaged in recombination, and utilized the general mechanisms of meiosis to achieve appropriate segregation. A marker locus in a region of structural homology borne on the Y chromosome could be transferred to the X chromosome and thereby would not exhibit strict father-to-son transmission. Similarly, a marker locus borne on the X chromosome that could be transferred to the Y chromosome would not exhibit the usual father-to-daughter transmission. Such a region of homology on the sex chromosomes that is known to engage in regular exchange is referred to as a pseudoautosomal region (PAR) because genes there would appear to have autosomal or partially sex-linked inheritance (Burgoyne, 1982; Rappold, 1993). What are the genetic properties of such regions and how do they fare in the evolution of the sex chromosomes?

In human, two PARs have been identified (Cooke *et al.*, 1985; Freije *et al.*, 1992; Simmler *et al.*, 1985). The larger, PAR1, is localized to the distal tips of the short arms of the sex chromosomes and the smaller, PAR2, is localized to the distal tips of the long arms of the sex chromosomes. From long-range restriction mapping and molecular cloning experiments, PAR1 is estimated to be 2600 kb in length (Brown, 1988; Petit *et al.*, 1988; Rappold and Lehrach, 1988) and PAR2 is 320 kb (Kvaløy *et al.*, 1994). In the mouse, there is a single PAR localized to the distal tips of the long arms of the sex chromosomes (Singh and Jones, 1982). Long-range restriction mapping of the mouse PAR is incomplete; however, its size is likely to be greater than 320 kb but not much larger than 2600 kb (Kipling *et al.*, 1996). In these examples, the PARs constitute less than 0.1% of the genome and are not microscopically visible. Eutherian sex chromosomes are extremely heteromorphic, but the PARs are submicroscopic homomorphic regions.

Genetic evidence suggests that enough exchange occurs between the sex chromosomes in the human PAR1 and the mouse PAR to ensure that virtually every XY bivalent at anaphase has at least one crossover; that is, a marker locus at the telomere segregates independently of the sex-specific regions (Cooke *et al.*, 1985; Keitges *et al.*, 1985). In human, marker loci proximal to the telomere exhibit different degrees of sex linkage forming a gradient of recombination from the telomere to the proximal limit of the PAR1 (Rouyer *et al.*, 1986), and the probability of genetic exchange is uniform throughout it (there are no hotspots). The relationship between genetic map distance and physical distance is 1% recombination for every 50 kb of genomic DNA, an amount 20-fold greater than the genome average (Petit *et al.*, 1988). Exchange in human PAR2 is detectable in only about 4% of meiotic products examined, indicating that recombination there is not necessary for segregation of the sex chromosomes, yet even in the PAR2 the ratio of genetic map distance and physical distance is elevated over the genome average (Kvaløy *et al.*, 1994). Non-disjunction of the X and Y chromosomes is associated with failure of exchange in the PAR1 in male meiosis (Hassold *et al.*, 1991). Moreover, in persons carrying a deletion of PAR1, spermatogenesis fails before the completion of meiosis I, and such persons are infertile (Gabriel-Robez *et al.*, 1990; Mohandas *et al.*, 1992). Thus, PAR1 is required for successful gamete production.

Genes in the PAR are present in two copies in males and two copies in females. In human, eight functional genes have been localized to the PAR1 (*Table 10.2*). Of the four genes in which it has been possible to test, all four escape X-inactivation. In mouse, a single gene, steroid sulphatase (*Sts*), has been localized to the PAR, and it too escapes X-chromosome inactivation (Keitges and Gartler, 1986; Salido *et al.*, 1996). Because these genes escape X-inactivation one might expect that they might also not be conserved during mammalian evolution.

10.5.1 The message from marsupials and monotremes

The grounding of Ohno's law in X-inactivation sets the conditions under

Table 10.2. Human and mouse pseudoautosomal genes

Gene symbol	Gene name	X-inactivated
Human PAR1		
CD39	Late lymphocyte activation antigen	
CSF2RA	Granulocyte–macrophage colony-stimulating factor receptor 2 α subunit	
IL3RA	Interleukin-3 receptor α subunit	
ANT3	Adenine nucleotide translocase 3	No
ASMT	Acetylserotonin methyltransferase	
XE7	XE7 product	No
MIC2R	*MIC2*-related (non-functional pseudogene)	N.A.
MIC2	CD99; the 12E7 antigen	No
XG	Xg blood group gene	No
Human PAR2		
IL9R	Interleukin-9 receptor	No
Mouse PAR		
STS	Steroid sulphatase	No

A blank indicates that the X-inactivation status is unknown. References for the genes in the tables are given in the text except for *CD39* (Miller *et al.*, 1995), *CSF2RA* (Gough *et al.*, 1990), *IL3RA* (Milatovich *et al.*, 1993), *ASMT* (Yi *et al.*, 1993), *XE7* (Ellison *et al.*, 1992), *MIC2R* (Smith and Goodfellow, 1994), *MIC2* (Goodfellow *et al.*, 1986) and *XG* (Ellis *et al.*, 1994). N.A., not applicable.

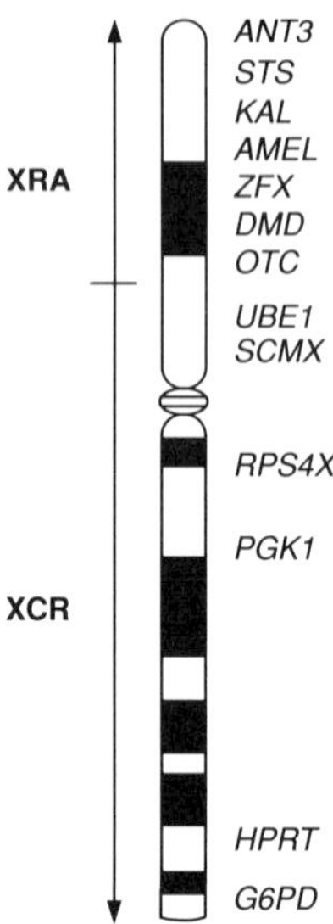

Figure 10.3. Model of the evolutionary origins of the eutherian X chromosome. The X conserved region (XRC) contains genes that were present on the ancestral mammalian X chromosome. The X recently added (XRA) region contains genes that were translocated to the X chromosome after the divergence of metatherian and eutherian lineages but before the radiation of eutherian lineages. The chromosome depicted here is an ideogram of the human X chromosome. Selected genes from each of the regions are shown. All the gene names are found in the text, except those for Duchenne muscular dystrophy (*DMD*), ornithine transcarbamylase (*OTC*), hypoxanthine phosphoribosyltransferase (*HPRT*) and glucose-6-phosphate dehydrogenase (*G6PD*).

which the law could be broken: a chromosomal segment can be added to the X chromosome provided it doesn't carry genes required in male meiosis. A chromosomal segment can be lost from the X chromosome if dosage imbalance of genes in the segment can be tolerated. Conservation of genes on the X chromosome indicates the presence of the ancestral sex chromosomes, whereas lack of conservation indicates a more recent origin. By comparison of widely converged mammalian species, these possibilities have been tested experimentally. A large sampling of genes from the eutherian X chromosome has been used as probes for *in situ* hybridization to metaphase chromosomes of marsupial and monotreme cells, and the following surprising picture has emerged.

Genes on human Xq and the proximal part of Xp are present on the metatherian and prototherian X chromosomes (McKay *et al.*, 1992; Spencer *et al.*, 1991a; Watson *et al.*, 1990) – evidence that these genes were on the sex chromosome pair ancestral to all mammals. In contrast, genes on the distal part of human Xp are in fact autosomal in both marsupials and monotremes (McKay *et al.*, 1992; Spencer *et al.*, 1991b; Watson *et al.*, 1991). The human Xq and proximal Xp has been defined as the X conserved region (XCR) whereas genes on the distal Xp have been viewed as recent additions to the X chromosome (XRA) (*Figure 10.3*; see Graves, 1995). Graves proposed that the XCR was present on the ancestral sex chromosomes. The genes in the XRA could have been added in one piece from a single ancestral autosomal segment or in a series of additions. Because XRA genes are found on several different autosomes in both metatherian and protherian lineages, it seems likely that genes were added in several successive events. All XRA genes are conserved on the X chromosome of virtually all eutherian mammals, indicating that these genes were added to the X chromosome after eutherian and metatherian lineages diverged, but before any eutherian lineages diverged from each other.

The XRA genes were either translocated to the X and Y chromosomes simultaneously or were exchanged between the sex chromosomes by recombination. Once shared between the sex chromosomes they could act as PARs in the eutherian lineage. Now, any inversion of the Y chromosome that included a break in such an ancient PAR would remove the sequences included in the inversion from the PAR and make them Y-specific. When sequences become Y-specific, they become subject to Muller's ratchet and selective sweeps. In the eutherian lineage, most of the XRA genes will, therefore, have accumulated mutations and been lost from the Y chromosome. Many of these genes have come under control of the X-inactivation system; however, some have not come under X-inactivation control (Disteche, 1995). Others, both functional and non-functional, are still present on the Y chromosome (see *Table 10.2*). The acquisition of newly added genes to form an expanded PAR was perhaps a chance event(s) but once new genes had been added successfully to the PAR, they then were subjected to genetic forces peculiar to the sex chromosomes – conservation on the one hand, and non-conservation on the other.

10.5.2 Recent pseudoautosomal evolution

Genes in the pseudoautosomal region are not subject to X-inactivation. Consequently, they need not be subject to Ohno's law. Cytogenetic analysis of the sex chromosomes in meiosis suggests that if marsupials maintain a PAR at all, it is restricted to the telomeres themselves (Sharp, 1982). Prototherian sex chromosomes, on the other hand, maintain a large segment of homology that is cytogenetically visible (Wrigley and Graves, 1988) (see *Figure 10.1*). Thus, PARs can be much smaller or much larger than the ones found in human and in mouse, implying that the genetic content of PARs is generally not conserved.

The analysis of eutherian *STS* shows how a pseudoautosomal gene can evolve into an X-linked gene. *STS* is part of the XRA region present in all eutherians. As stated above, *STS* is pseudoautosomal in mouse. Although human *STS* is X-linked, a non-functional copy of it, *STSP*, exists on Yq (Yen *et al.*, 1987). Data from Southern blotting experiments were used to propose that the *STS* gene is pseudoautosomal in prosimian species and became X-linked only more recently in higher primates (Yen *et al.*, 1988). In the lineage leading to higher primates, a pericentric inversion on the Y chromosome with one break in the PAR distal to *STS* and the other break near the centromere could have accomplished this change in gene order on the Y chromosome. Consistent with this proposal, other genes (*KALX* and *AMELX*) close to *STS* on the X chromosome are found close to *STS* on the Y chromosome (*KALY* and *AMELY*) (del Castillo *et al.*, 1992; Incerti *et al.*, 1992). The original model in this simple form, however, has been abandoned because additional Y-linked genes with X-linked homologues from the region around *STS* have been identified, and their order on the Y chromosome does not correspond to their order on the X chromosome. Thus, at least four inversions must be hypothesised (*Figure 10.4*).

There is a more serious problem with the model that must be addressed. Both the *STS* (X-linked in human) and *ANT3* (pseudoautosomal in human; Schiebel *et al.*, 1993) have been shown to be autosomal, not pseudo-autosomal, in prosimian species (Toder *et al.*, 1995), raising the possibility that the distal segment of the human X chromosome originated by a translocation from an autosome in the lineage leading to higher primates. But *STS* is pseudoautosomal in mouse and X-linked in rat (Salido *et al.*, 1996). If *STS* were autosomal in the lineage leading to all primates, then its appearance on the sex chromosomes in both the rodent and higher primate lineages had to occur independently, which is unlikely. Therefore, *STS* was translocated from the sex chromosomes to an autosome in the lineage leading to prosimian species.

The *ANT3* gene, on the other hand, could have been anywhere in the genome of our most primitive primate ancestors. Although hybridization of human *ANT3* does not identify a mouse *Ant3* (Ellison *et al.*, 1996), other genes in the human PAR have been identified in mouse and they are autosomal there (Disteche *et al.*, 1992; Ellison *et al.*, 1996; Miyajima *et al.*, 1995). Similarly, changes in the distal X genes have been detected in the genus *Mus*. *Clcn4*, an X-linked gene in human, rat, and the mouse species *Mus spretus*, is autosomal in *Mus musculus* (Palmer *et al.*, 1995; Rugarli *et al.*, 1995). The loss of *Clcn4* from the *M. musculus* X chromosome occurred since the divergence of lineages leading to *M. spretus* and *M. musculus* approximately 3 million years ago. *Clcn4* is subject to X-inactivation in *M. spretus*, indicating that the dosage imbalance of *Clcn4* did not deter its translocation to an autosome. The lesson is clear. Genes in the most distal part of the X chromosome, including the PAR, can come and go from there by chance genetic alterations, and the organism suffers no deleterious consequences. These genes are not conserved on the eutherian X chromosome.

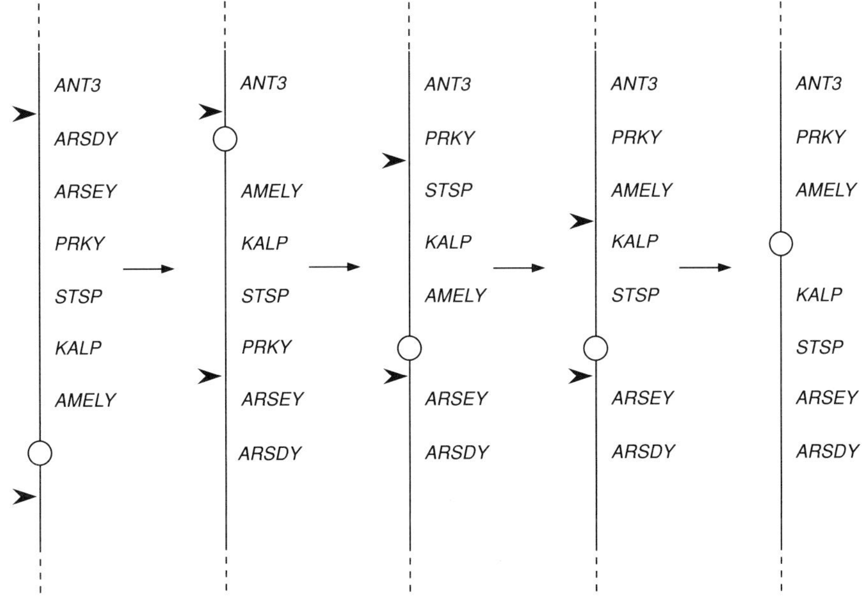

Figure 10.4. Inversion hypothesis for the evolution of the Y chromosome in higher primates. On the left, the Y chromosome present in the ancestral primate contained genes that were pseudoautosomal (*ANT3* to *AMELY*). The order of these genes is assumed to be the same as the order of the X-linked homologues found on the distal end of the short arm of the human X chromosome and the *ANT3* gene is assumed to have been on the ancestral Y chromosome. The first step toward the evolution of the Y chromosome found in humans is a pericentric inversion of the Y chromosome, the breakpoints of which are depicted by arrowheads. The inversion removes from the pseudoautosomal region all the genes proximal to the breakpoint on the short arm of the chromosome but leaves *ANT3* pseudoautosomal. To arrive at the order of genes known on the present-day human Y chromosome, a total of four inversions is required. The location of the *ARSEY* and *ARSDY* genes on the long arm of the human Y chromosome is unknown. In this model, they are assumed to be distal to *STSP*; if they are proximal to *STSP*, at least one additional inversion would be required.

10.5.3 Pseudoautosomal boundary

The pseudoautosomal boundary, *PABXY*, is defined as the transition between the pseudoautosomal sequences and the sex-chromosome-specific sequences. In human, the proximal limits of PAR1 and PAR2 are *PABXY1* and *PABXY2*, respectively. These *PABXY*s have been cloned and sequenced (Ellis *et al.*, 1989; Kvaløy *et al.*, 1994), and coincidentally they are defined by repeat sequence elements; an *Alu* repeat sequence at *PABXY1* and a LINE element at *PABXY2* (*Figure 10.5*). The *Alu* repeat element at *PABXY1* is present on the Y chromosome but absent from the X chromosome, whereas the LINE repeat element at *PABXY2* is present on both sex chromosomes. In

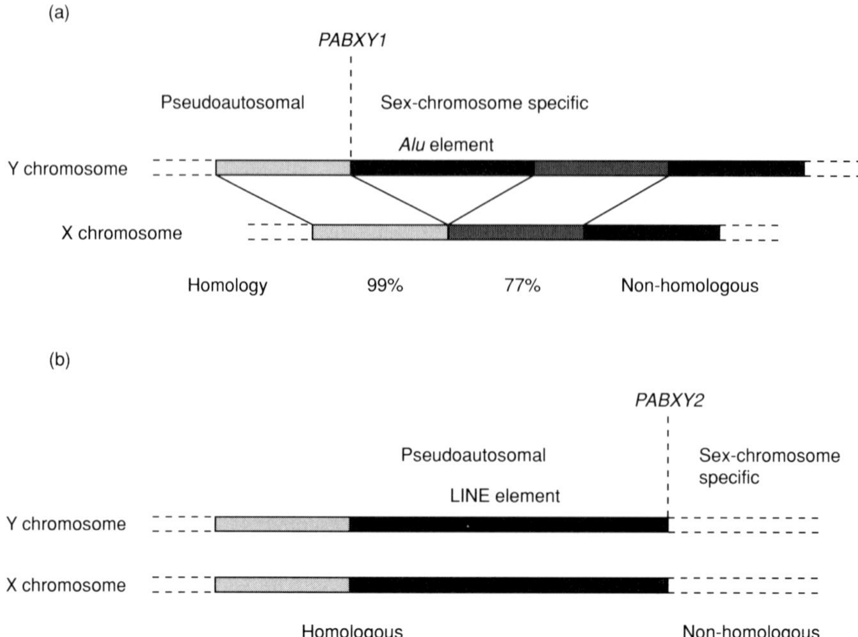

Figure 10.5. Structure of the human pseudoautosomal boundaries (a) on the short arms (*PABXY1*) and (b) on the long arms (*PABXY2*) of the X and Y chromosomes. (a) *PABXY1* is defined by a 303 bp *Alu* repeat element inserted on the Y chromosome but not present on the X chromosome. Distal to the *Alu* insertion site, the X- and Y-sequences are homologous. Proximal to the *Alu* insertion site, there exists a 220 bp segment in which the homology between the sex chromosomes is reduced. Proximal to this small segment, the X- and Y-sequences are non-homologous. (b) *PABXY2* is defined by a LINE repeat element present on both the X and Y chromosomes. Distal to and including the LINE repeat element, the X- and Y-sequences are homologous. Proximal to the LINE repeat element, the X- and Y-sequences are non-homologous.

both cases, the boundaries are formed by an abrupt transition in sequence homology. Sequences distal to the repeat elements are homologous and exchange regularly. With a minor reservation explained below, sequences proximal to the repeat elements are non-homologous. The repeat elements define the position proximal to which genetic exchange between the sex chromosomes does not occur regularly.

At *PABXY1*, there is a 220 bp segment of reduced homology (77%) between the sex chromosomes on the proximal side of the *Alu* insertion site (*Figure 10. 5*). The existence of a short segment of reduced homology implies that these sequences were at one time highly homologous. This observation lends itself to the hypothesis that the boundary once was localized to the proximal end of the short segment of reduced homology, and it later moved 220 bp distally (Ellis *et al.*, 1989). On the basis of percentage divergence in great apes and Old World monkeys, determined by DNA sequencing, the X and Y sequences in the 220 bp segment had separated and begun to diverge roughly 35 million years ago (Ellis *et al.*, 1990). This 'old' boundary probably was formed before the simian and higher primate lineages diverged. The divergence between the X and Y sequences at the boundary coincides with the divergence of the *STSP*, suggesting that the postulated inversion event that made *STSP* Y-specific was also the event that formed the old boundary. Insertion of the *Alu* repeat element at the boundary was not the cause of the distal movement of the 'old' boundary to the 'new' boundary because it is absent from Old World monkey Y-boundaries and the percentage divergence between different *Alu* repeat elements in the great apes is consistent with a insertion date after Old World monkey and great ape lineages diverged. Instead, the boundary probably moved distally by a gradual process of attrition. The attrition was the result of a small imbalance between two forces: the mutation rate of sequences very close to the boundary, causing them to diverge, and the rate of recombination between the sex chromosomes in these same sequences, causing them to remain the same. Attrition is taking place in each of the primate lineages independently, guided by the stochastic nature of mutation and recombination.

The evolution of PAR2 and its boundary is a recent event, because the human PAR2 is not present in other great apes (Bickmore and Cooke, 1987). It has been hypothesized that *PABXY2* was formed by an exchange event between the sex chromosomes at non-homologous LINE repeat elements (Kvaløy *et al.*, 1994). The exchange event resulted in transfer of distal X-specific chromatin to the Y chromosome and in loss of distal Y-specific chromatin. By this model, no essential functions were present in the Y-chromatin (it was dispensable) and the newly formed PAR2 might have been advantageous if the recombination there, even though it is infrequent, provided extra stability to the XY bivalent and hence more reliable segregation of the sex chromosomes. One gene has been identified in PAR2, *IL9R*, which escapes X-inactivation, and in mouse is localized to chromosome 11 (Vermeesch *et al.*, 1995). *IL9R* is yet another gene that escaped Ohno's law.

10.6 Conclusions

Two major genetic mechanisms act on the sex chromosomes in opposing manners. The action of Muller's ratchet caused the genetic isolation of the Y chromosome, which is now the least conserved element in the mammalian genome. As the Y chromosome degenerated, the X-inactivation system evolved which set stringent conditions for a locus to enter or leave the X chromosome. As a result, the X chromosome is now the most conserved element in the mammalian genome. That said, the X chromosome is not an immutable element and the Y chromosome is not so free that no genes at all are conserved there. Since the divergence of metatherians and eutherians, a large segment of genes has been added to form the eutherian X chromosome.

By and large, the composition of this X chromosome is conserved in eutherian mammals; however, exceptions to this conservation have been demonstrated for genes that are close to the X chromosomes' most distal ends. The Y chromosome is rapidly evolving, gaining and losing new sequences; but there are genes on the Y chromosome that have been there for over 150 million years. *SRY*, the testis-determining gene, was probably on the sex chromosomes ancestral to all mammals. Some of the other Y-conserved genes may have been kept on the Y chromosome because for biological reasons they cannot easily come under the control of the X-inactivation system. Once they do, however, they are lost from the Y chromosome, or on rare occasions they evolve novel functions which are often specific to the testis.

In the classical model of sex chromosome evolution, recombination suppression arose to keep together a haplotypic arrangement of mutations at two loci; together, the two alleles were an advantage to the males carrying them. This initial genetic system set the Y chromosome off on its degenerate path. *SRY* was most likely one of the elements of this system. Once the Y chromosome degenerated to the point where YY individuals were not viable and a dosage compensation mechanism had evolved to normalize gene products between males and females, the *SRY* sex-determining mechanism was stabilized on the Y chromosome. This is manifest because a portion of the X chromosome is conserved in all mammals. Generally, no other sex-determining mechanism has successfully invaded and superseded the present SRY-based system, however, in a few extant mammalian species such an invasion process may have occurred (Fredga, 1988; Just *et al.*, 1995).

The other element in the initial genetic system may or may not still remain on the Y chromosome, because the selective conditions that operated to favour that pair of loci may have changed. As mentioned above, new functions relating to spermatogenesis often evolve on the Y chromosome. But other traits can be selected for on the Y chromosome. For example, a gene that leads to greater stature has been mapped to the Y chromosome, and this gene may have been sexually selected. Thus, different species may possess different sexually selected traits whose genes localize to the Y chromosome.

Acknowledgements

The author expresses his gratitude to A. Ballabio, H. Cooke, B. Franco, J. Graves, D. Kipling, G. Rappold, L. Shapiro and S. Wilcot for making available unpublished information.

References

Adler DA, Bressler SL, Chapman VM, Page DC, Disteche CM. (1991) Inactivation of the *Zfx* gene on the mouse X chromosome. *Proc. Natl Acad. Sci. USA* **88:** 4592–4595.

Agulnik AI, Mitchell MJ, Lerner JL, Woods DR, Bishop CE. (1994a) A mouse Y chromosome gene encoded by a region essential for spermatogenesis and expression of a male-specific minor histocompatibility antigens. *Hum. Mol. Genet.* **3:** 873–878.

Agulnik AI, Mitchell MJ, Mattei MG, Borsani G, Avner PA, Lerner JL, Bishop CE. (1994b) A novel X gene with a widely transcribed Y-linked homologue escapes X-inactivation in mouse and human. *Hum. Mol. Genet.* **3:** 879–884.

Ariel M, Robinson E, McCarrey JR, Cedar H. (1995) Gamete-specific methylation correlates with imprinting on the murine *Xist* gene. *Nature Genetics* **9:** 315–315.

Arnemann J, Epplen J, Cooke H, Sauermann U, Engel W, Schmidtke J. (1987) A human Y-chromosomal DNA sequence expressed in testicular tissue. *Nucleic Acids Res.* **15:** 8713–8724.

Ashworth A, Rastan R, Lovell-Badge R, Kay G. (1991) X-chromosome inactivation may explain the difference in viability of XO humans and mice. *Nature* **351:** 406–408.

Baverstock PR, Adams M, Polkinghorne RW, Gelder M. (1982) Sex linked enzymes in birds: Z-chromosome conservation but no dosage compensation. *Nature* **296:** 7633–7636.

Berta P, Hawkins JR, Sinclair AH, Taylor A, Griffiths BL, Goodfellow PN, Fellous M. (1990) Genetic evidence equating SRY and the testis-determining factor. *Nature* **348:** 448–450.

Bickmore WA, Cooke HJ. (1987) Evolution of homologous sequences on the human X and Y chromosome, outside of the meiotic pairing segment. *Nucleic Acids Res.* **15:** 6261–6271.

Boer PH, Adra CN, Lau YF, McBurney MW. (1987) The testis-specific phosphoglycerate kinase gene Pgk-2 is a recruited retroposon. *Mol. Cell Biol.* **7:** 3107–3112.

Borsani G, Tonlorenzi R, Simmler, MC *et al.* (1991) Characterization of a murine gene expressed from the inactive X chromosome. *Nature* **351:** 325–329.

Brockdorff N, Ashworth A, Kay GF *et al.* (1991) Conservation of position and exclusive expression of mouse Xist from the inactive X chromosome. *Nature* **351:** 329–331.

Brockdorff N, Ashworth A, Kay GF, McCabe VM, Norris DP, Cooper PJ, Swift S, Rastan S. (1992) The product of the mouse *Xist* gene is a 15 kb inactive X-specific transcript containing no conserved ORF and located in the nucleus. *Cell* **71:** 515–526.

Brown WRA. (1988) A physical map of the human pseudosutosomal region. *EMBO J.* **7:** 2377–2385.

Brown CJ, Willard HF. (1994) The human X-inactivation center is not required for maintenance of X-chromosome inactivation. *Nature* **368:** 154–156.

Brown CJ, Lafreniere RG, Powers VE *et al.* (1991a) Localization of the X-inactivation centre on the human X chromosome to Xq13. *Nature* **349:** 82–84.

Brown CJ, Ballabio A, Rupert JL, Lafreniere RG, Grompe M, Tonlorenzi R, Willard HF. (1991b) A gene from the region of the human X inactivation centre is expressed exclusively from the inactive X chromosome. *Nature* **349:** 38–44.

Brown CJ, Hendrich BD, Rupert JL, Lafreniere RG, Xing Y, Lawrence J, Willard HF. (1992) The human XIST gene: analysis of a 17 kb inactive X-specific RNA that contains conserved repeats and is highly localized within the nucleus. *Cell* **71:** 527–542.

Bull JJ. (1983) *Evolution of Sex Determining Mechanisms*. Benjamin/Cummings, Don Mills, Ontario.

Burd CG, Dreyfuss G. (1994) Conserved structures and diversity of functions of RNA-binding proteins. *Science* **265:** 615–621.

Burgoyne PS. (1982). Genetic homology and crossing over in the X and Y chromosomes of mammals. *Hum. Genet.* **61:** 85–90.

del Castillo I, Cohen-Salmon M, Blanchard S, Lutfalla G, Petit C. (1992) Structure of the X-linked Kallmann syndrome gene and its homologous pseudogene on the Y chromosome. *Nature Genetics* **2:** 305–310.

Chapman V, Keitz B, Disteche C, Lau E, Snead M. (1991) Linkage of amelogenin to the distal portion of the mouse X chromosome. *Genomics* **10:** 23–28.

Charlesworth B. (1978) Model for evolution of Y chromosomes and dosage compensation. *Proc. Natl Acad. Sci. USA* **75:** 5618–5622.

Charlesworth B. (1991) The evolution of sex chromosomes. *Science* **251:** 1030–1033.
Christidis L. (1989) Chromosomal repatterning and systematics in the passeriformes (songbirds). In: *Chromosomes Today,* Vol. 10 (eds K Fredga, BA Kihlman and MD Bennett). Unwin Hyman, Boston, pp. 279–294.
Cooke HJ, Brown WRA, Rappold GA. (1985) Hypervariable telomeric sequences from the human sex chromosomes are pseudoautosomal. *Nature* **317:** 687–692.
Cooper DW. (1995) Sex differentiation is different down under. *Trends Genet.* **11:** 385.
Cooper DW, Johnston PG, Graves JAM, Watson JM. (1993) X-chromosome inactivation in marsupials and monotremes. *Sem. Dev. Biol.* **4:** 117–129.
Disteche CM. (1995) Escape from X inactivation in human and mouse. *Trends Genet.* **11:** 17–22.
Disteche CM, Brannan CI, Larsen A, Adler DA, Schorderet DF, Gearing D, Copeland NG, Jenkins NA, Park LS. (1992) The human pseudoautosomal GM-CSF receptor α subunit gene is autosomal in mouse. *Nature Genetics* **1:** 333–336.
Dominguez-Steglich M, Meng G, Bettecken T, Muller CR, Schmid M. (1990) The dystrophin gene is autosomally located on a microchromosome in chicken. *Genomics* **8:** 536–540.
Dorit RL, Akashi H, Gilbert W. (1995) Absence of polymorphism at the ZFY locus on the human Y chromosome. *Science* **268:** 1183–1185.
Ellis NA, Goodfellow PJ, Pym B, Smith M, Palmer M, Frischauf AM, Goodfellow PN. (1989) The pseudoautosomal boundary in man is defined by an *Alu* repeat sequence inserted on the Y chromosome. *Nature* **337:** 81–84.
Ellis N, Yen P, Neiswanger K, Shapiro LJ, Goodfellow PN. (1990) Evolution of the pseudoautosomal boundary in Old World monkeys and great apes. *Cell* **63:** 977–986.
Ellis NA, Tippett P, Petty A *et al.* (1994) *PBDX* is the *XG* blood group gene. *Nature Genetics* **8:** 285–290.
Ellison JW, Ramos C, Yen PH, Shapiro LJ. (1992) Structure and expression of the human pseudoautosomal gene XE7. *Hum. Mol. Genet.* **9:** 691–696.
Ellison JW, Li X, Frankel U, Shapiro LJ. (1996) Rapid evolution of pseudoautosomal genes and their mouse homologs. *Mamm. Genome* **7:** 25–30.
Felsenstein J. (1974) The evolutionary advantage of recombination. *Genetics* **78:** 737–756.
Ferrari S, Harley VR, Pontiggia A, Goodfellow PN, Lovell-Badge R, Bianchi ME. (1992) SRY, like HMG1, recognizes sharp angles in DNA. *EMBO J.* **11**: 4497–4506.
Fisher RA. (1935) The sheltering of lethals. *Am. Natural.* **69:** 446–455.
Fisher EMC, Beer-Romero P, Brown LG, Ridley A, McNeil JA, Lawrence JB, Willard HF, Bieber FR, Page DC. (1990) Homologous ribosomal protein genes on the human X and Y chromosomes: escape from X inactivation and possible implications for Turner syndrome. *Cell* **63:** 1205–1218.
Foster JW, Brennan FE, Hampikian GK *et al.* (1992) Evolution of sex determination and the Y chromosome SRY-related sequences in marsupials. *Nature* **359:** 531–533.
Foster JW, Graves JAM. (1994) A SRY-related sequence on the marsupial X chromosome – implications for the evolution of the mammalian testis determining gene. *Proc. Natl Acad. Sci. USA* **91:** 1927–1931.
Franco B, Meroni G, Parenti G *et al.* (1995) A cluster of sulfatase genes on Xp22.3: mutations in chondrodysplasia punctata (CDPX) and implication for warfarin embryopathy. *Cell* **81:** 15–25.
Freije D, Helms C, Watson MS, Donis-Keller H. (1992) Identification of a second pseudoautosomal region near the Xq and Yq telomeres. *Science* **258:** 1784–1787.
Fredga K. (1988) Aberrant chromosomal sex-determining mechanisms in mammals, with special reference to species with XY females. *Phil. Trans. R. Soc. Lond.* **322:** 83–95.
Ferguson-Smith MA. (1966) X–Y chromosomal interchange in the aetiology of true hermaphroditism and of XX Klinefelter's syndrome. *Lancet* **2:** 475–476.
Gabriel-Robez O, Rumpler Y, Ratomponirina C, Petit C, Levilliers J, Croquette MF, Couturier J. (1990) Deletion of the pseudoautosomal region and lack of sex-chromosome

pairing at pachytene in two infertile men carrying an X;Y translocation. *Cytogenet. Cell Genet.* **54:** 38–42.

Gartler SM, Dyer KA, Goldman MA. (1992) Mammalian X inactivation. *Mol. Genet. Med.* **2:** 121–160.

Goodfellow PJ, Darling SM, Thomas NS, Goodfellow PN. (1986) A pseudoautosomal gene in man. *Science* **234:** 740–743.

Gough NM, Gaering DP, Nicola NA, Baker E, Pritchard M, Callen DF, Sutherland GR. (1990) Localisation of the human GM-CSF receptor gene to the X–Y pseudoautosomal region. *Nature* **345:** 734–736.

Graves JAM. (1995) The origin and function of the mammalian Y chromosome and Y-borne genes – an evolving understanding. *BioEssays* **17:** 311–321.

Grumbach MM, Morishima A, Taylor JH. (1963) Human sex chromosome abnormalities in relation to DNA replication and heterochromatinization. *Proc. Natl Acad. Sci. USA* **49:** 581–589.

Gubbay J, Collignon J, Koopman P, Capel B, Econonmou A, Munsterberg A, Vivian N, Goodfellow PN, Lovell-Badge R. (1990) A gene mapping to the sex-determining region of the mouse Y chromosome is a member of a novel family of embryonically expressed genes. *Nature* **346:** 245–250.

Hamilton WD, Axelrod R, Tanese R. (1990) Sexual reproduction as an adaptation to resist parasites (a review). *Proc. Natl Acad. Sci. USA* **87:** 3566–3573.

Hammer MF. (1995) A recent common ancestry for human Y chromosomes. *Nature* **378:** 376–378.

Harley VR, Jackson DI, Hextall PJ, Hawkins JR, Berkovitz GD, Sockanathan S, Lovell-Badge R, Goodfellow PN. (1992) DNA binding activity of recombinant SRY from normal males and XY females. *Science* **255:** 453–456.

Harley VR, Lovell-Badge R, Goodfellow PN. (1994) Definition of a consensus DNA binding site for SRY. *Nucleic Acids Res.* **22:** 1500–1501.

Hassold TJ, Sherman SL, Pettay D, Page DC, Jacobs PA. (1991) XY chromosome nondisjunction in man is associated with diminished recombination in the pseudoautosomal region. *Am. J. Hum. Genet.* 49**:** 253–260.

Henderson SA. (1964) RNA synthesis during male meiosis and spermiogenesis. *Chromosoma* **15:** 345–366.

Hope RM, Cooper S, Waiwright B. (1990) Globin macromolecular sequences in marsupials and monotremes. In: *Mammals from Pouches and Eggs: Genetics, Breeding and Evolution of Marsupials and Monotremes* (eds JAM Graves, RM Hope and DW Cooper). CSIRO Press, Melbourne, pp. 147–172.

Incerti B, Guioli S, Pragliola A *et al.* (1992) Kallmann syndrome gene on the X and Y chromosomes: implications for evolutionary divergence of human sex chromosomes. *Nature Genetics* **2:** 311–314.

Jaafar H, Gariel-Robez O, Rumppler Y. (1993) Chromosomal anomalies and disturbance of transcriptional activity at the pachytene stage of meiosis: relationship to male sterility. *Cytogenet. Cell Genet.* **64:** 273–280.

Jager RJ, Anvret M, Hall K, Scherer G. (1990) A human XY female with a frameshift mutation in the candidate testis-determining gene SRY. *Nature* **348:** 452–453.

Jost A, Vigier B, Prepin J, Perchellet JP. (1973) Studies on sex differentiation in mammals. *Recent Prog. Horm. Res.* **29:** 1–41.

Just W, Rau W, Vogel W, Akhverdian M, Fredga K, Graves JA, Lyapunova E. (1995) Absence of *Sry* in species of the vole *Ellobius*. *Nature Genetics* **11:** 117–118.

Kay GF, Ashworth A, Penny GD, Dunlop M, Swift S, Brockdorff N, Rastran S. (1991) A candidate spermatogenesis gene on the mouse Y chromosome is homologous to ubiquitin-activating enzyme E1. *Nature* **354:** 486–489.

Kay GF, Penny GD, Patel D, Ashworth A, Brockdorff N, Rastan S. (1993) Expression of Xist during mouse development suggests a role in the initiation of X chromosome inactivation. *Cell* **72:** 171–182.

Kay GF, Barton SC, Surani MA, Rastan S. (1994) Imprinting and X chromosome counting mechanisms determine Xist expression in early mouse development. *Cell* 77: 639–650.

Keitges E, Rivest M, Siniscalco M, Gartler SM. (1985) X linkage of steroid sulfatase in mouse is evidence for a functional Y linked allele. *Nature* **315:** 226–227.

Keitges E, Gartler SM. (1986) Dosage of the *Sts* gene in mouse. *Am. J. Hum. Genet.* **39:** 470–476.

Kipling D, Salido EC, Shapiro LJ, Cooke HJ. (1996) High frequency *de novo* alterations in the long-range genomic structure of the mouse pseudoautosomal region. *Nature Genetics* **13:** 78–82.

Klink A, Schiebel K, Winkelmann M, Rao E, Horsthemke B, Ludecke HJ, Claussen U, Scherer G, Rappold G. (1995) The human protein kinase gene PKX1 on Xp22.3 displays Xp/Yp homology and is a site of chromosomal instability. *Hum. Mol. Genet.* **4:** 869–878.

Koopman P, Gubbay J, Vivian N, Goodfellow PN, Lovell-Badge R. (1991) Male development of chromosomally female transgenic mice for Sry. *Nature* **351:** 117–121.

Kratzer PG, Chapman V, Lambert H, Evens RW, Liskay RM. (1983) Differences in the DNA of inactive X chromosomes of fetal and extraembryonic tissues of mice. *Cell* **33:** 37–42.

Kvaløy K, Galvagni F, Brown WRA. (1994) The sequence organization of the long arm pseudoautosomal region of the human sex chromosomes. *Hum. Mol. Genet.* **3:** 771–778.

Kwaja O, Jones M, Sargent C, Affara N. (1994) A novel XY homologous gene. *Cytogenet. Cell Genet.* **67:** 399.

Lifschytz E, Lindsley DL. (1972) The role of X-chromosome inactivation during spermatogenesis. *Proc. Natl Acad. Sci. USA* **69:** 182–186.

Lucotte G, Ngo NY. (1985) A highly polymorphic probe that detects Taq1 RFLPs on the human Y chromosome. *Nucleic Acids Res.* **13:** 8285.

Ma K, Inglis JD, Sharkey A *et al.* (1993) A Y chromosome gene family with RNA-binding protein homology: candidates for the azoospermia factor AZF controlling human spermatogenesis. *Cell* 75: 1287–1295.

Mahadevaiah SK, Lovell-Badge R, Burgoyne PS. (1993) Tdy-negative XY, XXY, and XYY female mice: breeding data and synaptonemal complex analysis. *J. Reprod. Fertil.* **97:** 151–160.

Mardon G, Mosher R, Disteche CM, Nishioka Y, McLaren A, Page DC. (1989) Duplication, deletion, and polymorphism in the sex-determining region of the mouse Y chromosome. *Science* **243:** 78–80.

Maynard Smith J. (1978) *The Evolution of Sex.* Cambridge University Press, Cambridge.

McCarrey JR, Thomas K. (1987) Human testis-specific PGK gene lacks introns and possesses characteristics of a processed gene. *Nature* **326:** 501–505.

McCarrey JR, Dilworth DD. (1992) Expression of Xist in mouse germ cells correlates with X-chromosome inactivation. *Nature Genetics* **2:** 200–203.

McKay LM, Watson JM, Graves JAM. (1992) Mapping human X-linked genes in the phalangerid marsupial *Trichosurus vulpecula. Genomics* **14:** 302–308.

Migeon BR. (1994) X-chromosome inactivation: molecular mechanisms and genetic consequences. *Trends Genet.* **10:** 230–235.

Migeon BR, Shapiro LJ, Norum RA, Mohandas T, Axelman J, Dabora RL. (1982) Differential expression of steroid sulfatase locus on active and inactive human X chromosome. *Nature* **299:** 838–840.

Migeon, BR, Wolf SF, Axelman J, Kaslow DC, Schmidt M. (1985) Incomplete X chromosome dosage in chorionic villi of human placenta. *Proc. Natl Acad. Sci. USA* **82:** 3390–3394.

Miltovich A, Kitamura T, Miyajima A, Francke U. (1993) Gene for the α-subunit of the human interleukin-3 receptor (IL3RA) localized to the X–Y pseudoautosomal region. *Am. J. Hum. Genet.* **53:** 1146–1153

Miller AP, Eble B, Voland JR, Schlessinger D, Willard HF. (1995) Poster at the Sixth International Workshop on Human X Chromosome Mapping 1995. Reported in *Cytogenet. Cell Genet.* **71:** 338.

Mitchell MJ, Woods DR, Tucker PK, Opp JS, Bishop CE. (1991) Homology of a candidate spermatogenic gene from the mouse Y chromosome to the ubiquitin-activating enzyme E1. *Nature* **354:** 483–486.

Mitchell MJ, Woods DR, Wilcox SA, Graves JAM, Bishop CE. (1992) Marsupial Y chromosome encodes a homologue of the mouse Y-linked candidate spermatogenesis gene Ube1y. *Nature* **359:** 528–531.

Miyajima I, Levitt L, Hara T, Bedell MA, Copeland NG, Jenkins NA, Miyajima A. (1995) The murine interleukin-3 receptor α subunit gene: chromosomal localization, genomic structure, and promoter function. *Blood* **85:** 1246–1253.

Mohandas TK, Speed RM, Passage MB, Yen PH, Chandley AC, Shapiro LJ. (1992) Role of the psuedoautosomal region in sex-chromosome pairing during male meiosis: meiotic studies in a man with a deletion of distal Xp. *Am. J. Hum. Genet.* **51:** 526–533.

Muller HJ. (1964) The relation of recombination to mutational advance. *Mutat. Res.* **1:** 2–9.

Nagamine CM, Chan K, Kozak CA, Lau YF. (1989) Chromosome mapping and expression of a putative testis-determining gene in mouse. *Science* **243:** 80–83.

Nakahori Y, Takenaka O, Nakagome Y. (1991) A human X–Y homologous region encodes 'amelogenin'. *Genomics* **9:** 264–269.

Ohno S. (1967) *Sex Chromosomes and Sex-Linked Genes*. Springer, New York.

Page DC, Mosher R, Simpson EM, Fisher EMC, Mardon G, Pollack J, McGillivray B, de la Chapelle A, Brown LG. (1987) The sex-determining region of the human Y chromosome encodes a zinc finger protein. *Cell* **51:** 1091–1104.

Palmer MS, Sinclair AH, Berta P, Ellis NA, Goodfellow PN, Abbas NE, Fellous M. (1989) Genetic evidence that ZFY is not the testis-determining factor. *Nature* **342:** 937–939.

Palmer MS, Berta P, Sinclair AH, Pym B, Goodfellow PN. (1990) Comparison of human ZFY and ZFX transcripts. *Proc. Natl Acad. Sci. USA* **87:** 1681–1685.

Palmer S, Perry J, Ashworth A. (1995) A contravention of Ohno's law in mice. *Nature Genetics* **10**: 472–476.

Penny GD, Kay GF, Sheardown SA, Rastan S, Brockdorff N. (1996) Requirement for Xist in X chromosome inactivation. *Nature* **379:** 131–137.

Petit C, Levilliers J, Weissenbach J. (1988) Physical mapping of the human pseudoautosomal region; comparison with the genetic linkage map. *EMBO J.* **7:** 2369–2376.

Rappold GA. (1993) The pseudoautosomal region of the human sex chromosomes. *Hum. Genet.* **92:** 315–324.

Rappold GA, Lehrach HA. (1988) Long range restriction map of the pseudoautosomal region by partial digest PFGE analysis from the telomere. *Nucleic Acids Res.* **16:** 5361–5377.

Rastan S. (1994) X chromosome inactivation and the XIST gene. *Curr Opin. Genet. Dev.* **4:** 292–297.

Reijo R, Lee TY, Salo P *et al.* (1995) Diverse spermatogenic defects in humans caused by Y chromosome deletions encompassing a novel RNA-binding protein gene. *Nature Genetics* **10:** 383–393.

Rice WR. (1994) Degeneration of a nonrecombining chromosome. *Science* **263**: 230–232.

Richler C, Soreq H, Wahrman J. (1992) X inactivation in mammalian testis is correlated with inactive X-specific transcription. *Nature Genetics* **2:** 192–195.

Rouyer F, Simmler MC, Vergnaud G, Johnsson C, Cooke HJ, Weissenbach J. (1986) A gradient of sex linkage in the pseudoautosomal region of the human sex chromosomes. *Nature* **319:** 291–295.

Rugarli EI, Adler DA, Borsani G, Tsuchiya K, Franco B, Hauge X, Disteche C, Chapman V, Ballabio A. (1995) Different chromosomal localization of the Clcn4 gene in *Mus spretus* and C57BL/6J mice. *Nature Genetics* **10:** 466–471.

Salido EC, Yen PH, Mohandas TK, Shapiro LJ. (1992) Expression of the X-inactivation-associated gene XIST during spermatogenesis. *Nature Genetics* **2:** 196–199.

Salido EC, Li XM, Yen PH, Martin N, Mohandas TK, Shapiro LJ. (1996) Cloning and expression of the mouse pseudoautosomal steroid sulfatase gene (Sts). *Nature Genetics* **13:** 83–86.

Sandler L, Hecht F. (1973) Genetic effects of aneuploidy. *Am. J. Hum. Genet.* **25:** 332–339.
Schmid M, Nanda I, Steinlein C, Epplen JT. (1989) Evolution of sex chromosomes in Amphibia. In: *Chromosomes Today*, Vol. 10 (eds K Fredga, BA Kihlman and MD Bennett). Unwin Hyman, Boston, pp. 199–222.
Schneider-Gädicke A, Beer-Romero P, Brown LG, Nussbaum R, Page DC. (1989) ZFX has a gene structure similar to ZFY, the putative human sex determinant, and escapes X inactivation. *Cell* **57:** 1247–1258
Scott DM, Ehrmann IE, Ellis PS, Bishop CE, Agulnik AI, Simpson E, Mitchell MJ. (1995) Identification of a mouse male specific transplantation antigen, H-Y. *Nature* **376:** 695–698.
Sharp P. (1982) Sex chromosome pairing during male meiosis in marsupials. *Chromosoma* **86:** 27–47.
Shimmin LC, Chang BHJ, Li WH.. (1993) Male-driven evolution of DNA sequences. *Nature* **362:** 745–747.
Simmler MC, Rouyer F, Vergnaud G, Nystrom-Lahti M, Ngo KY, de la Chapelle A, Weissenbach J. (1985) Pseudoautosomal DNA sequences in the pairing region of the human sex chromosomes. *Nature* **317:** 692–697.
Sinclair AH, Foster J, Spencer J, Page DC, Palmer MS, Goodfellow PN, Graves JAM. (1988) Sequences homologous to ZFY, a candidate human sex-determining gene, are autosomal in marsupials. *Nature* **336:** 780–783.
Sinclair AH, Berta P, Palmer MS *et al.* (1990) A gene from the human sex-determining region encodes a protein with homology to a conserved DNA-binding motif. *Nature* **346:** 240–244.
Singh L, Jones KW. (1982) Sex reversal in the mouse (*Mus musculus*) is caused by a recurrent nonreciprocal crossover involving the X and an aberrant Y chromosome. *Cell* **28:** 205–216.
Smith MJ, Goodfellow PN. (1994) MIC2R: a transcribed MIC2–related sequence associated with a CpG island in the human pseudoautosomal region. *Hum. Mol. Genet.* **2:** 1575–1582.
Sola L, Cataudella S, Capanna E. (1981) New developments in vertebrate cytotaxonomy III. Karyology of bony fishes: a review. *Genetica* **54:** 285–328.
Spencer JA, Watson JM, Graves JAM. (1991a) The X chromosome of marsupials shares a highly conserved region with eutherians. *Genomics* **9:** 598–604.
Spencer JA, Sinclair AH, Watson JM, Graves JAM. (1991b) Genes on the short arm of the human X chromosome are not shared with the marsupial X. *Genomics* **11:** 339–345.
Stevanovic M, Lovell-Badge R, Collignon J, Goodfellow PN. (1993) SOX3 is an X-linked gene related to SRY. *Hum. Mol. Genet.* **2:** 2013–2018.
Sutcliffe MJ, Burgoyne PS. (1989) Analysis in the testes of H-Y negative XO Sxr^b mice suggests that the spermatogenesis gene (Spy) acts during the differentiation of the A spermatogonia. *Development* **107:** 373–380.
Takagi N, Sasaki M. (1975) Preferential inactivation of the paternally derived X chromosome in the extraembryonic membranes of the mouse. *Nature* **256:** 640–643.
Tiepolo L, Zuffardi O. (1976) Localization of factors controlling spermatogenesis in the nonfluorescent portion of the human Y chromosome long arm. *Hum. Genet.* **34:** 119–124.
Toder R, Rappold G, Schiebel K, Schempp W. (1995) ANT3 and STS are autosomal in prosimian lemurs: implications for the evolution of the pseudoautosomal region. *Hum. Genet.* **95:** 22–28.
Tres LL. (1975) RNA synthesis in meiotic prophase spermatocytes in the human testis. *Chromosoma* **53:** 141–151.
Tucker PK, Lundrigan BL. (1993) Rapid evolution of the sex determining locus in Old World mice and rats. *Nature* **364:** 715–717.
VandeBerg JL, Robinson ES, Samollow PB, Johnston PG. (1987) X-linked gene expression and X-chromosome inactivation: marsupial, mouse, and man compared. *Isozymes: Curr. Top. Biol. Med. Res.* **15:** 225–253.
Vergnaud G, Page DC, Simmler MC, Brown L, Rouyer F, Noel B, Botstein D, de la Chapelle A, Weissenbach J. (1986) A deletion map of the human Y chromosome based on DNA hybridization. *Am. J. Hum. Genet.* **38:** 109–124.

Vermeesch JR, Kermmouni A, Renauld JC, Maryen P. (1996) The IL9 receptor gene is located in the pseudoautosomal region of the long arm of the sex chromosomes, escapes X inactivation and its murine homologue is located on an autosome. Reported in *Cytogenet. Cell Genet.*, in press.

Wang W, Meadows LR, den Haan JM *et al.* (1995) Human H-Y: a male-specific histocompatibility antigen derived from the SMCY protein. *Science* **269:** 1588–1590.

Watanabe M, Zinn AR, Page DC, Nishimoto T. (1993) Functional equivalence of human X- and Y-encoded isoforms of ribosomal protein S4 consistent with a role in Turner syndrome. *Nature Genetics* **4:** 268–271.

Watson JM, Spencer JA, Riggs AD, Graves JAM. (1990) The X chromosome of monotremes shares a highly conserved region with the eutherian and marsupial X chromosomes, despite the absence of X inactivation. *Proc. Natl Acad. Sci. USA* **87:** 7125–7129.

Watson JM, Spencer JA, Riggs AD, Graves JAM. (1991) Sex chromosome evolution: platypus gene mapping suggests that part of the human X chromosome was originally autosomal. *Proc. Natl Acad. Sci. USA* **88:** 11256–11260.

Whitfield LS, Lovell-Badge R, Goodfellow PN. (1993) Rapid sequence evolution of the mammalian sex-determining gene SRY. *Nature* **364:** 713–715.

Whitfield LS, Sulston JE, Goodfellow PN. (1995) Sequence variation of the human Y chromosome. *Nature* **378:** 379–380.

Whitkop CJ. (1967) Partial expression of sex-linked recessive amelogenesis imperfecta in females compatible with the Lyon hypothesis. *Oral Surg. Oral Med. Oral Pathol.* **23:** 174–182.

Winge O. (1932) The nature of sex chromosomes. *Proc. 6th Int. Congr. Genet.* **1:** 343–355.

Wrigley JM, Graves JAM. (1988) Sex chromosome homology and incomplete, tissue-specific X-inactivation suggest that monotremes represent an intermediate stage of mammalian sex chromosome evolution. *J. Hered.* **79:** 115–118.

Yen PH, Allen E, Marsh B, Mohandas T, Wong N, Taggart RT, Shapiro LJ. (1987) Cloning and expression of steroid sulfatase cDNA and the frequent occurrence of deletion in STS deficiency: implications for X–Y interchange. *Cell* **49:** 443–454.

Yen PH, Marsh B, Allen E, Tsai SP, Ellison J, Connolly L, Neiswanger K, Shapiro LJ. (1988) The human X-linked steroid sulfatase gene and Y-encoded pseudogene: evidence for an inversion of the Y-chromosome during primate evolution. *Cell* **55:** 1123–1135.

Yi H, Donohue SJ, Klein DC, McBride OW. (1993) Localization of the hydroxyindole-O-methyltransferase gene to the pseudoautosomal region: implication for mapping of psychiatric disorders. *Hum. Mol. Genet.* **2:** 127–131.

Zinn AR, Bressler SL, Beer-Romero P, Adler DA, Chapman VM, Page DC, Disteche CM. (1991) Inactivation of the Rps4 gene on the mouse X chromosome. *Genomics* **11:** 1097–1101.

Zuccotti M, Monk M. (1995) Methylation of the mouse Xist gene in sperm and eggs correlated with imprinted Xist expression and paternal X-inactivation. *Nature Genetics* **9:** 316–320.

11

Mitochondrial DNA variation and human evolution

Mark Stoneking

11.1 Introduction

The study of recent human evolution, or the origin of modern humans, is traditionally based on fossil and archaeological evidence. However, genetic evidence, especially analyses of mitochondrial DNA (mtDNA) variation, are increasingly supplanting the fossil and archaeological record as the major source of new insights (and arguments) concerning human evolution. This chapter is an update of previous reviews of the mtDNA evidence for human origins (Stoneking, 1993, 1994), which has also been recently reviewed by others (Rogers and Jorde, 1995; Wallace, 1995).

11.2 Theories of modern human origins

Historically, there are four theories of modern human origins that can be distinguished. All four take as their starting point the observation that hominids evolved in Africa and spread from Africa to the rest of the Old World some 1–1.5 million years ago with a subsequent transformation to our species, anatomically modern humans. While all of these theories were originally based on observations from the fossil record, they are all in essence predictions of the genetic relationships of African and non-African populations. They can, therefore, be distinguished on the basis of the predicted contribution of African genes to non-African populations (*Figure 11.1*).

11.2.1 The candelabra hypothesis

At one extreme is the candelabra hypothesis, which holds that the transforma-

Human Genome Evolution, edited by M. Jackson, T. Strachan and G. Dover.

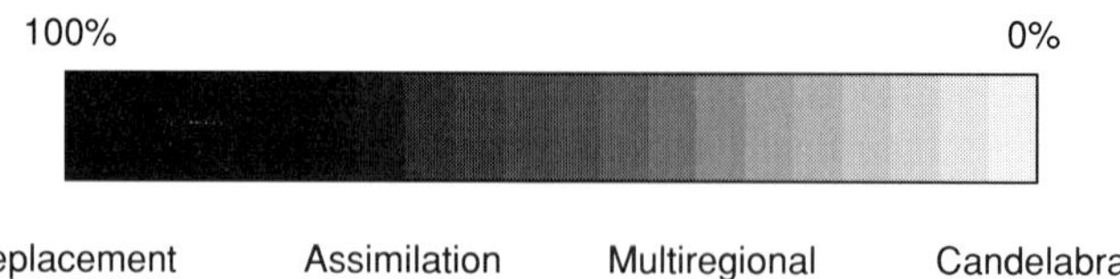

Figure 11.1. Four theories of modern human origins, distinguished by predictions for the contribution of African genes to non-African populations. The replacement and candelabra hypotheses hold that the African contribution is 100% and 0%, respectively, for all genes, while the assimilation and multiregional hypotheses would predict varying contributions for different genes.

tion to anatomically modern humans occurred completely independently in different regions of the Old World (Coon, 1962). Hence, modern Africans evolved from ancient Africans, modern Asians from ancient Asians, modern Europeans from ancient Europeans, and so on, with no genetic contributions whatsoever from outside a particular region. The candelabra hypothesis thus predicts that no African genes were contributed to non-African populations. This hypothesis has been discredited on the grounds that it invokes completely independent evolution of modern humans many times, which is extremely unlikely.

11.2.2 Multiregional evolution

Multiregional evolution is similar to the candelabra hypothesis in holding that the main genetic contributions were within each region of the Old World, but allows for a single origin of various modern human traits by invoking gene flow to spread these traits around the Old World (Wolpoff, 1989; Wolpoff *et al.*, 1984). Hence, while modern Asians and Europeans are primarily descended from ancient Asians and Europeans, respectively, they have received some African genes by gene flow. The contribution of African genes to non-African populations would, therefore, be expected to vary from gene to gene (*Figure 11.1*). The principal evidence cited in favour of multiregional evolution is the observation of regional continuity in the fossil record; that is, features of modern Asian skeletons that occur only in ancient Asian skeletons, features of modern European skeletons that occur only in ancient European skeletons, and so on.

11.2.3 The recent African origin hypothesis

The recent African origin hypothesis (or replacement hypothesis) differs from the first two in proposing that there was a single origin of modern humans in Africa about 100 000 years ago with subsequent dispersal throughout the rest of the world, and complete replacement of the non-modern populations (Stringer, 1992; Stringer and Andrews, 1988). Hence, all of the genes in non-

African populations would have a recent African origin (*Figure 11.1*). Proponents of this view of the fossil record dispute the evidence for regional continuity, and instead point to growing evidence for the earliest appearance of fossils of anatomically modern humans in Africa about 100 000 years ago.

11.2.4 Assimilation hypotheses

Assimilation hypotheses, of which there are several varieties (Bräuer, 1989; Smith *et al.*, 1989), are an attempt to reconcile the fossil evidence for the earliest appearance of anatomically modern humans in Africa with the evidence for regional continuity. These hypotheses agree with the recent African origin hypothesis in proposing a single recent origin of modern humans in Africa followed by dispersal from Africa, but differ in allowing for some degree of admixture between the modern African and the archaic non-African populations to account for regional continuity. Hence, as with multiregional evolution, these hypotheses predict that there will be varying contributions of African and non-African genes to non-African populations (*Figure 11.1*).

11.3 Properties of mtDNA

The human mtDNA genome consists of about 16 500 base pairs of DNA, or about 0.0006% of the human nuclear DNA genome. Such a small component of the human genome has, nevertheless, several properties which make it particularly suitable for some evolutionary studies (Avise, 1986; Stoneking, 1993; Wallace, 1995; Wilson *et al.*, 1985).

First, mtDNA is present in a high copy number in human cells. The average somatic cell has just two copies of any given nuclear gene or DNA segment, but hundreds to thousands of copies of mtDNA (Robin and Wong, 1988). This property, along with the extra-nuclear, cytoplasmic location of mtDNA, makes it easier to obtain mtDNA for analysis, and also makes mtDNA the molecule of choice for analysing ancient DNA.

Secondly, mtDNA is perhaps the best known eukaryotic genome: the complete sequence and gene organization (*Figure 11.2*) is known for humans (Anderson *et al.*, 1981) and many other organisms. Knowledge of the complete sequence made it possible to construct detailed restriction site maps (Cann *et al.*, 1984, 1987; Horai and Matsunaga, 1986; Johnson *et al.*, 1983; Merriweather *et al.*, 1991; Stoneking *et al.*, 1990) by comparing observed restriction enzyme fragment patterns with those predicted from the published sequence. More recently, the polymerase chain reaction (PCR; Arnheim *et al.*, 1990; Erlich and Arnheim, 1992) has facilitated the rapid analysis of any desired region of the mtDNA genome. Currently, most laboratories engaged in studies of human mtDNA variation use PCR to amplify and sequence hypervariable segments of the non-coding control region (DiRienzo and Wilson, 1991; Lum *et al.*, 1994; Mountain *et al.*, 1995; Redd *et al.*, 1995;

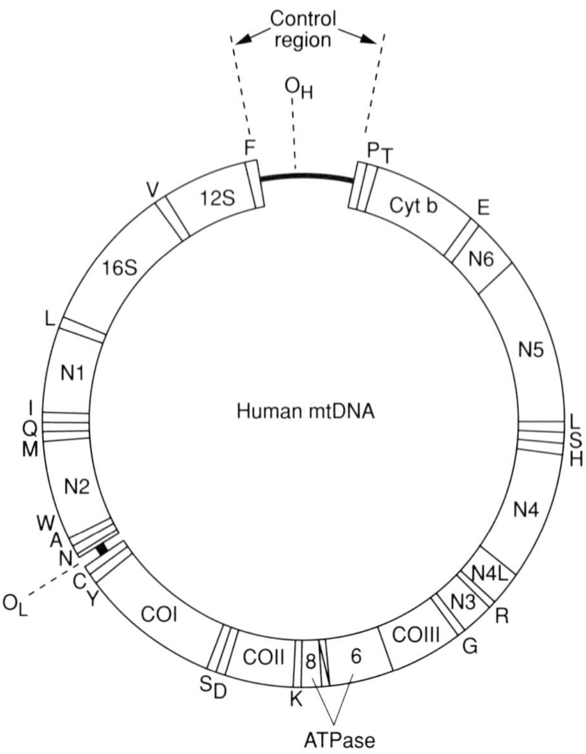

Figure 11.2. The human mtDNA genome, with functional regions indicated as follows: two ribosomal RNA genes (12S and 16S); 13 protein-coding genes, including three cytochrome oxidase subunits (COI–COIII), seven NADH-dehydrogenase subunits (N1–N6, N4L), two ATPase subunits (6 and 8), and cytochrome *b* (Cyt b); and 22 transfer RNA genes (designated by the standard single letter code). The thin solid bars indicate the major noncoding regions, including the origin of replication of the heavy (O_H) and light (O_L) strands. Derived from Stoneking (1993).

Stoneking *et al.*, 1992a; Vigilant *et al.*, 1989, 1991; Ward *et al.*, 1991), but another approach combines PCR with restriction enzyme analysis by amplifying the entire mtDNA genome in overlapping segments and constructing detailed restriction site maps of each segment (Ballinger *et al.*, 1992; Chen *et al.*, 1995; Schurr *et al.*, 1990; Torroni *et al.*, 1992, 1993, 1994). More recently, a rapid procedure for determining mtDNA types based on hybridization of sequence-specific oligonucleotide probes to known polymorphic sites in the mtDNA control region has been developed (Stoneking *et al.*, 1991) and is being implemented in studies of human mtDNA variation (Melton *et al.*, 1995; Soodyall *et al.*, 1995).

The third unusual property of mtDNA is that it is strictly maternally inherited, with no paternal contribution or recombination between mitochondrial genomes (Giles *et al.*, 1980). However, strict maternal inheritance of mtDNA has been called into question by studies showing low levels of paternal mtDNA inheritance in fruit flies (Kondo *et al.*, 1990), mice (Gyllensten *et al.*, 1991) and mussels (Zouros *et al.*, 1992). Mussels differ from humans in that extensive mtDNA heteroplasmy (presence of multiple mtDNA genomes, implying biparental inheritance of mitochondria) is common. The validity of using data from mussel species to draw conclusions about human mtDNA inheritance is, therefore, debatable. The other studies used repeated backcrosses involving interspecific hybrids in order to build up detectable levels of paternal mtDNA that could be differentiated from the maternal mtDNA. This raises the question of whether the observed low levels of paternal inheritance were actually a peculiarity associated with the hybrids rather than a phenomenon within species. Recently, it has been shown that in intraspecific crosses of mice, paternal mtDNA is preferentially eliminated shortly after fertilization, while it persists in the zygote of interspecific crosses (Kaneda *et al.*, 1995). Hence, it is likely that the previously observed paternal inheritance is a consequence of using interspecific hybrids and that strict maternal inheritance occurs within species.

The final property of mtDNA which makes it a unique tool to investigate human evolution is that it evolves rapidly, some 5–10 times faster than an average segment of nuclear DNA (Brown *et al.*, 1979). This means that mtDNA will be most useful for comparisons involving closely related species or populations of the same species, as a sufficient number of mutations will have occurred even over a short evolutionary timespan to permit an assessment of species or population relationships. The rate of mtDNA evolution is a function of two processes: the rate at which new mutations arise and the probability that a newly arisen mutation will become fixed. Plausible reasons for believing that both of these processes are elevated for mtDNA have been detailed elsewhere (Gray, 1989; Wilson *et al.*, 1985) and include such factors as the expanded codon recognition pattern in mitochondria, exposure of mtDNA to oxidative damage, absence of DNA repair functions, an error-prone DNA replication system, an inability to remove mildly deleterious mutations by natural selection (due to the lack of recombination) and a rapid turnover of mitochondria and mtDNA.

Initial studies of human mtDNA variation, benefitting from the properties of mtDNA outlined above, led to what has become popularly known as the 'African Eve' hypothesis (Cann *et al.*, 1987); subsequent work has largely supported and extended this hypothesis (Chen *et al.*, 1995; Horai and Hayasaka, 1990; Horai *et al.*, 1995; Kocher and Wilson, 1991; Merriweather *et al.*, 1991; Ruvolo *et al.*, 1993; Stoneking and Cann, 1989; Stoneking *et al.*, 1992a; Vigilant *et al.*, 1989, 1991). There are three aspects to this hypothesis: (i) all mtDNA types in contemporary populations trace back to a single ancestor; (ii) this ancestor probably lived in Africa; and (iii) this ancestor probably lived about 200 000 years ago. Each of these three aspects will be considered in turn.

11.4 A single mtDNA ancestor

This seemingly straightforward conclusion follows directly from principles of biological evolution. Nonetheless, it has been responsible for more consternation and misinterpretation (particularly among the popular press) than any other aspect of the hypothesis. Simply put, assuming that there was a single origin of life on this planet and assuming that all living things are descended from this single point of origin, then it must follow that all of the variation in *any* segment of DNA (mitochondrial or nuclear) in the present generation must ultimately trace back to just one ancestor in some previous generation.

Suppose, for example, that we survey a population today and find *n* different mtDNA types. If we could go back in time and survey this population one generation ago, we would find (at most) *n* of these mtDNA types (ignoring the additional types we might find that are not represented in the population today). If we keep going back in time, generation by generation, at some point we will find only *n–1* of the mtDNA types present today. This is because a new mutation in this generation will have produced a new mtDNA type; conversely, two mtDNA types present in the subsequent generation will have coalesced (Hudson, 1990) in this generation. This process will repeat itself as we move back in time, until ultimately all *n* of the mtDNA types in the present generation will have coalesced to one ancestor – by definition, the ancestor of all contemporary human mtDNA types.

Figure 11.3 illustrates this principle, as well as some additional concepts that follow directly. First, the mtDNA ancestor was female, since mtDNA is maternally inherited. Secondly, the mtDNA ancestor was not the only individual alive; she was a member of a population, but the mtDNA types of her contemporaries ultimately became extinct because they, or their descendants, left either no offspring or only male offspring. Third, she was not the first woman to have appeared on this planet; along with everyone else, she had ancestors that belonged to past populations, but she represents the point at which all contemporary mtDNA types coalesce. Fourth, she need not have contributed any other genes or DNA segments to contemporary populations. While every gene or DNA segment must have a common ancestor, each ancestor could be a different individual living at a different time. In fact, there is no *a priori* requirement that the mtDNA ancestor (or any other DNA ancestor) was necessarily even human; for example, alleles of certain genes of the human major histocompatability complex (HLA) are shared between humans and non-human primates (Gyllensten and Erlich, 1989; Figueroa *et al.*, 1988; Lawlor *et al.*, 1988; see also Chapter 3), indicating that the origin of these alleles predates the speciation event(s) which separated humans from non-human primates. Therefore, the principle that all contemporary mtDNA types trace back to a single ancestor should not be of any particular interest or concern. Rather, the interesting part of the story is determining where and when she lived.

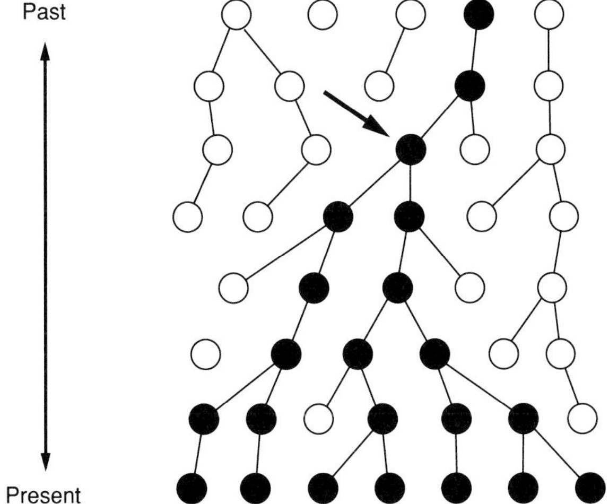

Figure 11.3. An illustration of the principle that all contemporary mtDNA types must trace back to a single ancestor. The solid circles indicate the path of descent from the ancestor (arrow) to the present generation (or, alternatively, the path of coalescence from the present population of mtDNA types to the ancestor); empty circles represent mtDNA types that became extinct. While the contemporary mtDNA types ultimately trace back to a single ancestor, note that other individuals co-existed with the mtDNA ancestor, and that the mtDNA ancestor also had ancestors. Derived from Stoneking (1993).

11.5 An African origin for the human mtDNA ancestor

There are two types of data that have been interpreted as supporting an African origin for the human mtDNA ancestor: phylogenetic analysis and relative amounts of mtDNA sequence divergence.

11.5.1 Phylogenetic analysis

The maternal, haploid inheritance of human mtDNA means that phylogenetic analysis is relatively straightforward. Since in the absence of recombination the only source of new variation is mutation, the number of mutations separating two mtDNA types reflects how closely related they are: two mtDNA types differing by a single mutation are (on average) more closely related to each other (i.e. share a more recent common ancestor) than either would be to a third mtDNA type that differs by two mutations. Trees depicting the phylogenetic relationship of individual mtDNA types can therefore be readily constructed by a variety of methods (Swofford and Olsen, 1990).

Initially, such trees (Cann *et al.*, 1987; Vigilant *et al.*, 1991) invariably contained two primary branches, one consisting solely of African mtDNAs, the

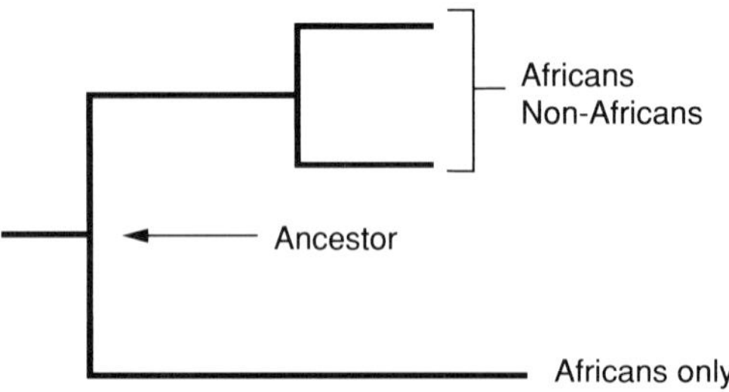

Figure 11.4. A simplified version of the results of phylogenetic analysis of mtDNA types in human populations. Invariably such trees consist of two primary branches, with only Africans represented on both branches. This sort of pattern is most easily explained by assuming that the ancestor was African. Derived from Stoneking (1993).

other consisting of both African and all non-African mtDNAs (*Figure 11.4*). While other explanations are possible, the simplest explanation for this pattern is to suppose that the ancestor was indeed African. However, subsequent re-analyses of the data led to the conclusion that there were better-fitting trees that did not fit the pattern depicted in *Figure 11.4* (Hedges *et al.*, 1992; Maddison, 1991; Maddison *et al.*, 1992; Templeton, 1992, 1993), thereby casting doubt on the claim that phylogenetic analysis of human mtDNA types supported an African origin. Instead, these trees indicated that no reliable information on the geographical origin of the human mtDNA ancestor was revealed by phylogenetic analysis of human mtDNA types.

The debate over the adequacy and merits of various methods of phylogenetic analysis is beyond the scope of this review; the above papers and some additional comments (Stoneking, 1993, 1994; Stoneking *et al.*, 1992b) should be consulted by the interested reader. While the issue is far from settled, a number of additional phylogenetic analyses of various types have been carried out on various datasets (Hedges *et al.*, 1992; Merriweather *et al.*, 1991; Penny *et al.*, 1995; Tamura and Nei, 1993) and the evidence currently does seem to favour trees of the form depicted in *Figure 11.4*.

11.5.2 Sequence divergence

All studies of worldwide human mtDNA variation have found that, on average, African populations have the greatest mtDNA sequence divergence, followed by Asian and European populations (Cann *et al.*, 1987; Chen *et al.*, 1995; Merriweather *et al.*, 1991; Vigilant *et al.*, 1991); an example is illustrated in *Figure 11.5*. It is important to note that this measure of mtDNA variation is not a simple measure of variation based on gene frequency, such as the average heterozygosity that is typically used to measure variation for blood group, serum protein, red cell enzymes, or other autosomal genes. Rather, the mean pairwise mtDNA sequence divergence directly reflects the

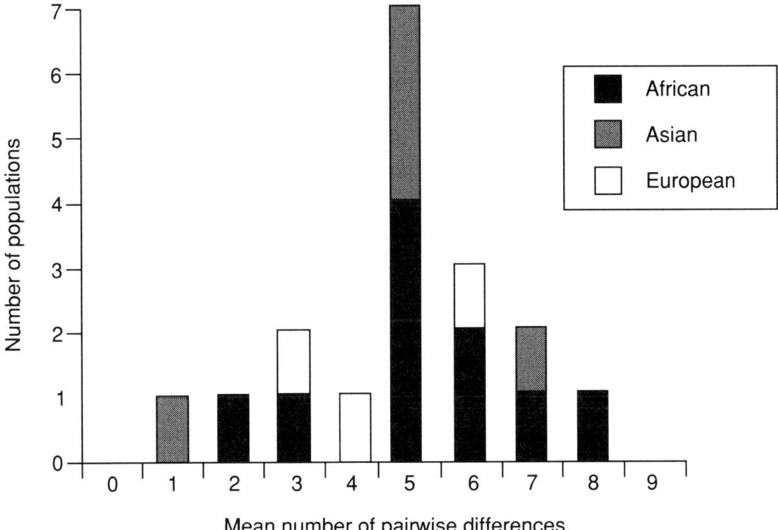

Figure 11.5. Histogram of the average number of pairwise differences for 18 populations for which sequences of the first hypervariable segment of the mtDNA control region were obtained. Data include published studies (DiRienzo and Wilson, 1991; Horai and Hayasaka, 1990; Vigilant *et al.*, 1991; Ward *et al.*, 1991) and unpublished data from my laboratory (M. Stoneking). African populations clearly have the most mtDNA mutations, a finding which has been replicated in every comprehensive study of human mtDNA variation (Cann *et al.*, 1987; Chen *et al.*, 1995; Merriweather *et al.*, 1991; Vigilant *et al.*, 1991). Derived from Stoneking (1993).

average number of mtDNA mutations that have accumulated within populations. The fact that Africans have the largest mtDNA sequence divergence indicates that they have accumulated the most mtDNA mutations, and the argument for an African origin states that the population with the most diversity is likely to be the ancestral population (Cann *et al.*, 1987; Stoneking and Cann, 1989). Not only do non-African populations have, on average, much less mtDNA diversity but they also appear to have a subset of the diversity present in Africa; that is, the mtDNA type diversity in Africa appears to encompass all of the diversity found outside of Africa, again in keeping with an African origin.

However, while all studies agree that Africans have the most mtDNA mutations, not everyone agrees that this necessarily indicates an African origin. This interpretation assumes that the rate of mtDNA evolution has been constant across populations; if Africans have experienced a higher rate of mtDNA evolution, then they would have more mtDNA mutations as well. However, a comparison of the average number of substitutions between human and chimpanzee mtDNAs does not reveal any differences among populations (Vigilant, 1990), arguing against different rates of evolution in different populations.

Another possible explanation is that selection might have altered patterns of mtDNA variation to give a false indication of an African origin. It is difficult to dismiss selection, since one can place the geographical origin of human mtDNA anywhere in the world by merely arguing that selection has been operating to distort patterns of mtDNA variation. There are ways to test whether the observed frequency distribution of mtDNA types in a population is consistent with what would be predicted from neutral equilibrium theory (Ewens, 1972; Tajima, 1989; Watterson, 1978). Application of these tests to human mtDNA data do indicate some instances in which the observed distributions differ significantly from the expected neutral distributions (Excoffier, 1990; Merriweather *et al.*, 1991; Whittam *et al.*, 1986). However, it is important to keep in mind that these tests assume that populations are at equilibrium; the significant departures from neutral expectations are generally considered to reflect deviations from equilibrium such as population agglomeration and recent population expansion, and not necessarily selection (Merriweather *et al.*, 1991; Stoneking, 1990; Whittam *et al.*, 1986).

Finally, the greater mtDNA diversity in Africa might reflect ancient population size rather than population age (Relethford and Harpending, 1994). If African populations have indeed been larger than non-African populations, then they would be expected to have more mtDNA diversity, since the neutral theory predicts that diversity is directly proportional to the effective population size (Kimura, 1983). However, even if African populations have been larger, this does not mean they could not be the source population (indeed, the presumably random mutations leading to anatomically modern humans would be expected to have had the greatest chance of occurring in the largest population). Furthermore, the patterns of mtDNA diversity, in which non-African populations appear to possess a subset of the diversity in Africa, suggest an African origin. In summary, the combined observations of more mtDNA diversity and more divergent mtDNA lineages in Africans are most easily reconciled with an African origin of human mtDNA.

11.6 Age of the human mtDNA ancestor

To distinguish between the recent African and multiregional evolution hypotheses, establishing an African origin of human mtDNA is not as important as the age of the human mtDNA ancestor. This is because both hypotheses acknowledge that hominids did arise in Africa and that the earliest hominid migrations out of Africa probably began about 1–1.5 million years ago, involving members of the genus *Homo* (Stringer and Andrews, 1988; Wolpoff, 1989). Therefore, if the human mtDNA ancestor lived about 1–1.5 million years ago, the migrations out of Africa inferred from the mtDNA data would correspond to these first hominid migrations, and would not throw light on putative subsequent migrations involving anatomically modern humans. Previous estimates of the age of the human mtDNA ancestor have centred around 200 000 years ago (Cann *et al.*, 1987; Stoneking and Cann, 1989; Vigilant *et al.*, 1991), implying that the migrations out of Africa inferred from mtDNA data are more recent than the earliest migrations of the genus *Homo*, and hence presumably involve anatomically modern humans. However, these studies were unable to estimate standard errors, which means that a much older age might not be statistically ruled out.

Thus, there has been considerable attention focused on this problem. There are two aspects to estimating the age of the human mtDNA ancestor: (i) determining the amount of sequence divergence that has accumulated since the existence of the mtDNA ancestor; and (ii) determining the rate of human mtDNA sequence divergence. The amount of sequence divergence that has accumulated since the mtDNA ancestor lived is easily determined from the phylogenetic trees; even with the controversy over the adequacy of phylogenetic analysis for the purpose of inferring geographical origin, it is unlikely that the average amount of sequence divergence traced back to the ancestor will be greatly influenced. For example, parsimony, NJ (neighbour joining), and UPGMA (unweighted pair-group method using arithmetic averages) analysis of data (Vigilant *et al.*, 1991) yield different trees but virtually identical estimates of the amount of sequence divergence since the ancestor (M. Stoneking and S.T. Sherry, unpublished).

Of more concern, then, is determining the rate of human mtDNA sequence divergence and the associated standard error. The usual method of determining the rate of mtDNA divergence is to calculate the average amount of sequence divergence between human mtDNA and our closest non-human relative, the chimpanzee, and divide this by the time taken for the human and chimpanzee separation to get the rate of divergence. While this procedure looks simple enough, it is complicated by the fact that the observed amount of mtDNA sequence divergence between human and chimpanzee must be corrected for multiple substitutions at the same nucleotide site. The dynamics of mtDNA sequence evolution (in particular the rapid rate of change, the greatly elevated frequency of transition mutations and the decidedly non-uniform distribution of mutational sites across the mtDNA genome) combined with the relatively long period of evolutionary time separating the human and chimpanzee make this correction for multiple substitutions particularly troublesome. The standard methods of correction, with their assumption of an equal probability of all types of mutations at all sites, are simply inadequate (Hasegawa and Horai, 1991; Kocher and Wilson, 1991).

To complicate matters further, there are multiple sources of variance that must be considered when estimating a standard error for the age of the human mtDNA ancestor (Nei, 1987; Tajima, 1983). These include variances due to the sampling of individuals, the sampling of mtDNA diversity and the inherent stochastic nature of mtDNA evolution itself. Nevertheless, a variety of methods have recently been developed for estimating the age of

Table 11.1. Estimates of the age of the human mtDNA ancestor, and associated 95% confidence intervals (CI)

Study (reference)	Ancestor age (years)	95% CI
Hasegawa and Horai (1991)	280 000	180 000–380 000
Pesole *et al.* (1992)	400 000	200 000–600 000
Nei (1992)	207 000	110 000–504 000
Stoneking *et al.* (1992a)	135 000	63 000–386 000
Tamura and Nei (1993)	160 000	80 000–480 000
Hasegawa *et al.* (1993)[a]	211 000	0–433 000
Hasegawa *et al.* (1993)[b]	101 000	0–205 000
Templeton (1993)	213 000	102 000–389 000
Ruvolo *et al.* (1993)	298 000	129 000–536 000
Horai *et al.* (1995)	143 000	107 000–179 000

[a] Based on control region sequences.
[b] Based on coding sequences.

the human ancestor and the associated standard error (Hasegawa and Horai, 1991; Hasegawa *et al.*, 1993; Nei, 1992; Pesole *et al.*, 1992; Stoneking *et al.*, 1992a; Tamura and Nei, 1993; Templeton, 1993). While the details of these methods are beyond the scope of this review, they include various approaches to the problems identified above (including Markov models, maximum likelihood methods, and estimates based on the ratio of transition mutations to transversion mutations). The results of applying these methods to various human mtDNA datasets are summarized in *Table 11.1* in the form of approximate 95% confidence intervals for the age of the human mtDNA ancestor. Despite the variety of methods and datasets used, almost all of these confidence intervals place the age of the human mtDNA ancestor within the range of 50 000 to 500 000 years, as predicted from conjectures concerning the likely magnitude of the standard error (Stoneking and Cann, 1989).

In considering the relevance of the age of the human mtDNA ancestor for comparing the various theories of recent human origins, it is important to realize that the age of the mtDNA ancestor marks a genetic divergence, not a population divergence (Nei, 1987). Placing the age of the mtDNA ancestor at 500 000 years does not mean that modern humans arose 500 000 years ago, since there is no necessary requirement that the ancestor was anatomically modern (Stoneking and Cann, 1989). However, while it is true that the mtDNA ancestor does mark a genetic divergence, it is also true that genetic divergence tends to precede population divergence (Nei, 1987). This is because at the time of population divergence, the ancestral population is usually polymorphic for any particular gene, so the coalescence for that polymorphism must predate the population divergence (*Figure 11.6*). Therefore, if the human mtDNA ancestor really did live not more than 500 000 years ago then the corresponding inference is that non-African human populations are no more than 500 000 years old. This is clearly incompatible with multiregional evolution,

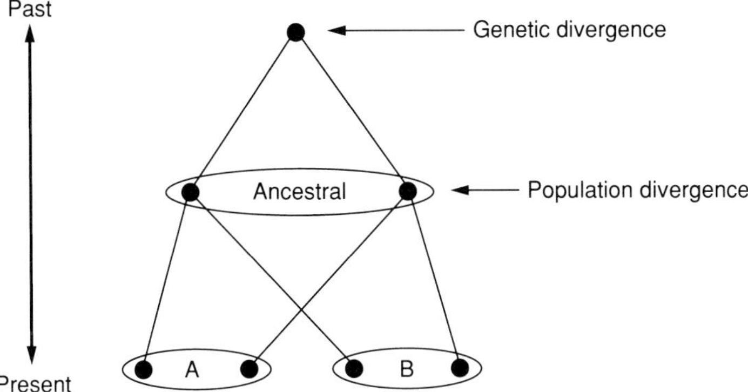

Figure 11.6. An illustration of the principle that genetic divergence will generally precede population divergence. An ancestral population is depicted that gave rise to two descendant populations (A and B). The ancestral population contained two mtDNA types, which therefore had to have diverged before population divergence occurred. In general, if the ancestral population is polymorphic (which must be the case, unless a severe bottleneck occurred), then the genetic divergence (or coalescence) must precede population divergence. The importance of the age of the human mtDNA ancestor is that this age is an upper bound for human population divergence. Therefore, human population divergences cannot be older than the age of the human mtDNA ancestor; they could, however, be considerably younger. Derived from Stoneking (1993).

which holds that human population differences should be 1–1.5 million years old, and hence genetic divergences should be even older.

It has also been pointed out recently that the age of the human mtDNA ancestor is primarily a property of the effective population size of the ancestral human population (Harpending *et al.*, 1993; Rogers and Jorde, 1995). The average amount of sequence divergence can be used to estimate the effective population size (Tajima, 1983); for mtDNA the resulting estimate is about 6000 females (Wilson *et al.*, 1985) which is in good agreement with estimates from nuclear DNA of about 10 000 males and females (Nei and Graur, 1984; Takahata, 1993). This is too small to make multiregional evolution feasible since it does not seem possible that only 6000 females can be distributed from South Africa to northern Asia and Europe in populations that are big enough to survive and close enough to maintain the level of gene flow required by multiregional evolution (Harpending *et al.*, 1993; Rogers and Jorde, 1995).

11.7 Human origins reconsidered

The available evidence is most consistent with placing the mtDNA ancestor in Africa about 200 000 years (and not more than 500 000 years) ago. This is easily reconciled with a recent origin of modern humans in Africa, some

100 000 years ago, with subsequent spread of modern humans throughout the Old World. Furthermore, the relatively recent coalescent time argues strongly against multiregional evolution and the candelabra theory. What about the assimilation theories?

Since non-African populations that pre-dated the mtDNA ancestor apparently did not contribute mtDNAs to contemporary human populations, it would appear that the spread of modern humans was accomplished with little or no admixture with the resident non-African populations; if it were otherwise, then there should be evidence of much more divergent mtDNA types in contemporary human populations (Cann, 1988; Stoneking and Cann, 1989).

Thus, while the mtDNA evidence appears to argue strongly for replacement and against assimilation, there are alternative explanations to be considered. First, the lack of extremely divergent, non-African mtDNA types could reflect sampling; it may be that such mtDNA types do exist in contemporary human populations, but they have not yet been studied. This explanation appears unlikely, since at the time of this review mtDNA types have been determined from around 10 000 people, by a number of laboratories from around the world using a variety of techniques, and no such extremely divergent mtDNA type has been reported. If these 10 000 people are a random sample of the world's population, then we can be 95% confident that the frequency of any unreported mtDNA type must be less than 0.0003. As more individuals are surveyed, eventually either extremely divergent mtDNA types will be found or the likelihood of finding such a type will drop to essentially zero.

Secondly, the fact that extremely divergent mtDNA types are not present in contemporary populations does not necessarily imply that they never contributed to the ancestral human population. Perhaps non-African mtDNAs made a contribution (as in the assimilation hypothesis), but were lost in subsequent generations by the process of random extinction depicted in *Figure 11.3*. However, random extinction is by definition without bias, whereas this argument supposes preferential loss of the non-African mtDNA types. Furthermore, the probability of random extinction is influenced by changes in population size and a population that is increasing in size (even very slowly) tends to retain most of its mtDNA variation (Avise *et al.*, 1984). No doubt there have been fluctuations in human population size over time but the overall trend has been one of growth. While this is a qualitative response to a problem that should be addressed quantitatively, it does not seem very likely that any substantial contribution of non-African mtDNA types could have disappeared simply via random extinction.

Another possible explanation for the disappearance of non-African mtDNA types from the ancestral modern human gene pool involves selection. If a selectively favoured mutation had arisen anywhere in the human mtDNA genome around, say, 200 000 years ago, then the mtDNA data would be reflecting the spread and fixation of that advantageous mutation, not human population movements *per se.* As discussed previously, selection largely does not appear to be influencing patterns of mtDNA variation in contemporary human populations. However, this does not rule out a previous selective event involving human mtDNA. In fact, even if future studies prove beyond any statistical doubt that there was a recent African origin for human mtDNA, there is always the possibility that this reflects something unique about the history of human mtDNA, and not necessarily the history of our species.

This is because, when all is said and done, mtDNA is only a single genetic locus (albeit a highly interesting and informative one). The danger of relying on a single genetic locus to reconstruct human population history is that the pattern of variation at this locus might, by chance or by selection, differ drastically from other loci and from the true picture. To reconstruct human population history accurately, many genetic loci must be examined. Since humans unfortunately have only one mtDNA locus, for these other loci we must rely on studies of nuclear DNA, including both Y-chromosomal ([Y-DNA], see Chapter 12) and autosomal DNA variation. While a review of the growing body of nuclear DNA data is outside the scope of this chapter, studies of a wide variety of genes, including classical markers (Cavalli-Sforza *et al.*, 1988; Nei and Roychoudhury, 1993), restriction fragment length polymorphism (RFLP) loci (Mountain and Cavalli-Sforza, 1994; Mountain *et al.*, 1992; Wainscoat *et al.*, 1986), short tandem repeat loci (Bowcock *et al.*, 1994; Goldstein *et al.*, 1995), *Alu* insertion polymorphisms (Batzer *et al.*, 1994), and Y-DNA (Dorit *et al.*, 1995; see also Chapter 12), are all consistent with a recent and/or African origin of modern humans. The issue is far from settled but all of the currently available genetic data favour a recent African origin of our species.

References

Anderson S, Bankier AT, Barrell BG *et al.* (1981) Sequence and organization of the human mitochondrial genome. *Nature* **290**: 457–465.

Arnheim N, White T, Rainey WE. (1990) Application of PCR: organismal and population biology. *BioScience* **40**: 174–182.

Avise JC. (1986) Mitochondrial DNA and the evolutionary genetics of higher animals. *Phil. Trans. R. Soc. Lond. Ser. B* **312**: 325–342.

Avise JC, Neigel JE, Arnold J. (1984) Demographic influences on mitochondrial DNA lineage survivorship in animal populations. *J. Mol. Evol.* **20**: 99–105.

Ballinger SW, Schurr TG, Torroni A, Gan YY, Hodge JA, Hassan K, Chen K-H, Wallace DC. (1992) Southeast Asian mitochondrial DNA analysis reveals genetic continuity of ancient mongoloid migrations. *Genetics* **130**: 139–152.

Batzer MA, Stoneking M, Alegria-Hartman M *et al.* (1994) African origin of human-specific polymorphic Alu insertions. *Proc. Natl Acad. Sci. USA* **91**: 12288–12292.

Bowcock AM, Ruiz-Linares A, Tomfohrde J, Minch E, Kidd JR, Cavalli-Sforza LL. (1994) High resolution of human evolutionary trees with polymorphic microsatellites. *Nature* **368**: 455–457.

Bräuer G. (1989) The evolution of modern humans: a comparison of the African and non-African evidence. In: *The Human Revolution* (eds P Mellars and CB Stringer). Edinburgh University Press, Edinburgh, pp. 123–154.

Brown EM, George M, Wilson AC. (1979) Rapid evolution of animal mitochondrial DNA. *Proc. Natl Acad. Sci. USA* **76**: 1967–1971.

Cann RL. (1988) DNA and human origins. *Annu. Rev. Anthropol.* **17**: 127–143.

Cann RL, Brown WM, Wilson AC. (1984) Polymorphic sites and the mechanism of evolution in human mitochondrial DNA. *Genetics* **106**: 479–499.

Cann RL, Stoneking M, Wilson AC. (1987) Mitochondrial DNA and human evolution. *Nature* **325**: 31–36.

Cavalli-Sforza LL, Piazza A, Menozzi P, Mountain J. (1988) Reconstruction of human evolution: bringing together genetic, archaeological, and linguistic data. *Proc. Natl Acad. Sci. USA* **85**: 6002–6006.

Chen Y-S, Torroni A, Excoffier E, Santachiara-Benerecetti A, Wallace DC. (1995) Analysis of mtDNA variation in African populations reveals the most ancient of all human continent-specific haplogroups. *Am. J. Hum. Genet.* **57**: 133–149.

Coon CS. (1962) *The Origin of Races.* Knopf, New York.

DiRienzo A, Wilson AC. (1991) Branching pattern in the evolutionary tree for human mitochondrial DNA. *Proc. Natl Acad. Sci. USA* **88**: 1597–1601.

Dorit RL, Akashi H, Gilbert W. (1995) Absence of polymorphism at the *ZFY* locus on the human Y chromosome. *Science* **268**: 1183–1185.

Erlich HA, Arnheim N. (1992) Genetic analysis using the polymerase chain reaction. *Annu. Rev. Genet.* **26**: 479–506.

Ewens WJ. (1972) The sampling theory of selectively neutral alleles. *Theor. Popul. Biol.* **3**: 87–112.

Excoffier L. (1990) Evolution of human mitochondrial DNA: evidence for departure from a pure neutral model of populations at equilibrium. *J. Mol. Evol.* **30**: 125–139.

Figueroa F, Gunther E, Klein J. (1988) MHC polymorphism pre-dating speciation. *Nature* **335**: 265–268.

Giles RE, Blanc H, Cann HM, Wallace DC. (1980) Maternal inheritance of human mitochondrial DNA. *Proc. Natl Acad. Sci. USA* **77**: 6715–6719.

Goldstein DB, Linares AR, Cavalli-Sforza LL, Feldman MW. (1995) Genetic absolute dating based on microsatellites and the origin of modern humans. *Proc. Natl Acad. Sci. USA* **92**: 6723–6727.

Gray, MW. (1989) Origin and evolution of mitochondrial DNA. *Annu. Rev. Cell Biol.* **5**: 25–50.

Gyllensten UB, Erlich HA. (1989) Ancient roots for polymorphism of the HLA-DQα locus in primates. *Proc. Natl Acad. Sci. USA* **86**: 9986–9990.

Gyllensten UB, Wharton D, Joseffson A, Wilson AC. (1991) Paternal inheritance of mitochondrial DNA in mice. *Nature* **352**: 255–257.

Harpending HC, Sherry ST, Rogers AR, Stoneking M. (1993) The genetic structure of ancient human populations. *Curr. Anthropol.* **34**: 483–496.

Hasegawa M, Horai S. (1991) Time of the deepest root for polymorphism in human mitochondrial DNA. *J. Mol. Evol.* **32**: 37–42.

Hasegawa M, Di Rienzo A, Kocher TD, Wilson AC. (1993) Toward a more accurate time scale for the human mitochondrial DNA tree. *J. Mol. Evol.* **37**: 347–354.

Hedges SB, Kumar S, Tamura K, Stoneking M. (1992) Human origins and the analysis of mitochondrial DNA sequences. *Science* **255**: 737–739.

Horai S, Hayasaka K. (1990) Intraspecific nucleotide sequence differences in the major noncoding region of human mitochondrial DNA. *Am. J. Hum. Genet.* **46**: 828–842.

Horai S, Matsunaga E. (1986) Mitochondrial DNA polymorphism in Japanese. *Hum. Genet.* **72**: 105–117.

Horai S, Hayasaka K, Kondo R, Tsugane K, Takahata N. (1995) Recent African origin of modern humans revealed by complete sequences of hominoid mitochondrial DNAs. *Proc. Natl Acad. Sci. USA* **92**: 532–536.

Hudson RR. (1990) Gene genealogies and the coalescent process. *Oxf. Surv. Evol. Biol.* **7**: 1–44.

Johnson MJ, Wallace DC, Ferris SD, Rattazzi MC, Cavalli-Sforza LL. (1983) Radiation of human mitochondria DNA types analyzed by restriction endonuclease cleavage patterns. *J. Mol. Evol.* **19**: 255–271.

Kaneda H, Hayashi J-I, Takahama S, Taya C, Lindahl KF, Yonekawa H. (1995) Elimination of paternal mitochondrial DNA in intraspecific crosses during early mouse embryogenesis. *Proc. Natl. Acad. Sci. USA* **92**: 4542–4546.

Kimura M. (1983) *The Neutral Theory of Molecular Evolution.* Cambridge University Press, Cambridge.

Kocher TD, Wilson AC. (1991) Sequence evolution of mitochondrial DNA in humans and chimpanzees: control region and a protein-coding region. In: *Evolution of Life: Fossils, Molecules and Culture* (eds S Osawa and T Honjo). Springer-Verlag, Tokyo, pp. 391–413.

Kondo R, Satta Y, Matsuura ET, Ishiwa H, Takahata N, Chigusa SI. (1990) Incomplete maternal transmission of mitochondrial DNA in *Drosophila. Genetics* **126**: 657–663.

Lawlor DA, Zenmour J, Ennis PP, Parham P. (1988) HLA-A and B polymorphisms predated the divergence of humans and chimpanzees. *Nature* **335**: 268–271.

Lum JK, Rickards O, Ching C, Cann RL. (1994) Polynesian mitochondrial DNAs reveal three deep maternal lineage clusters. *Hum. Biol.* **66**: 567–590.

Maddison DR. (1991) African origin of human mitochondrial DNA reexamined. *Syst. Zool.* **40**: 355–363.

Maddison DR, Ruvolo M, Swofford DL. (1992) Geographic origins of human mitochondrial DNA: phylogenetic evidence from control region sequences. *Syst. Biol.* **41**: 111–124.

Melton T, Peterson R, Redd AJ, Saha N, Sofro ASM, Martinson J, Stoneking M. (1995) Polynesian genetic affinities with Southeast Asian populations as identified by mtDNA analysis. *Am. J. Hum. Genet.* **57**: 403–414.

Merriweather DA, Clark AG, Ballinger SW, Schurr TG, Soodyall H, Jenkins T, Sherry ST, Wallace DW. (1991) The structure of human mitochondrial DNA variation. *J. Mol. Evol.* **33**: 543–555.

Mountain JL, Cavalli-Sforza LL. (1994) Inference of human evolution through cladistic analysis of nuclear DNA restriction polymorphisms. *Proc. Natl Acad. Sci. USA* **91**: 6515–6519.

Mountain JL, Lin AA, Bowcock AM, Cavalli-Sforza LL. (1992) Evolution of modern humans: evidence from nuclear DNA polymorphism. *Phil. Trans. R. Soc. Lond. Ser. B* **337**: 159–165.

Mountain JL, Hebert JM, Bhattacharya S, Underhill PA, Ottolenghi C, Gadgil M, Cavalli-Sforza LL. (1995) Demographic history of India and mtDNA-sequence diversity. *Am. J. Hum. Genet.* **56**: 979–992.

Nei M. (1987) *Molecular Evolutionary Genetics.* Columbia University Press, New York.

Nei M. (1992) Age of the common ancestor of human mitochondrial DNA. *Mol. Biol. Evol.* **9**: 1176–1178.

Nei M, Graur D. (1984) Extent of protein polymorphism and the neutral mutation theory. *Evol. Biol.* **17**: 73–118.

Nei M, Roychoudhury AK. (1993) Evolutionary relationships of human populations on a global scale. *Mol. Biol. Evol.* **10**: 927–943.

Penny D, Steel M, Waddell PJ, Hendy MD. (1995) Improved analyses of human mtDNA sequences support a recent African origin for *Homo sapiens. Mol. Biol. Evol.* **12**: 863–882.

Pesole G, Sbisa E, Preparata G, Saccone C. (1992) The evolution of the mitochondrial D-loop region and the origin of modern man. *Mol. Biol. Evol.* **9**: 587–598.

Redd AJ, Takezaki N, Sherry ST, McGarvey ST, Sofro ASM, Stoneking M. (1995) Evolutionary history of the COII/tRNA-lys intergenic 9 base pair deletion in human mitochondrial DNAs from the Pacific. *Mol. Biol. Evol.* **12**: 604–615.

Relethford JH, Harpending HC. (1994) Craniometric variation, genetic theory, and modern human origins. *Am. J. Phys. Anthropol.* **95**: 249–270.

Robin ED, Wong R. (1988) Mitochondrial DNA molecules and virtual number of mitochondria per cell in mammalian cells. *J. Cell. Physiol.* **136**: 507–513.

Rogers AR, Jorde LB. (1995) Genetic evidence on modern human origins. *Hum. Biol.* **67**: 1–36.

Ruvolo M, Zehr S, von Dornum M, Pan D, Chang B, Lin J. (1993) Mitochondrial COII sequences and modern human origins. *Mol. Biol. Evol.* **10**: 1115–1135.

Schurr TG, Ballinger SW, Gan Y-Y, Hodge JA, Merriweather DA, Lawrence DN, Knowler WC, Weiss KM, Wallace DC. (1990) Amerindian mitochondrial DNAs have rare Asian mutations at high frequencies, suggesting they derived from four primary maternal lineages. *Am. J. Hum. Genet.* **46**: 613–623.

Smith FH, Falsetti AB, Donnelly SM. (1989) Modern human origins. *Yrbk Phys. Anthropol.* **32**: 35–68.

Soodyall H, Jenkins T, Stoneking M. (1995) 'Polynesian' mtDNA in the Malagasy. *Nature Genetics* **10**: 377–378.

Stoneking M. (1990) Departure of human mitochondrial DNA variation from neutral expectations: an alternative explanation. *J. Mol. Evol.* **31**: 343–346.

Stoneking M. (1993) DNA and recent human evolution. *Evol. Anthropol.* **2**: 60–73.

Stoneking M. (1994) Mitochondrial DNA and human evolution. *J. Bioenerget. Biomembr.* **26**: 251–259.

Stoneking M, Cann RL. (1989) African origin of human mitochondrial DNA. In: *The Human Revolution* (eds P Mellars and CB Stringer). Edinburgh University Press, Edinburgh. pp. 17–30.

Stoneking M, Jorde LB, Bhatia K, Wilson AC. (1990) Geographic variation in human mitochondrial DNA from Papua New Guinea. *Genetics* **124**: 717–733.

Stoneking M, Hedgecock D, Higuchi RG, Vigilant L, Erlich HA. (1991) Population variation of human mtDNA control region sequences detected by enzymatic amplification and sequence-specific oligonucleotide probes. *Am. J. Hum. Genet.* **48**: 370–382.

Stoneking M, Sherry ST, Redd AJ, Vigilant L. (1992a) New approaches to dating suggest a recent age for the human mtDNA ancestor. *Phil. Trans. R. Soc. Lond. Ser. B* **337**: 167–175.

Stoneking M, Sherry ST, Vigilant L. (1992b) Geographic origin of human mitochondrial DNA revisited. *Syst. Biol.* **41**: 384–391.

Stringer CB. (1992) Replacement, continuity and the origin of *Homo sapiens*. In: *Continuity or Replacement Controversies in* Homo sapiens *Evolution* (eds G Bräuer and FH Smith). Balkema, Rotterdam, pp. 9–24.

Stringer CB, Andrews P. (1988) Genetic and fossil evidence for the origin of modern humans. *Science* **239**: 1263–1268.

Swofford DL, Olsen GJ. (1990) Phylogeny reconstruction. In: *Molecular Systematics* (eds DM Hillis and G Moritz). Sinauer, Sunderland, MA, pp. 411–501.

Tajima F. (1983) Evolutionary relationship of DNA sequences in finite populations. *Genetics* **105**: 437–460.

Tajima F. (1989) Statistical method for testing the neutral mutation hypothesis by DNA polymorphism. *Genetics* **123**: 585–595.

Takahata N. (1993) Allelic genealogy and human evolution. *Mol. Biol. Evol.* **10**: 2–22.

Tamura K, Nei M. (1993) Estimation of the number of nucleotide substitutions in the control region of mitochondrial DNA in humans and chimpanzees. *Mol. Biol. Evol.* **10**: 512–526.

Templeton AR. (1992) Human origins and analysis of mitochondrial DNA sequences. *Science* **255**: 737.

Templeton AR. (1993) The 'Eve' hypotheses: a genetic critique and reanalysis. *Am. Anthropol.* **95**: 51–72.

Torroni A, Schurr TG, Yang C-C *et al.* (1992) Native American mitochondrial DNA analysis indicates that the Amerind and Nadene populations were founded by two independent migrations. *Genetics* **130**: 153–162.

Torroni AS, Schurr TG, Cabell MF, Brown MD, Neel JV, Larsen M, Smith DG, Vullo CM, Wallace DC. (1993) Asian affinities and continental radiation of the four founding Native American mtDNAs. *Am. J. Hum. Genet.* **53**: 563–590.

Torroni A, Chen Y-S, Semino O, Santachiara-Beneceretti AS, Scott CR, Lott MT, Winter M, Wallace DC. (1994) MtDNA and Y-chromosome polymorphisms in four native American populations from southern Mexico. *Am. J. Hum. Genet.* **54**: 303–318.

Vigilant L. (1990) *Control Region Sequences from African Populations and the Evolution of Human Mitochondrial DNA*. PhD Thesis, University of California, Berkeley, CA.
Vigilant L, Pennington R, Harpending H, Kocher TD, Wilson AC. (1989) Mitochondrial DNA sequences in single hairs from a southern African population. *Proc. Natl Acad. Sci. USA* **86**: 9350–9354.
Vigilant L, Stoneking M, Harpending H, Hawkes K, Wilson AC. (1991) African populations and the evolution of human mitochondrial DNA. *Science* **253**: 1503–1507.
Wainscoat JS, Hill AVS, Boyce AL *et al.* (1986) Evolutionary relationships of human populations from an analysis of nuclear DNA polymorphisms. *Nature* **319**: 491–493.
Wallace DC. (1995) Mitochondrial DNA variation in human evolution, degenerative disease, and aging. *Am. J. Hum. Genet.* **57**: 201–223.
Ward RH, Frazier BL, Dew-Jager K, Pääbo S. (1991) Extensive mitochondrial diversity within a single Amerindian tribe. *Proc. Natl Acad. Sci. USA* **88**: 8720–8724.
Watterson GA. (1978) The homozygosity test of neutrality. *Genetics* **88**: 405–417.
Whittam TS, Clark AG, Stoneking M, Cann RL, Wilson AC. (1986) Allelic variation in human mitochondrial genes based on patterns of restriction site polymorphism. *Proc. Natl Acad. Sci. USA* **83**: 9611–9615.
Wilson AC, Cann RL, Carr SM *et al.* (1985) Mitochondrial DNA and two perspectives on evolutionary genetics. *Biol. J. Linn. Soc.* **26**: 375–400.
Wolpoff MH. (1989) Multiregional evolution: the fossil alternative to Eden. In: *The Human Revolution* (eds P Mellars and C Stringer). Edinburgh University Press, Edinburgh, pp. 62–108.
Wolpoff MH, Wu XZ, Thorne AG. (1984) Modern *Homo sapiens* origins: a general theory of hominid evolution involving the fossil evidence from East Asia. In: *The Origins of Modern Humans* (eds FH Smith and F Spencer). Alan R. Liss, New York, pp. 411–483.
Zouros E, Freeman KR, Ball AO, Pogson GH. (1992) Direct evidence for extensive paternal mitochondrial DNA inheritance in the marine mussel *Mytilus*. *Nature* **359**: 412–414

12

Y chromosome sequences as a tool for studying human evolution

Mark Jobling

12.1 Introduction

Within the human nuclear genome there is one section of DNA, the Y chromosome, which has a particularly unusual evolutionary history. In a diploid sexual organism it is unique in being constitutively haploid and uniparentally inherited, largely escaping from the processes of recombination which reshuffle the other chromosomes in each generation. This haploidy has important consequences for the genes which lie on the Y chromosome, for the mutation processes which can act upon it, and for its behaviour in populations. The evolutionary history and population genetics of the Y are, in addition, profoundly influenced by the cultural effects of the cardinal phenotype which it confers: maleness.

The purpose of this chapter is to review human Y chromosome diversity in the context of sequence evolution, and to ask how studying this can illuminate our understanding of the recent evolution of the human species.

12.2 The Y chromosome as an evolutionary tool

The diversity of modern human DNA results from the accumulation of mutations as it passes down to us from our ancestors. This diversity, in the form of DNA polymorphisms, can be analysed and used to investigate our genetic history and relatedness – a method of studying human evolution which complements the more traditional approaches of archaeology, palaeontology and linguistics.

Human Genome Evolution, edited by M. Jackson, T. Strachan and G. Dover.

Because of the reshuffling which occurs through recombination, the history of X chromosomes and autosomes is complex, and each has multiple ancestors. The uniparental inheritance and lack of recombination of the Y chromosome, on the other hand, means that its history is comparatively simple, and that modern Ys have a single paternal ancestor – a man sometimes referred to as 'Y-chromosomal Adam'. Unlike the Biblical account, this does not mean that there was ever only one single male in the population, or that there was anything unusual about him (see Chapter 11, Section 4, for a discussion of the concept of a single ancestor).

12.2.1 Analogy with mtDNA

In its mode of inheritance and lack of recombination, the Y chromosome is analogous to another part of the genetic material, the mitochondrial genome (mtDNA). This is maternally inherited because the sperm contributes no mtDNA to the zygote, and has been used extensively as a tool to study the evolution of human maternal lineages from a common maternal ancestor, 'mitochondrial Eve' (Cann *et al.*, 1987; Stoneking, 1993; see also Chapter 11). The aims of Y chromosome and mtDNA research in human evolution are the same: demonstration of the true relationships of modern Ys or mtDNAs in a tree which has a root and dated branchpoints, together with an estimate of the relatedness of human populations by the measurement of the frequencies of different Y or mtDNA types in these populations – the more closely related two populations are, the more similar these frequencies are expected to be.

These are several advantages of mtDNA which make it technically easy to analyse. It is a small (16.5 kb) circular molecule whose entire sequence is known and it has a base-substitutional mutation rate about ten times higher than that of nuclear DNA, possibly because of differences in repair systems (reviewed by Gray, 1989). This gives modern mtDNAs enormous diversity, which can be assayed simply by methods such as DNA sequencing or restriction analysis of PCR-amplified segments (PCR–RFLPs).

The Y chromosome does not have these advantages: it is large, with much higher sequence complexity than mtDNA, and little sequence information is available. It does not have a higher mutation rate than the rest of the genome. However, these properties of complexity and low mutation rate should ultimately allow the Y to yield more informative data than mtDNA. Its greater complexity allows it to carry a far more diverse range of polymorphisms than the compact, gene-rich and simple mtDNA. Rearrangements such as insertions, deletions, duplications, inversions and polymorphic tandemly repeated loci such as satellites, minisatellites and microsatellites are all likely to occur, in addition to base substitutions. These different kinds of polymorphic loci have different mutation rates, which should allow the choice of appropriate markers for studying evolution on different timescales. Although the high mutation rate of mtDNA is responsible for its high diversity, it also precludes the determination of the ancestral states of individual bases, since these will have undergone multiple substitutions in the lineages connecting humans and other primates. Recurrence and reversion within human lineages is also a problem and complicates the construction of trees. On the Y, in contrast, the rate is so low that individual mutations are very unlikely to recur or revert, and can thus be regarded as unique events defining groups of chromosomes related by descent; their ancestral states can be determined by analysis of DNA from other primate species. Binary polymorphisms of this kind can be used to draw a tree, simply and without computer assistance, using the principle of parsimony (an assumption of a minimum number of mutations).

12.3 Y chromosome genetics and cytogenetics

The human sex chromosomes are morphologically distinct (*Figure 12.1*). While the X chromosome is an average-sized (150 Mb) chromosome to which many genes have been genetically and physically mapped, the Y is small (60 Mb) and very poor in genes (Affara *et al.*, 1994). Although the Y is haploid, pairing and recombination between the X and the Y chromosomes is necessary for proper segregation in male meiosis, and occurs within the shared pseudoautosomal region (PAR1: 2.6 Mb) at the tips of their short arms. Markers within PAR1 show a gradient of sex linkage from 50% at the telo-

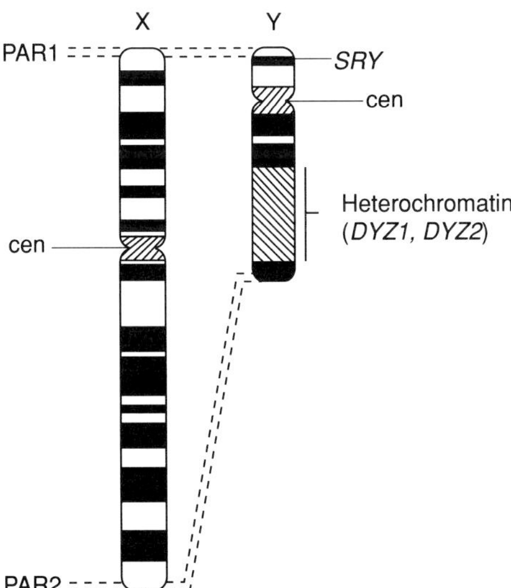

Figure 12.1. The human sex chromosomes. Ideograms of G-banded human X and Y chromosomes, showing the positions of the pseudoautosomal regions (PAR1, PAR2), the sex-determining gene (*SRY*), the centromere (cen), and the Y-chromosomal non-centromeric heterochromatin composed of the repeat sequences *DYZ1* and *DYZ2*. Material on the Y between PAR1 and PAR2 is haploid, male-specific and exempt from recombination.

mere to zero at the boundary with sex-specific DNA (Rouyer *et al.*, 1986). A second, smaller pseudoautosomal region (PAR2: 320 kb) exists at the termini of the long arms of X and Y, but recombination here is not obligate (Kvaløy *et al.*, 1994). Between these two regions, although areas of ancestral X–Y homology exist, recombination does not occur on the Y chromosome.

12.3.1 Cytogenetic diversity among the primates

Just 5 kb proximal to the PAR1 boundary lies *SRY* (sex-determining region Y), the testis-determining gene (Sinclair *et al.*, 1990), which can justly be regarded as the *raison d'être* of the Y chromosome. In theories of sex chromosome evolution (see Chapter 10) it is the existence of this locus which is responsible for the evolution of a barrier to recombination, followed by the accumulation of mutations and rearrangements which lead to the genetic degeneration of the sex-determining chromosome. This tendency to accumulate mutations has certainly led to great diversity of Y chromosome morphology within closely related groups of species such as the primates (*Figure 12.2a*): comparative cytogenetics can be used to construct chromosome phylogenies for all primate chromosomes except the Y (Dutrillaux, 1979). Molecular studies have confirmed this diversity (Lambson *et al.*, 1992). For example, the Y chromosome of one of our closest relatives, the chimpanzee, lacks material which in the human lineage has transposed from Xq21.1–21.3 to Yp11.1 (Geldwerth *et al.*, 1985; Page *et al.*, 1984), and also lacks PAR2 (Bickmore and Cooke, 1987; Kvaløy *et al.*, 1994).

12.3.2 Cytogenetic diversity among humans

Within the human species, too, cytogenetic analysis has shown the Y chromosome to be highly heteromorphic (*Figure 12.2b*). A characteristic terminal quinacrine-fluorescent segment of heterochromatin, composed of the repeat sequences *DYZ1* and *DYZ2*, lies towards the end of the long arm. Studies of normal males have demonstrated enormous variation in the length of this segment (Bobrow *et al.*, 1971; Verma *et al.*, 1978), from complete absence to a length greater than that of the whole of the rest of the chromosome. The non-fluorescent segment also shows heteromorphisms, in the form of length variations (Verma *et al.*, 1978) and pericentric inversions (Bernstein *et al.*, 1986; Verma *et al.*, 1982). Although rarer than these, satellited Y chromosomes, in which an active nucleolus organizer region is translocated to the long arm of the Y from the short arm of an acrocentric autosome, are also regarded as heteromorphisms since they have no clinical consequences and can exist in families for many generations (Schmid *et al.*, 1984).

Together, these observations of between- and within-species variability, and the considerations from theories of sex chromosome evolution, might be expected to imply high sequence diversity of human Y chromosomes. This expectation is reinforced by evidence that mutation rates are higher in the

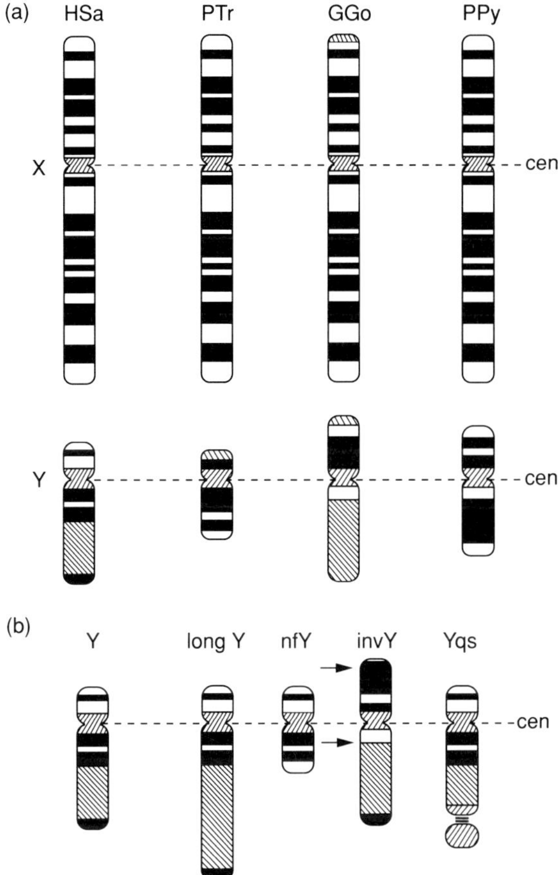

Figure 12.2. Inter- and intraspecies cytogenetic diversity of Y chromosomes. (a) Ideograms of G-banded sex chromosomes of four primate species: HSa, human; PTr, common chimpanzee; GGo, gorilla; PPy, orangutan. While the X chromosomes are almost identical, the Y chromosomes of these species are highly diverged. Heterochromatin (both centromeric and non-centromeric) is shown as hatched areas.'cen' marks the position of the centromere. Banding patterns are taken from Müller and Schempp (1991) and Weber *et al.* (1988). (b) Idiograms of G-banded heteromorphic human Y chromosomes. Long Y and non-fluorescent (nf)Y show length polymorphisms of the fluorescent heterochromatin (Verma *et al.*, 1978); invY, pericentrically inverted Y, with breakpoints indicated by arrows; Yqs, satellited Y, carrying short arm material from an acrocentric chromosome. invY and Yqs banding patterns are taken from Weber *et al.* (1988).

male than in the female germline, perhaps because of differences in the number of cell divisions and concomitant DNA replication errors (see Charlesworth, 1993, for a review).

12.4 Reduced sequence diversity of the Y chromosome

The first Y chromosome DNA polymorphisms were described 10 years ago, amid optimism about the use of the Y as an evolutionary tool (Casanova *et al.*, 1985; Lucotte and Ngo, 1985). The lack of progress since then has been due to the difficulty of finding polymorphisms on the chromosome, and the fact that many of those which have been identified are difficult to use and interpret. A number of serious searches have been made for RFLPs, by hybridizing many Y-chromosomal probes against DNAs from many males digested with many restriction enzymes (Jakubiczka *et al.*, 1989; Jobling, 1994; Malaspina *et al.*, 1990; Spurdle and Jenkins, 1992). Using 22 different Y chromosomes one study screened 833 bp (Jakubiczka *et al.*, 1989) and another 2215 bp (Malaspina *et al.*, 1990) per chromosome, and found between them only three polymorphisms. Searches using DNA sequencing have been similarly unsuccessful. A single polymorphic base substitution was found in a study which sequenced 1357 bp in 12–16 chromosomes from three continents (Seielstad *et al.*, 1994), and no variation whatever was found in a segment of 729 bp sequenced in 38 very ethnically diverse chromosomes (Dorit *et al.*, 1995). The base substitution frequency in non-coding regions of autosomes is in striking contrast to this, at about one in 235 bp for 12–20 chromosomes, all Caucasian in origin (Nickerson *et al.*, 1992).

This has seemed puzzling in view of the considerations discussed above (Section 12.3.2): how can a chromosome which is free to accumulate mutations, and which is known to be cytogenetically diverse both within the human species and between humans and other primates, exhibit such restricted sequence diversity? There are a number of possible explanations. The most basic is a population genetic explanation: for each Y chromosome there are three X chromosomes and four of each autosome. This reduced population size of Y chromosomes is expected to lead to correspondingly reduced diversity (Kimura, 1983) – one-quarter of that of autosomes. The effective population size is further reduced because of cultural practices governing mating behaviour: not all men have sons, but a few men have many sons, a factor which may be particularly important in some modern, and many ancient societies. A different kind of explanation relies on the lack of recombination on the Y. Consider an advantageous mutation arising on a Y chromosome, affecting a phenotype such as fecundity – this could be as small as a single base change. This mutation would spread through the population, carrying with it all 60 Mb of the chromosome to which it is permanently linked; this is the hitch-hiking effect (Maynard-Smith and Haigh, 1974). The observation of exceptionally high interspecies divergence of the *SRY* gene (Whitfield *et al.*,

1993) has led to the suggestion that strong selection on this particular locus, in selective sweeps, might be responsible for reduced Y diversity.

How can the low sequence diversity be reconciled with high cytogenetic diversity? Length heteromorphisms probably reflect the presence of rapidly evolving blocks of repeated sequences on the Y (see Section 12.5.5), while other rearrangements may be common because this haploid chromosome is unconstrained by the requirements of pairing along its length or by a high gene density, and so may be freer to accumulate such rearrangements than other chromosomes.

12.5 Classes of Y polymorphism

The polymorphism searches described above (Section 12.4) have yielded a number of conventional RFLPs. In addition, long-range polymorphisms detectable by pulsed-field gel electrophoresis (PFGE) are relatively abundant. More usefully, various kinds of polymorphisms detectable by PCR are now becoming available. The different classes of polymorphism, together with examples, are described in this section.

For the purposes of phylogenetic tree construction, the most useful markers are binary polymorphisms which represent unique events and whose ancestral states are known. For practical reasons it should be possible to type these by PCR, which allows screening to be done rapidly, using small amounts of DNA. These ideal polymorphisms are few and far between. The other classes of marker found on Y have practical and theoretical disadvantages.

Tandem repeat polymorphisms are not of great use in the construction of phylogenies. Ancestral states cannot be determined, and conventional tree-building methodologies are not suitable for these kinds of multiallelic data. However, these loci display great variability because of their high mutation rates, and can be used to distinguish among closely related Y chromosomes. For many loci, alleles associate non-randomly with the states of simpler polymorphic markers, reflecting divergence from founding alleles. If mutation rates can be measured, some of these rapidly evolving loci may be useful in dating the branches of a Y-chromosomal tree.

12.5.1 Base substitutions

As discussed above (Section 12.4), efforts to identify base substitutions directly in sequencing studies have not met with much success, although one study, which has not yet been fully reported, has identified three within a region of 2.6 kb (Hammer and Bonner, 1994). The ancestral state at such a locus can be determined simply by sequencing in a number of chimpanzees and gorillas. The single Y-specific substitution described to date, sY81 (*DYS271*; Seielstad *et al.*, 1994), is an A to G transition, and can be detected easily by PCR–RFLP analysis, since it creates an *Nla*III site within a PCR product.

RFLPs which are detectable using only one restriction enzyme are also likely to represent base substitutions. However, these have the disadvantage that the ancestral state cannot always be determined, as well as the practical drawbacks associated with non-PCR-based detection methods – usually Southern blot hybridization – which require several micrograms of DNA. The probe 47z (*DXYS5*) detects loci on both sex chromosomes, lying in the block of homology described in Section 12.3.1 (Xq21.1–21.3 and Yp11.1; Page *et al.*, 1984). Two different Y alleles, of 17 kb and 5.3 kb, can be detected in *Stu*I digests only (Nakahori *et al.*, 1989). The ancestral state can in this case be deduced, based on knowledge about the origin of this block of homology (transposition from X to Y): since the fragment detected on the X chromosome is of 17 kb, it is probable that the 17 kb Y allele is ancestral to the 5.3 kb allele.

For some other RFLPs which probably represent base substitutions, such as 92R7/*Hin*dIII and M911/*Xba*I (Mathias *et al.*, 1994), ancestral states cannot be determined because the hybridization pattern of these probes to chimpanzee DNA is completely different to that of human DNA. This problem can in principle be solved by detailed molecular analysis, which should also allow the conversion of all of these markers to PCR-based assays.

Powerful technologies developed for the efficient detection of mutations in human genetic disorders, together with large-scale sequencing projects (Whitfield *et al.*, 1995), promise to yield more Y-chromosomal base substitutions in the near future.

12.5.2 Insertions, deletions and duplications

These simple rearrangements are varied in their usefulness. Those that have been described to date have been identified by observing an identical change in fragment size detected with more than one restriction enzyme, but more may be identified in the course of sequencing projects.

The most useful marker in this class is YAP (*DYS287*), the Y *Alu* polymorphic element (Hammer, 1994). This is a recent insertion of a member of the *Alu* family of repeated DNA elements into the long arm of the Y. *Alu* elements spread through genomes by retrotransposition events (see Chapter 9), but there is no mechanism which specifically removes them; they are thus powerful lineage markers and have been used in studies of autosomal diversity as well as Y diversity (Batzer *et al.*, 1994). Analysis of primate DNAs indicates that the YAP insertion occurred after the human–chimpanzee divergence, and the ancestral state is therefore YAP-negative. Sequencing of several Y chromosomes has shown that the insertion lies between the same two base pairs in all cases. This, then, is a unique insertion event with a known ancestral state and a PCR assay available to type it (Hammer and Horai, 1995). Additional polymorphism, in the form of variation of the length of a poly(A) tail, is also associated with the YAP element (Hammer and Bonner, 1994).

One of the first Y polymorphisms to be described, detected by the probe 12f2 (*DYS11*; Casanova *et al.*, 1985), is probably also a unique insertion or deletion event but more obscure in origin than YAP. It was identified as a 2 kb change in the size of a hybridizing fragment in both *Eco*RI and *Taq*I digests, and no direct evidence exists as to its ancestral state.

More complex than this is the duplication polymorphism detected by pDP31 (*DXYS1Y*) in *Eco*RI digests (Page *et al.*, 1982). All chromosomes have a fragment of 4.5 kb, while some have an additional band of 5.2 kb, which is interpreted to arise from a local tandem duplication. In some cases the larger band is increased in intensity, which appears to result from a triplication event. Analysis of chromosomes which are typed for other markers suggests that these rearrangements have occurred more than once in independent lineages and so are unlikely to be very useful for evolutionary studies (Spurdle *et al.*, 1994b).

12.5.3 Inversions

As has already been discussed (see Section 12.3.2; *Figure 12.2b*), there is cytogenetic evidence for large-scale pericentric inversion polymorphisms of the Y chromosome (Bernstein *et al.*, 1986; Verma *et al.*, 1982). Molecular evidence for paracentric inversions of the short arm has come from deletion mapping of aberrant Y chromosomes (Affara *et al.*, 1986; Nielsen *et al.*, 1988; Page, 1986). In these studies, Y chromosome probes were being used to investigate the breakpoints in interchange XY females and XX males: these are individuals who have either lost a segment of Yp, including the testis-determining locus, or had a similar segment transferred to one of their X chromosomes by an aberrant exchange event in paternal meiosis. Most of these individuals are internally consistent in the map order of markers which they define, but some are not: some XY females possess distal markers and apparently lack more proximal ones while some XX males lack distal but possess more proximal markers. This can most easily be explained by assuming that the Y chromosomes from which they derive have short arm inversions, since alternative explanations require multiple independent deletion events. These inversions might prove to be useful lineage markers but no means yet exist to identify them by molecular methods in normal Y chromosomes.

12.5.4 Complex rearrangements

Complex rearrangements are the most common class of polymorphism which emerge from conventional RFLP searches on the Y chromosome. The best example of this class is the polymorphism detected by the probe 49f [*DYS1* (also 49a); Lucotte and Ngo, 1985; Ngo *et al.*, 1986]. This has been the most popular marker for population studies because of its ability to detect more than 100 different Y-chromosomal haplotypes (Jobling, 1994; Ngo *et al.*, 1986; Persichetti *et al.*, 1992; Spurdle and Jenkins, 1991; Spurdle *et al.*, 1994a,b; Torroni *et al.*, 1990, 1994). These are usually determined by hybridization of the 49f probe to *Taq*I-digested DNAs, although variable hap-

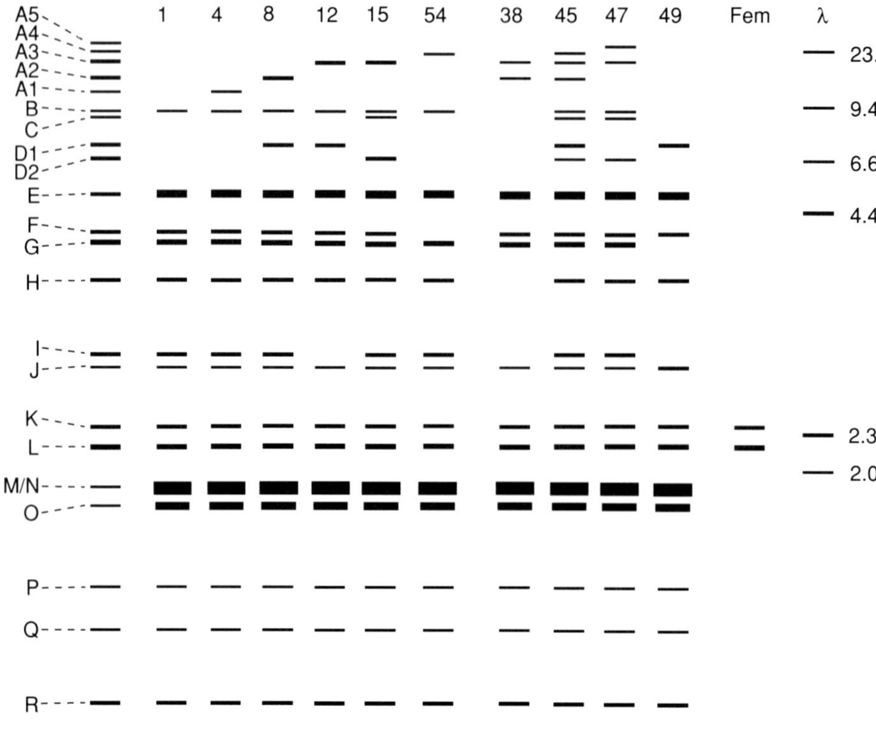

Figure 12.3. Polymorphic *Taq*I haplotypes of the Y chromosome detected with probe 49f (*DYS1*). Schematic representation of patterns obtained by hybridizing 49f (*DYS1*) to *Taq*I digests of unrelated males. Bands are identified by a letter from A to R (at left). The commonly variable bands are A, C, D, F and I. Band A has five possible sizes or can be absent (A0). Band D has two possible sizes, or can be absent (D0). Bands C, F and I can be present or absent. Bands K and L are autosomal, and present in the female (Fem). Haplotype numbers (from Spurdle and Jenkins, 1991) are given at the top. The haplotypes on the left illustrate polymorphism of the commonly variable bands. The four haplotypes on the right show the presence of two or more bands from a single 'allelic series' (haplotypes 38, 45, 47), and polymorphism of bands B, G and H. The extreme right hand lane illustrates phage lambda/*Hin*dIII marker fragments (λ) with sizes in kb.

lotypes are also revealed in digests with several other enzymes (Hammer and Horai, 1995; Spurdle and Jenkins, 1991). The kinds of *Taq*I hybridization patterns observed, and the way in which haplotypes are defined, are illustrated schematically in *Figure 12.3*. Two autosomal and about 15 Y-specific *Taq*I fragments are normally detected, and eight of these Y-specific bands (A, B, C, D, F, G, H and I) can be present, absent, or variable in size. These polymorphic bands were originally considered to represent independent loci ('alleles'), and the different allele sizes were proposed to result from C to T base substitutions within *Taq*I sites (Ngo *et al.*, 1986). These contain the CpG dinucleotide, which is hypermutable, and, it is argued, will therefore tend to be lost during evolution; small *Taq*I fragments are thus viewed as being ancestral to large ones. Based on these assumptions, an attempt has been made to deduce the evolutionary relationships between observed haplotypes (Hazout and Lucotte, 1986), and comparisons of human and chimpanzee hybridization patterns have been used to draw conclusions about the ancestral states of *Taq*I fragments (Abbas *et al.*, 1988).

There are problems with these studies, however. The underlying assumptions about the mechanisms responsible for haplotype diversity can be questioned: *Taq*I sites are likely to have been created as well as lost. Other Y-chromosomal probes detect large numbers of *Taq*I fragments but show no polymorphism (Jobling, 1994): If the mechanism proposed for 49f mutations is correct, then there must be something special about the base substitutional mutation process in this region of the chromosome – it seems more likely that a local rearrangement process is at work. This is supported by the finding of more than one member of each 'allelic series' in some individuals (Jobling, 1994; Spurdle and Jenkins, 1991; Torroni *et al.*, 1990), a finding which also undermines the assumption that each variable band represents a distinct 'allele'. Comparison of 49f haplotypes with the state of a unique-event marker, YAP, indicates that the A, D, F, G and I 'allelic series' have certainly arisen more than once in human evolution (Spurdle *et al.*, 1994b). In addition, geographical considerations indicate that some entire haplotypes have probably arisen more than once (Jobling, 1994; Spurdle and Jenkins, 1991).

The 49f probe certainly reveals an enormous amount of Y chromosome variability. However, a proper interpretation of this variability awaits detailed molecular characterization of the locus. In the meantime, 49f haplotypes undoubtedly have uses in estimating population relatedness by frequency analysis, but are of little phylogenetic utility.

12.5.5 Tandem repeat polymorphisms: satellites

As is the case with other human chromosomes, the size of the block of alphoid satellite DNA at the centromere of the Y chromosome (*DYZ3*) is hypervariable (Oakey and Tyler-Smith, 1990; see also Chapter 5). Using a restriction enzyme (such as *Bgl*II) which does not cut in the alphoid repeat unit and PFGE, the sizes of alphoid blocks can be measured. These range from about 200 kb to over 1600 kb (Mathias *et al.*, 1994). As well as length polymorphism, the alphoid DNA exhibits other polymorphisms: some alphoid blocks contain sites for the enzymes *Ava*II and *Eco*O109I, perhaps representing internal point mutations, and some blocks contain a minority of alphoid units which differ in length from the majority 5.7 kb units and which can be detected by conventional gel electrophoresis and filter hybridization. Two other hypervariable polymorphic satellites were initially described on the Y: an array of the repeat Y-190 (*DYZ5*) on the short arm; and a hypervariable *Bgl*II fragment extending into the adjacent heterochromatin, detected by the single-copy probe poxY1 (Oakey and Tyler-Smith, 1990). The way in which the states of these different polymorphisms relate together will be discussed in the following sections.

A systematic survey for long-range DNA polymorphisms revealed a further five hypervariable loci whose polymorphism is most easily explained by variation in the number of tandem repeats in an array (Jobling, 1994). It would seem that these satellite polymorphisms are common on the Y chromosome; this is consistent with the observation of cytogenetic length heteromorphisms.

Although these polymorphisms reveal a great deal of variability, their usefulness is severely restricted by the need to use PFGE, which is time-consuming and requires fresh blood or cell lines as its starting material.

12.5.6 Tandem repeat polymorphisms: minisatellites

Minisatellites are loci consisting of tandemly arranged, short (generally 10–50 bp), repeated sequences and show variation not only in the number of these repeats but also in their sequence. This additional level of variability can be accessed by the technique of minisatellite variant repeat PCR (MVR–PCR) to give extremely variable digital DNA codes (Jeffreys *et al.*, 1991; Chapter 7). Minisatellites are common in PAR1 but only one is known in the Y-specific part of the chromosome. MVR–PCR codes at this A+T-rich locus, MSY1 (*DYF155S1*; Jobling *et al.*, 1994), are extremely diverse but fall into groups consistent with other haplotypic information. In principle, mutation rates and processes can be directly observed at this locus by analysis of new-length alleles in pedigrees and sperm DNA (Jeffreys *et al.*, 1994); this information will provide a firm foundation for diversity studies and may be useful in dating branches of a tree constructed using simpler polymorphisms.

12.5.7 Tandem repeat polymorphisms: microsatellites

Microsatellites, tandem arrays of repeat sequences 2–5 bp in length, are abundant elsewhere in the nuclear genome and are also likely to be plentiful on the Y chromosome. Several have already been described and shown to be polymorphic (Mathias *et al.*, 1994; Roewer *et al.*, 1992). They are of limited variability and alleles of particular lengths within populations are likely to have multiple origins (isoallelism); however, their likely abundance and ease of analysis will certainly make these loci useful. A study using 30 autosomal microsatellites has shown that phylogenetic information can be extracted by statistical methods from these data (Bowcock *et al.*, 1994), and methods for multiplex PCR analysis with fluorescent detection (Urquhart *et al.*, 1995) will increase their power in population and forensic studies.

12.5.8 Mutation processes at tandemly repeated loci

The variability of tandemly repeated loci on the Y chromosome must arise through haploid mutation processes – the Y provides a system for the study of these processes, isolated from interallelic mechanisms which act on other chromosomes. The major mechanism causing variation at satellite loci is probably unequal sister chromatid exchange (USCE; Chapter 5); at microsatellites replication slippage is thought to predominate (Chapter 8). Slippage is also the most likely process acting at MSY1; the observed allele structures indicate that, if USCE is occurring, it is physically constrained. Most minisatellites are G+C-rich, and undergo mutation primarily through events resembling interallelic gene conversion (Jeffreys *et al.*, 1994), which clearly would be precluded on the Y. It is interesting that no polymorphic minisatellites have been found on the Y, even though they have been sought (N. Fretwell and M.A. Jobling, unpublished). The only known Y-specific G+C-rich minisatellites are the RU2 elements which lie within the two copies of the CRI-S232 (*DYS136*) locus in Yq11.1. Such loci are monomorphic, despite the existence of homologues in Xp22.3 which are hypervariable (Li *et al.*, 1992). These observations suggest that, for this class of locus at least, a location on the Y chromosome is incompatible with polymorphism.

12.5.9 Compound haplotyping

Because of the haploidy of much of the Y, the states of many polymorphisms located in different parts of the chromosome can be combined to give highly informative compound haplotypes. Compound haplotyping was originally done using the three satellite polymorphisms discussed above (see Section 12.5.5; Oakey and Tyler-Smith, 1990) and other polymorphisms of the alphoid block; this analysis allowed 33 compound haplotypes, each defining a paternal lineage, to be identified among 42 Y chromosomes. Among these chromosomes, two major alphoid array size groups were seen, one peaking around 300 kb and one around 900 kb. The distribution of the other polymorphisms in these two groups was non-random: the 'small' alphoid block was associated with the absence of internal sites for *Ava*II and *Eco*O109I and the absence of 6.0 kb repeat units, as well as the absence of a poxY1 *Bgl*II fragment greater in length than 1300 kb; in contrast, the 'large' alphoid block was associated with the presence of internal sites and 6.0 kb units and a small poxY1 *Bgl*II fragment. The simplest explanation for these associations is that the Y chromosomes in each group (known respectively as 1 and 2) are descended from a single male ancestor, with alphoid block sizes of about 300 kb and 900 kb, respectively, associated with the other characteristic polymorphic states. In contrast, the distribution of *DYZ5* array sizes in each group was similar. This could be because both founders had similar *DYZ5* array sizes or because the mutation processes at this locus lead to very large size changes which tend to blur the record of different founding sizes.

Subsequent studies (Jobling, 1994; Mathias *et al.*, 1994) have extended the definition of the two groups, added two more and identified many chromosomes which do not fall into these groups but which probably represent fur-

Table 12.1. Defining characteristics of Y-chromosomal haplotypic groups 1–4

Predominant geographical origin	Group 1 Europe	Group 2 Diverse	Group 3 India	Group 4 Africa
Alphoid:				
Array size (kb)	200–450	280–1050	460–680	430–1620
6.0 kb unit	0	1	0	1
*Ava*II sites	0	1	0	1
*Eco*O109I sites	0	1	1	0
poxY1 size (kb)	>1300	≤1300	>1300	<1300
YAP	0	0	0	1
92R7	0	1	0	1
M911	0	1	0	1
Predominant 49f *Taq*I ht	ht 15	diverse	ht 11	ht 4

States of multiple and diverse polymorphisms which define four haplotypic groups. The sizes of alphoid arrays and poxY1 fragment were measured from *Bgl*II digests. Other polymorphisms are: presence (1) or absence (0) of 6.0 kb alphoid repeat units; presence (1) or absence (0) of internal alphoid restriction sites; presence (1) or absence (0) of the YAP element; 6.7 kb (0) or 4.6 kb (1) 92R7 *Hin*dIII fragment; 33 kb (0) or 55 kb (1) M911 *Hin*dIII fragment. The most common 49f *Taq*I haplotype (ht) for each group is also shown (nomenclature from Spurdle and Jenkins, 1991). Data are from Jobling, 1994; Mathias *et al.*, 1994; Spurdle *et al.*, 1994b.

ther groups in the population. The defining characteristics of groups 1–4 are shown in *Table 12.1.* Groups 1 and 3 are rather closely related and each is quite homogeneous haplotypically as well as ethnically: all but one of the group 1 males are European Caucasians and most group 3 males are Indian in origin. Although group 4 shows more heterogeneity, most members are sub-Saharan Africans and it appears to be genuinely monophyletic, corresponding to the YAP-positive class of chromosomes (Tyler-Smith and Hammer, 1995). Group 2 is the most heterogeneous, including males from Europe, Asia and Oceania and certainly contains chromosomes with derived unique-event mutations such as 12f2 (*DYS11*) and 47z (*DXYS5*).

12.6 Studying human evolution using Y chromosome polymorphisms

There are many issues in human evolution to which Y chromosome research can contribute. As well as the question of the origin of modern humans, there are questions about the spread of agriculture in Europe over 10 000 years ago, the colonization of the Americas and the Pacific islands and more general issues such as population admixture. Y chromosome histories will differ from mtDNA histories because of differences in male and female behaviour in colonizations, conflicts and migrations, and, of course, in mating practices; these differences are conferred by the Y chromosome itself.

12.6.1 Tree-building

As has already been discussed, construction of Y-chromosomal trees using parsimony should not require the complex methodologies needed for mtDNA tree construction, provided suitable polymorphisms can be found. Attempts have already been made to draw such trees, using markers which are assumed to represent unique events (Mathias *et al.*, 1994; Tyler-Smith and Hammer, 1995). Single, simple trees do emerge from this analysis: an example is shown in *Figure 12.4* which relates together six haplotypic groups including the four groups described in *Table 12.1*. As the number of appropriate polymorphisms increases, the complexity of these trees will also increase.

12.6.2 Population studies

Many qualitative population studies have been done using Y-chromosomal markers, chiefly 49f (e.g. Spurdle and Jenkins, 1991; Torroni *et al.*, 1990,

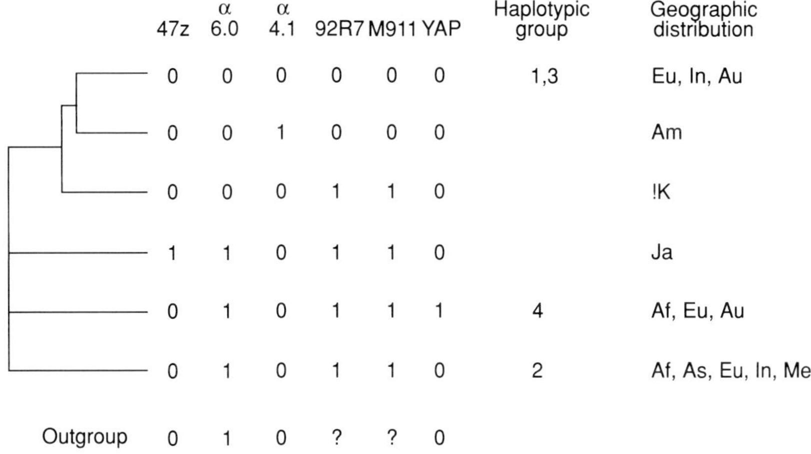

Figure 12.4. Evolutionary tree of Y chromosome haplotypes. Six haplotypes constructed using six polymorphisms are related in a maximum parsimony tree. The six polymorphisms are assumed to represent unique events, and the tree is rooted using assumed ancestral states at four of these loci in the chimpanzee outgroup. The six haplotypes include the four haplotypic groups defined in *Table 12.1*. Geographical distributions of each haplotype: Eu, Europe; In, India; Au, Australia; Am, America; !K, !Kung ('bushmen' of the Kalahari desert in southwestern Africa; the ! represents a clicking sound used in their language); Ja, Japan; Af, Africa; As, Asia; Me, Melanesia; 0 and 1 represent alternative alleles at individual loci as follows. For 47z (*DXYS5Y*), 0 = 17 kb and 1 = 5.3 kb *Stu*I fragment. For α6.0 and α4.1,0 = absence and 1 = presence of 6.0 kb or 4.1 kb alphoid repeat units. For 92R7; 0 = 6.7 kb and 1 = 4.6 kb *Hin*dIII fragment. For M911, 0 = 33 kb and 1 = 55 kb *Hin*dIII fragment. For YAP, 0 = absence and 1 = presence of the YAP element. Redrawn from Tyler-Smith and Hammer (1995) with permission from the *Y Chromosome Consortium Newsletter*.

1994). More recent studies have used 49f in conjunction with other markers such as YAP, 47z, and the microsatellite *DYS19* (Hammer and Horai, 1995; Spurdle *et al.*, 1994a, b). These have been useful in illuminating the relationships between these different polymorphisms, and have allowed population trees to be drawn by, for example, neighbour-joining analysis using allele frequency data (Hammer, 1994; Hammer and Horai, 1995). In a study on Japanese populations, genetic distances have also been calculated and are in agreement with distances calculated using mtDNA and autosomal markers (Hammer and Horai, 1995). The resolution of these studies is also expected to increase as more Y-chromosomal markers become available.

12.6.3 The future

Progress in Y chromosome research has undoubtedly been slow and the advances that have been made have failed to impress those outside the field: the Y occupied less than one of the 1032 pages of a recent and authoritative book on the subject of genetics and human evolution (Cavalli-Sforza *et al.*, 1994). However, technological advances promise to change this situation over the next few years. With suitable Y chromosome DNA polymorphisms and population samples, questions about human origins can be addressed from a Y chromosome perspective. Most geneticists support the 'Out of Africa' hypothesis, which holds that there was a single origin for modern humans in Africa less than 200 000 years ago, followed by dispersal and no admixture with existing populations such as the Neanderthals (Lahr and Foley, 1994). If a Y chromosome tree predicted a recent African ancestor, it would provide support for this hypothesis. The low levels of diversity of modern Y chromosomes do suggest that the common ancestor was indeed recent, but at present we cannot tell where this ancestor lived. mtDNA and Y chromosome trees show the relationships of mtDNAs and Y chromosomes, not of people. Genome history, complex though it is, will always be less complex than the real histories of populations; however, if we are ever to understand it, and to understand what it implies for population history, we will need to use information from all of the genome, not just selected and specialized parts. Y chromosome ancestors need not have been the ancestors of any other sequences in the genome, so it is possible that the Y ancestor lived at a different time in a different place to the mtDNA ancestor: Adam may never have met Eve.

Acknowledgements

The author acknowledges support through an MRC Training Fellowship, and thanks Chris Tyler-Smith for comments on the manuscript.

References

Abbas N, Guérin P, Bessières P, Ruffié J, Lucotte G. (1988) Séquences Y-chromosome humain spécifiques présente chez les singes Anthropoïdes. *Biochem. Syst. Ecol.* **16**: 105–109.

Affara NA, Ferguson-Smith MA, Tolmie JL, Kwok K, Mitchell M, Jamieson D, Cooke A, Florentin L. (1986) Variable transfer of Y-specific sequences in XX males. *Nucleic Acids Res.* **14**: 5375–5387.

Affara NA, Lau Y-FC, Briggs H, Davey P, Jones MH, Khwaja O, Mitchell M, Sargent C. (1994) Report of the First International Workshop on Y Chromosome Mapping 1994. *Cytogenet. Cell Genet.* **67**: 360–387.

Batzer MA, Stoneking M, Alegria-Hartman M *et al.* (1994) African origin of human-specific polymorphic Alu insertions. *Proc. Natl Acad. Sci. USA* **91**: 12288–12292.

Bernstein R, Wadee A, Rosendorff J, Wessels A, Jenkins T. (1986) Inverted Y chromosome polymorphism in the Gujerati Muslim Indian population of South Africa. *Hum. Genet.* **74**: 223–229.

Bickmore WA, Cooke HJ. (1987) Evolution of homologous sequences on the human X and Y chromosomes, outside of the meiotic pairing segment. *Nucleic Acids Res.* **14**: 6261–6271.

Bobrow M, Pearson PL, Pike MC, El-Alfi OS. (1971) Length variation in the quinacrine banding segment of human Y chromosomes of different sizes. *Cytogenetics* **10**: 190–198.

Bowcock AM, Ruiz-Linares A, Tomfohrde J, Minch E, Kidd JR, Cavalli-Sforza LL. (1994) High resolution of human evolutionary trees with polymorphic microsatellites. *Nature* **368**: 455–457.

Cann RL, Stoneking M, Wilson AC. (1987) Mitochondrial DNA and human evolution. *Nature* **325**: 31–36.

Casanova M, Leroy P, Boucekkine C, Weissenbach J, Bishop C, Fellous M, Purrello M, Fiori G, Siniscalco M. (1985) A human Y-linked DNA polymorphism and its potential for estimating genetic and evolutionary distance. *Science,* **230**: 1403–1406.

Cavalli-Sforza LL, Menozzi P, Piazza A. (1994) *The History and Geography of Human Genes.* Princeton University Press, Princeton, NJ.

Charlesworth B. (1993) More mutations in males? *Curr. Biol.* **3**: 466–467.

Dorit RL, Akashi H, Gilbert W. (1995) Absence of polymorphism at the ZFY locus on the human Y chromosome. *Science* **268**: 1183–1185.

Dutrillaux B. (1979) Chromosomal evolution in primates: tentative phylogeny from *Microcebus murinus* (Prosimian) to man. *Hum. Genet.* **48**: 251–314.

Geldwerth D, Bishop C, Guelläen G, Koenig M, Vergnaud G, Mandel J-L, Weissenbach J. (1985) Extensive DNA sequence homologies between the human Y and the long arm of the X chromosome. *EMBO J.* **4**: 1739-1743.

Gray MW. (1989) Origin and evolution of mitochondrial DNA. *Annu. Rev. Cell Biol.* **5**: 25–50.

Hammer MF. (1994) A recent insertion of an Alu element on the Y chromosome is a useful marker for human population studies. *Mol. Biol. Evol.* **11**: 749–761.

Hammer MF, Bonner MR. (1994) Evolutionary analysis of polymorphisms in the YAP region of the human Y chromosome. *Cytogenet. Cell Genet.* **67**: 395–396.

Hammer MF, Horai S. (1995) Y chromosomal DNA variation and the peopling of Japan. *Am. J. Hum. Genet.* **56**: 951–962.

Hazout S, Lucotte G. (1986) Vers une généalogie du chromosome Y. *Ann. Génét.* **29**: 246–252.

Jakubiczka S, Arnemann J, Cooke HJ, Krawczak M, Schmidtke J. (1989) A search for restriction fragment length polymorphism of the human Y chromosome. *Hum. Genet.* **84**: 86–88.

Jeffreys AJ, MacLeod A, Tamaki K, Neil DL, Monckton DG. (1991) Minisatellite repeat coding as a digital approach to DNA typing. *Nature* **354**: 204–209.

Jeffreys AJ, Tamaki K, MacLeod A, Monckton DG, Neil DL, Armour JAL. (1994) Complex gene conversion events in germline mutation at human minisatellites. *Nature Genetics* **6**: 136–145.

Jobling MA. (1994) A survey of long-range DNA polymorphisms on the human Y chromosome. *Hum. Mol. Genet.* **3**: 107–114.

Jobling MA, Fretwell N, Dover GA, Jeffreys AJ. (1994) Digital coding of human Y chromosomes: MVR-PCR at Y-specific minisatellites. *Cytogenet. Cell Genet.* **67**: 390.

Kimura M. (1983) *The Neutral Theory of Molecular Evolution.* Cambridge University Press, Cambridge, pp. 40–50.

Kvaløy K, Galvagni F, Brown WRA. (1994) The sequence organization of the long arm pseudoautosomal region of the human sex chromosomes. *Hum. Mol. Genet.* **3**: 771–778.

Lahr MM, Foley R. (1994) Multiple dispersals and human origins. *Evol. Anthropol.* **3**: 48–60.

Lambson B, Affara NA, Mitchell M, Ferguson-Smith MA. (1992) Evolution of DNA sequence homologies between the sex chromosomes in primate species. *Genomics* **14**: 1032–1040.

Li X-M, Yen PH, Shapiro LJ. (1992) Characterization of a low copy repetitive element S232 involved in the generation of frequent deletions of the distal short arm of the human X chromosome. *Nucleic Acids Res.* **20**: 1117–1122.

Lucotte G, Ngo KY. (1985) p49f, a highly polymorphic probe that detects RFLPs on the human Y chromosome. *Nucleic Acids Res.* **13**: 8285.

Malaspina P, Persichetti F, Novelletto A, Iodice C, Terrenato L, Wolfe J, Ferraro M, Prantera G. (1990) The human Y chromosome shows a low level of DNA polymorphism. *Ann. Hum. Genet.* **54**: 297–305.

Mathias N, Bayés M, Tyler-Smith C. (1994) Highly informative compound haplotypes for the human Y chromosome. *Hum. Mol. Genet.* **3**: 115–123.

Maynard-Smith J, Haigh J. (1974) The hitch-hiking effect of a favourable gene. *Genet. Res.* **23**: 23–35.

Müller G, Schempp W. (1991) Comparative mapping of ZFY in the hominoid apes. *Hum. Genet.* **88**: 59–63.

Nakahori Y, Tamura T, Yamada M, Nakagome Y. (1989) Two 47z (DXYS5) RFLPs on the X and the Y chromosome. *Nucleic Acids Res.* **17**: 2152.

Ngo KY, Vergnaud G, Johnsson C, Lucotte G, Weissenbach J. (1986) A DNA probe detecting multiple haplotypes of the human Y chromosome. *Am. J. Hum. Genet.* **38**: 407–418.

Nickerson DA, Whitehurst C, Boysen C, Charmley P, Kaiser R, Hood L. (1992) Identification of clusters of biallelic polymorphic sequence-tagged sites (pSTSs) that generate highly informative and automatable markers for genetic linkage mapping. *Genomics* **12**: 377–387.

Nielsen KB, Schwartz M, Sardemann H. (1988) Investigation of three XX males by cytogenetic and DNA analyses. *Hum. Genet.* **78**: 179–182.

Oakey R, Tyler-Smith C. (1990) Y chromosome DNA haplotyping suggests that most European and Asian men are descended from one of two males. *Genomics* **7**: 325–330.

Page DC. (1986) Sex reversal: deletion mapping the male-determining function of the human Y chromosome. *Cold Spring Harbor. Symp. Quant. Biol.* **51**: 229–235.

Page DC, de Martinville B, Barker D, Wyman A, White R, Francke U, Botstein D. (1982) Single-copy sequence hybridizes to polymorphic and homologous loci on human X and Y chromosomes. *Proc. Natl Acad. Sci. USA* **79**: 5352–5356.

Page DC, Harper ME, Love J, Botstein D. (1984) Occurrence of a transposition from the X-chromosome long arm to the Y-chromosome short arm during human evolution. *Nature* **311**: 119–122.

Persichetti F, Blasi P, Hammer M, Malaspina P, Jodice C, Terrenato L, Novelletto A. (1992) Disequilibrium of multiple DNA markers on the human Y chromosome. *Ann. Hum. Genet.* **56**: 303–310.

Roewer L, Arnemann J, Spurr NK, Grzeschik K-H, Epplen JT. (1992) Simple repeat sequences on the human Y chromosome are equally polymorphic as their autosomal counterparts. *Hum. Genet.* **89**: 389–394.

Rouyer F, Simmler M-C, Johnsson C, Vergnaud G, Cooke HJ, Weissenbach J. (1986) A gradient of sex linkage in the pseudoautosomal region of the human sex chromosomes. *Nature* **319**: 291–295.

Schmid M, Haaf T, Solleder E, Schempp W, Leipoldt M, Heilbronner H. (1984) Satellited Y chromosomes: structure, origin and clinical significance. *Hum. Genet.* **67**: 72–85.

Seielstad MT, Hebert JM, Lin AA, Underhill PA, Ibrahim M, Vollrath D, Cavalli-Sforza LL. (1994) Construction of human Y-chromosomal haplotypes using a new polymorphic A to G transition. *Hum. Mol. Genet.* **3**: 2159–2161.

Sinclair AH, Berta P, Palmer MS *et al.* (1990) A gene from the human sex-determining region encodes a protein with homology to a conserved DNA-binding motif. *Nature,* **346**: 240–244.

Spurdle A, Jenkins T. (1991) Y chromosome probe p49a detects complex PvuII haplotypes and many new TaqI haplotypes in Southern African populations. *Am. J. Hum. Genet.* **50**: 107–125.

Spurdle AB, Jenkins T. (1992) The search for Y chromosome polymorphism is extended to negroids. *Hum. Mol. Genet.* **1**: 169–170.

Spurdle AB, Woodfield DG, Hammer MF, Jenkins T. (1994a) The genetic affinity of Polynesians: evidence from Y chromosome polymorphisms. *Ann. Hum. Genet.* **58**: 251–263.

Spurdle AB, Hammer MF, Jenkins T. (1994b) The Y *Alu* polymorphism in Southern African populations and its relationship to other Y-specific polymorphisms. *Am. J. Hum. Genet.* **54**: 319–330.

Stoneking M. (1993) DNA and recent human evolution. *Evol. Anthropol.* **2**: 60–73.

Torroni A, Semino O, Scozzari R, Sirugo G, Spedini G, Abbas N, Fellous M, Santachiara-Benerecetti AS. (1990) Y chromosome DNA polymorphisms in human populations: differences between Caucasoids and Africans detected by 49a and 49f probes. *Ann. Hum. Genet.* **54**: 287–296.

Torroni A, Chen Y-S, Semino O, Santachiara-Benerecetti AS, Scott CR, Lott MT, Winter M, Wallace DC. (1994) mtDNA and Y-chromosome polymorphisms in four native American populations from Southern Mexico. *Am. J. Hum. Genet.* **54**: 303–318.

Tyler-Smith C, Hammer MF. (1995) The relationships of Y chromosomes determined using five RFLPs and the YAP element. *Y Chrom. Consort. News.* **2**: 4–5.

Urquhart A, Oldroyd NJ, Kimpton CP, Gill P. (1995) Highly discriminating heptaplex short tandem repeat PCR system for forensic identification. *BioTechniques* **18**: 116–121.

Verma RS, Dosik H, Scharf T, Lubs HA. (1978) Length heteromorphisms of fluorescent (f) and non-fluorescent (nf) segments of human Y chromosome: classification, frequencies, and incidence in normal Caucasians. *J. Med. Genet.* **15**: 277–281.

Verma RS, Rodriguez J, Dosik H. (1982) The clinical significance of pericentric inversion of the human Y chromosome: a rare "third" type of heteromorphism. *J. Hered.* **73**: 236–238.

Weber B, Walz L, Schmid M, Schempp W. (1988) Homeologic aberrations in human and chimpanzee Y chromosomes: inverted and satellited Y chromosomes. *Cytogenet. Cell Genet.* **47**: 26–28.

Whitfield LS, Lovell-Badge R, Goodfellow PN. (1993) Rapid sequence evolution of the mammalian sex-determining gene *SRY*. *Nature* **364**: 713–715.

Whitfield LS, Hawkins TL, Goodfellow PN, Sulston J. (1995) 41 kilobases of analyzed sequence from the pseudoautosomal and sex-determining regions of the short arm of the human Y chromosome. *Genomics* **27**: 306–311.

Index

ORDERING DETAILS

Main address for orders

BIOS Scientific Publishers Ltd
9 Newtec Place, Magdalen Road,
Oxford OX4 1RE, UK
Tel: +44 1865 726286
Fax: +44 1865 246823

Australia and New Zealand
DA Information Services
648 Whitehorse Road, Mitcham, Victoria 3132, Australia
Tel: (03) 9210 7777
Fax: (03) 9210 7788

India
Viva Books Private Ltd
4325/3 Ansari Road, Daryaganj, New Delhi 110 002, India
Tel: 11 3283121
Fax: 11 3267224

Singapore and South East Asia
(Brunei, Hong Kong, Indonesia, Korea, Malaysia, the Philippines, Singapore, Taiwan, and Thailand)
Toppan Company (S) PTE Ltd
38 Liu Fang Road, Jurong, Singapore 2262
Tel: (265) 6666
Fax: (261) 7875

USA and Canada
BIOS Scientific Publishers
PO Box 605, Herndon, VA 20172-0605, USA
Tel: (703) 661 1500
Fax: (703) 661 1501

Payment can be made by cheque or credit card (Visa/Mastercard, quoting number and expiry date). Alternatively, a *pro forma* invoice can be sent.

Prepaid orders must include £2.50/US$5.00 to cover postage and packing (two or more books sent post free)